METHODS IN MOLECULAR BIOLOGY

Series Editor
John M. Walker
School of Life and Medical Sciences
University of Hertfordshire
Hatfield, Hertfordshire, UK

For further volumes:
http://www.springer.com/series/7651

For over 35 years, biological scientists have come to rely on the research protocols and methodologies in the critically acclaimed Methods in Molecular Biology series. The series was the first to introduce the step-by-step protocols approach that has become the standard in all biomedical protocol publishing. Each protocol is provided in readily-reproducible step-by step fashion, opening with an introductory overview, a list of the materials and reagents needed to complete the experiment, and followed by a detailed procedure that is supported with a helpful notes section offering tips and tricks of the trade as well as troubleshooting advice. These hallmark features were introduced by series editor Dr. John Walker and constitute the key ingredient in each and every volume of the Methods in Molecular Biology series. Tested and trusted, comprehensive and reliable, all protocols from the series are indexed in PubMed.

Manuscript Instructions https://drive.google.com/file/d/1K4Az5vpAFVzYadJ3jKx2-QSo7Dqa_SNA/view?usp=sharing

Biological Macromolecule Engineering Technologies

Edited by

Huan Peng

College of Life Science and Technology, Huazhong University of Science and Technology, Wuhan, Hubei, China

S. Brian Kim

Department of Chemistry, York University, Toronto, ON, Canada

Humana Press

Editors
Huan Peng
College of Life Science and Technology
Huazhong University of Science
and Technology
Wuhan, Hubei, China

S. Brian Kim
Department of Chemistry
York University
Toronto, ON, Canada

ISSN 1064-3745 ISSN 1940-6029 (electronic)
Methods in Molecular Biology
ISBN 978-1-0716-5256-5 ISBN 978-1-0716-5257-2 (eBook)
https://doi.org/10.1007/978-1-0716-5257-2

This work was supported by National Natural Science Foundation of China (32201100).

This Humana imprint is published by the registered company Springer Science+Business Media, LLC, part of Springer Nature.
The registered company address is: 1 New York Plaza, New York, NY 10004, U.S.A.

Preface

The field of biological engineering is undergoing a profound transformation. We have moved beyond simply observing and describing natural phenomena to actively writing the genetic code, designing protein functions, and programming cellular behaviors. This new era is defined by our ability to engineer biological macromolecules with precision and creativity, turning cells into living foundries for sustainable chemistry, advanced therapeutics, and intelligent biosensors.

This volume, *Biological Macromolecule Engineering Technologies*, captures the cutting edge of this revolution. The collected chapters represent a snapshot of the sophisticated toolkit now available to researchers, showcasing methods that leverage and extend the capabilities of synthetic biology. The protocols within are organized to guide the reader from foundational molecular manipulations to complex cellular applications.

Part I: Engineering Proteins and Enzymes begins with the direct design and evolution of nature's workhorses, including novel approaches to create enzymes for unprecedented photobiocatalytic reactions and systems for high-level protein production.

Part II: Engineering Viral Delivery Systems highlights the repurposing of viral architectures, from bacteriophages to virus-like particles, as programmable nanoscaffolds. These engineered systems demonstrate powerful applications in vaccine development, targeted in vivo delivery, and biomarker discovery.

Part III: CRISPR-Driven Genome Engineering details the latest advancements in precision genome editing. This section moves beyond basic gene editing to cover multiplexed base and prime editing, the creation of complex cell models, and the direct evolution of proteins using CRISPR-based diversity generation.

Part IV: Directed Evolution and Selection explores sophisticated methods that harness the power of continuous evolution, coupling phage, and CRISPR systems to artificially drive the selection of new functional macromolecules, from regulatory elements to bioactive peptides.

Finally, *Part V: Nucleic Acid Synthesis and Cellular Recording* looks to the future, covering the synthesis of functional genetic libraries and the groundbreaking technology of biological "memory," where a cell's history is permanently recorded into its DNA.

We are profoundly grateful to the contributing authors, whose expertise and innovative protocols form the core of this collection. The interdisciplinary spirit of this work, bridging chemical biology, protein engineering, genetics, and computational design, is its greatest strength. It is our hope that this book will serve as a valuable resource for scientists and students alike, providing the foundational methods to drive the next wave of discovery in biological macromolecule engineering. The potential to address some of humanity's most pressing challenges in health, energy, and manufacturing has never been greater.

Wuhan, China ***Huan Peng***
Toronto, Canada ***S. Brian Kim***

Contents

Contributors

MARYAM ALI • *Department of Chemistry, University of Toronto, Mississauga, ON, Canada*
ANDRESSA GODOY AMARAL • *Biochemistry Department, Institute of Chemistry, University of Sao Paulo, São Paulo, SP, Brazil*
DANIELLA SABINO BATTAGELLO • *Biochemistry Department, Institute of Chemistry, University of Sao Paulo, São Paulo, SP, Brazil*
MAARTEN BOON • *Laboratory of Gene Technology, Department of Biosystems, KU Leuven, Leuven, Belgium*
CEN CHEN • *Xiaogan Agricultural Technology Extension Center, Xiaogan, Hubei, China*
STEPHEN L. CRAIG • *Department of Chemistry, Duke University, Durham, NC, USA*
MASAD J. DAMHA • *Department of Chemistry, McGill University, Montreal, QC, Canada*
TYLER C. DANIEL • *Department of Chemical and Biomolecular Engineering, University of Pennsylvania, Philadelphia, PA, USA; Center for Precision Engineering for Health (CPE4H), University of Pennsylvania, Philadelphia, PA, USA*
RIET DEHAENE • *Laboratory of Gene Technology, Department of Biosystems, KU Leuven, Leuven, Belgium*
MATT C. DEMELO • *Department of Biosciences, Rice University, Houston, TX, USA*
JIE FENG • *State Key Laboratory and Institute of Elemento-Organic Chemistry, College of Chemistry, Frontiers Science Center for New Organic Matter, Nankai University, Tianjin, China*
JINGXIANG FU • *College of Life Science and Technology, Huazhong University of Science and Technology, Wuhan, China*
XUE GAO • *Department of Chemical and Biomolecular Engineering, University of Pennsylvania, Philadelphia, PA, USA; Center for Precision Engineering for Health (CPE4H), University of Pennsylvania, Philadelphia, PA, USA; Department of Bioengineering, University of Pennsylvania, Philadelphia, PA, USA*
RICARDO JOSÉ GIORDANO • *Biochemistry Department, Institute of Chemistry, University of Sao Paulo, São Paulo, SP, Brazil*
DEVIN A. GOLLA • *Department of Chemical and Biomolecular Engineering, University of Pennsylvania, Philadelphia, PA, USA; Center for Precision Engineering for Health (CPE4H), University of Pennsylvania, Philadelphia, PA, USA*
LOGAN HAUGH • *Department of Chemical and Biomolecular Engineering, University of Pennsylvania, Philadelphia, PA, USA; Center for Precision Engineering for Health (CPE4H), University of Pennsylvania, Philadelphia, PA, USA*
RYAN HILI • *Department of Chemistry, Centre for Research on Biomolecular Interactions, York University, Toronto, ON, Canada*
SEAN K. HSU • *Center for Nano-ImmunoEngineering, University of California, San Diego, La Jolla, CA, USA; Shu and K.C. Chien and Peter Farrell Collaboratory, University of California, San Diego, La Jolla, CA, USA; Department of Molecular Biology, University of California, La Jolla, CA, USA*

YUSEN HUANG • *State Key Laboratory of Virology and Biosafety, Wuhan Institute of Virology, Chinese Academy of Sciences, Wuhan, China; University of Chinese Academy of Sciences, Beijing, China*
BASIL P. HUBBARD • *Department of Pharmacology and Toxicology, University of Toronto, Toronto, ON, Canada*
ALISON KERREMANS • *Laboratory of Gene Technology, Department of Biosystems, KU Leuven, Leuven, Belgium*
NATALIE KHAMISSI • *Department of Chemistry, Centre for Research on Biomolecular Interactions, York University, Toronto, ON, Canada*
S. BRIAN KIM • *Department of Chemistry, York University, Toronto, ON, Canada; Centre for Research on Biomolecular Interactions (CRBI), Toronto, ON, Canada*
ROB LAVIGNE • *Laboratory of Gene Technology, Department of Biosystems, KU Leuven, Leuven, Belgium*
MAO-LIN LI • *State Key Laboratory and Institute of Elemento-Organic Chemistry, College of Chemistry, Frontiers Science Center for New Organic Matter, Nankai University, Tianjin, China*
MENGLING LI • *State Key Laboratory of Agricultural Microbiology, Huazhong Agricultural University, Wuhan, Hubei, China; Hubei Hongshan Lab, Wuhan, Hubei, China; Cooperative Innovation Center for Sustainable Pig Production, College of Veterinary Medicine, Huazhong Agricultural University, Wuhan, Hubei, China*
XINFENG LI • *State Key Laboratory of Virology and Biosafety, Wuhan Institute of Virology, Chinese Academy of Sciences, Wuhan, China*
THERESA B. LOVELESS • *Department of Biosciences, Rice University, Houston, TX, USA*
WEIXUAN LUO • *College of Life Science and Technology, Huazhong University of Science and Technology, Wuhan, China*
ZIDI LYU • *Department of Chemistry, McGill University, Montreal, QC, Canada*
YASAMAN MAHDAVI-AMIRI • *Department of Chemistry, York University, Toronto, ON, Canada*
ROBERTA AQUINO MARTINS • *Biochemistry Department, Institute of Chemistry, University of Sao Paulo, São Paulo, SP, Brazil*
JIA NIU • *Department of Chemistry, Boston College, Chestnut Hill, MA, USA*
JOSHUA O'GRADY • *Department of Chemistry, Ottawa, ON, Canada*
HUAN PENG • *College of Life Science and Technology, Huazhong University of Science and Technology, Wuhan, China*
JORIEN POPPELIERS • *Laboratory of Gene Technology, Department of Biosystems, KU Leuven, Leuven, Belgium*
LUCA RUSTICO • *Department of Chemistry, Centre for Research on Biomolecular Interactions, York University, Toronto, ON, Canada*
DAVID SABATINO • *Department of Chemistry, Ottawa, ON, Canada; Institute of Biochemistry, Carleton University, Ottawa, ON, Canada*
JUMI A. SHIN • *Department of Chemistry, University of Toronto, Mississauga, ON, Canada*
NICOLE F. STEINMETZ • *Department of NanoEngineering, University of California, San Diego, La Jolla, CA, USA; Center for Nano-ImmunoEngineering, University of California, San Diego, La Jolla, CA, USA; Moores Cancer Center, University of California, San Diego, La Jolla, CA, USA; Shu and K.C. Chien and Peter Farrell Collaboratory, University of California, San Diego, La Jolla, CA, USA; Department of Radiology, University of California, San Diego, La Jolla, CA, USA; Department of*

Bioengineering, University of California, San Diego, La Jolla, CA, USA; Institute for Materials Discovery and Design, University of California, San Diego, La Jolla, CA, USA; Center for Engineering in Cancer, Institute for Engineering in Medicine, University of California, San Diego, La Jolla, CA, USA

QIWEN SU-TOBON • *Department of Chemistry, Boston College, Chestnut Hill, MA, USA*

PAN TAO • *State Key Laboratory of Agricultural Microbiology, Huazhong Agricultural University, Wuhan, Hubei, China; Hubei Hongshan Lab, Wuhan, Hubei, China; Cooperative Innovation Center for Sustainable Pig Production, College of Veterinary Medicine, Huazhong Agricultural University, Wuhan, Hubei, China*

WENLI TONG • *College of Life Science and Technology, Huazhong University of Science and Technology, Wuhan, China*

LUÍS FELIPE GALVÃO VALDIVIA • *Biochemistry Department, Institute of Chemistry, University of Sao Paulo, São Paulo, SP, Brazil*

THANDAVA VANAPILLI NURSIMULU • *Department of Chemistry, University of Toronto, Mississauga, ON, Canada*

TINA WANG • *Department of Chemistry, University of Wisconsin–Madison, Madison, WI, USA*

XINYA WANG • *College of Life Science and Technology, Huazhong University of Science and Technology, Wuhan, China*

YIYAO WANG • *State Key Laboratory of Virology and Biosafety, Wuhan Institute of Virology, Chinese Academy of Sciences, Wuhan, China; University of Chinese Academy of Sciences, Beijing, China*

RYAN M. WOLOSCHUK • *Janelia Research Campus, Howard Hughes Medical Institute, Ashburn, VA, USA*

G. ANDREW WOOLLEY • *Department of Chemistry, University of Toronto, Toronto, ON, Canada*

ZHUOHONG WU • *Department of NanoEngineering, University of California, San Diego, La Jolla, CA, USA; Center for Nano-ImmunoEngineering, University of California, San Diego, La Jolla, CA, USA; Shu and K.C. Chien and Peter Farrell Collaboratory, University of California, San Diego, La Jolla, CA, USA*

YINGQI XU • *College of Life Science and Technology, Huazhong University of Science and Technology, Wuhan, China*

HANG YANG • *State Key Laboratory of Virology and Biosafety, Wuhan Institute of Virology, Chinese Academy of Sciences, Wuhan, China; University of Chinese Academy of Sciences, Beijing, China; Hubei Jiangxia Laboratory, Wuhan, China*

LINWEI YANG • *Department of Chemistry, University of Wisconsin–Madison, Madison, WI, USA*

SHUN ZHANG • *Department of Pharmacology and Toxicology, University of Toronto, Toronto, ON, Canada*

MINGYUE ZHONG • *State Key Laboratory of Virology and Biosafety, Wuhan Institute of Virology, Chinese Academy of Sciences, Wuhan, China*

CHANG-LIANG ZHU • *State Key Laboratory and Institute of Elemento-Organic Chemistry, College of Chemistry, Frontiers Science Center for New Organic Matter, Nankai University, Tianjin, China*

Part I

Engineering Proteins and Enzymes

Chapter 1

High-Level Protein Expression in *Pseudomonas putida* Using the Phage-Based pPUT Expression System

Riet Dehaene, Jorien Poppeliers, Rob Lavigne and Maarten Boon

Abstract

High-yield expression systems, such as the phage T7 RNAP-based pET system in *Escherichia coli* have typically been restricted to a selected chassis organisms due to inherent toxicity. Indeed, emerging production hosts, such as *Pseudomonas putida*, show that traditional T7-based pET systems are detrimental to growth, limiting stable recombinant protein production in this organism. The recently developed phage phi15-based pPUT expression system addresses this, as it is fully adapted to this host, giving access to exceptional expression levels in *P. putida*. The pPUT system is highly modular, with elements to tweak stringency or decouple growth from production to further enhance yields. This chapter presents how to achieve this phi15-based expression system in *P. putida*, either for a strain of interest or dedicated strains for optimal expression. It outlines the workflow starting from initial selection of the modules and their use case. Once a system is selected, the workflow encompasses cloning gene(s) of interest into the expression vector pPUT-ST via SEVAtile assembly, transforming it to *P. putida* and subsequently introducing the selected modules. Once the strains are prepared, a lab scale expression of the gene of interest is achieved through a series of incubation and induction steps. The expression system has demonstrated up to fivefold increased protein yields compared other widely used systems and is both modular and tunable. It, thus, makes high-yield recombinant protein production available in *P. putida* for a wide range of applications and use cases.

Keywords *Pseudomonas putida*, Recombinant protein production, Fermentation, Phage-based, Microbial SynBio

1 Introduction

Efficient recombinant protein expression systems have revolutionized molecular biology, enabling the production of recombinant proteins at scales suitable for research, industrial applications, and therapeutic development. The T7 RNA polymerase (RNAP)-based expression in *Escherichia coli* has been central to this transformation, with the pET system as a hallmark [1, 2]. It offers tight regulation, high expression levels, and rapid induction and has become the gold standard for protein production due to its simplicity, efficiency, and

Huan Peng and S. Brian Kim (eds.), *Biological Macromolecule Engineering Technologies*,
Methods in Molecular Biology, vol. 3032, https://doi.org/10.1007/978-1-0716-5257-2_1,

compatibility with a wide range of genetic constructs. Recently, it was expanded upon by the introduction of a growth decoupler, enabling a switch between growth phase to production using the phage T7 gp2 protein to inhibit host RNAP, further increasing flexibility and protein yields in *E. coli* [3].

However, the reliance on *E. coli* as a host organism also presents limitations. Certain proteins, particularly those that are membrane-bound, toxic, or require specific post-translational modifications, are poorly expressed or misfolded in *E. coli* [4]. Moreover, the T7 RNAP system, while powerful in *E. coli*, is often cytotoxic when transferred to non-model organisms, including *Pseudomonas putida* [5]. This has hindered the broader application of high-yield expression systems in alternative bacterial hosts that offer unique metabolic capabilities and stress tolerances.

Pseudomonas putida has emerged as a promising chassis for synthetic biology and metabolic engineering due to its robustness, metabolic versatility, and ability to thrive in harsh industrial environments. This makes it ideally suited for waste stream valorisation, production of chemicals or fermentation conditions (e.g., biphasic) that are too harsh for *E. coli* [6]. Yet, the lack of efficient, tightly regulated, and high-yield expression systems has been a major bottleneck to introduce this production chassis for cost-efficient, recombinant protein production.

This method describes the use of the recently developed pPUT expression platform based on the phi15 bacteriophage RNAP, which addresses the intrinsic pET toxicity issues [7]. The system mimics the modularity and strength of the T7 system while overcoming its toxicity in *P. putida*. The core of the pPUT platform consists of an orthogonal phi15 RNAP under control of a 3-methylbenzoate (3mBz)-inducible XylS/Pm promoter and an optimised weak RBS (Fig. 1a, b). Upon induction, the phi15 RNAP drives expression from its cognate phi15 promoter, independent of host transcription. This promoter is situated on the pPUT-ST plasmid, dedicated to optimal gene of interest (GOI) expression. pPUT-ST is a *Tn7*-based suicide plasmid that delivers the expression cassette site-specifically at the *glmS* locus of the genome. The single-copy genomic integration allows for consistent results and removes the need to select on antibiotics for plasmid maintenance. In addition to the phi15 promoter, pPUT-ST is equipped with an optimised strong bicistronic ribosomal binding site, enabling high-level expression regardless of sequence context from the GOI, and dedicated twin terminators to completely insulate the element from genomic context while boosting upstream GOI expression. While the combination of pPUT-ST with the phi15 RNAP enables the highest level of expression, it can have low levels of leaky expression that can be detrimental when expressing toxic proteins. For applications where stringency is crucial, the system can be used with an additional element encoding an engineered phi15 lysozyme

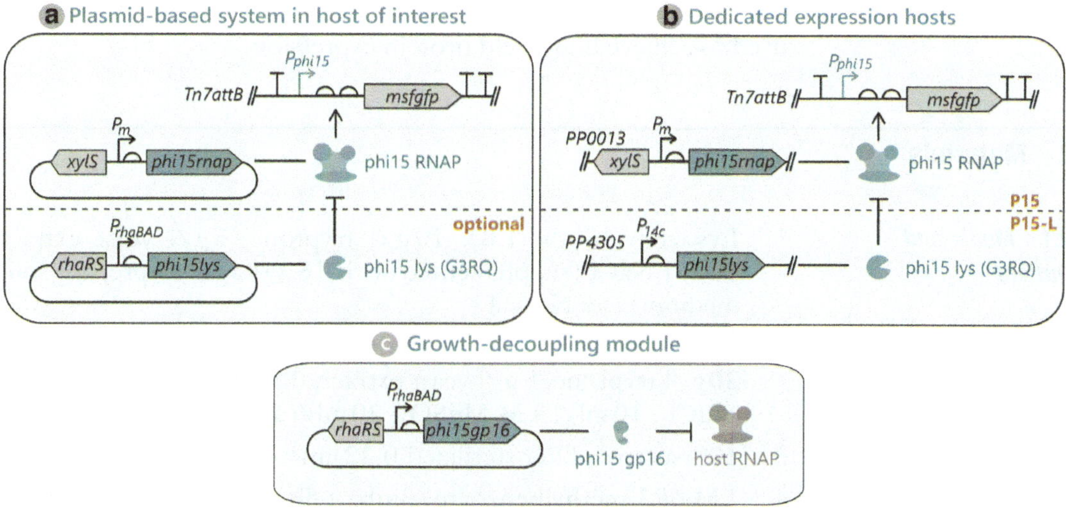

Fig. 1 Overview of the pPUT expression system. (**a**) The plasmid-based system is most suitable for expression in a *Pseudomonas* host of your choice. The simplest version consists of [1] the Tn7-based pPUT-ST plasmid, dedicated to optimal gene of interest (GOI) expression driven by the phi15 promoter, and [2] the pSTDesXa-phi15RNAP plasmid containing the orthogonal phi15 RNAP under control of a 3-methylbenzoate (3mBz)-inducible XylS/Pm promoter. When stringent regulation of expression is imperative, the optional lysozyme module [pSTDesR-p14c-BCD22-phi15lys(G3RQ)] can be included. (**b**) A set of dedicated expression hosts containing the different modules of the pPUT expression system readily integrated was developed in *P. putida* SEM11. Base strain P15 encodes the phi15-RNAP at locus PP0013, while strain P15-L has the additional phi15 lysozyme module at locus PP4305. (**c**) Both the plasmid-based system (without lysozyme module) and the dedicated expression hosts, can be supplemented with the optional growth-decoupling module (pSTDesR-phi15gp16). Upon induction, the phi15 gp16 inhibits the host RNAP, thereby arresting the growth phase and redirecting the cell to invest in expression of the gene of interest (production phase). (Figure adapted from [7])

(Fig. 1a, b). This lysozyme will inhibit low levels of phi15 RNAP that may be present in the non-induced cell due to leaky expression. This addition makes the system highly stringent, while minimizing impact on the maximum expression levels. Lastly, for applications requiring complete separation of the initial growth phase from the subsequent protein production phase (i.e., to siphon more cellular resources to protein production), a plasmid-based phi15 gp16 inhibitor of the host RNAP can be expressed from plasmid pSTDesR under rhamnose control (Fig. 1c). This inhibits host transcription of host genes at the time of interest, leaving only GOI transcription by the phi15 RNAP and resulting in higher yields of these expressed proteins per cell. All combined, the full pPUT expression system has demonstrated up to fivefold yield increases compared to widely adopted expression systems and can thus bring high-yield protein expression to *P. putida* [7].

This chapter outlines the implementation of this phi15-based expression system for GOI expression in *P. putida*. It includes considerations to be made in module selection, genetic construction, integration of systems in the host strain, and induction protocols used to achieve high-yield protein expression.

2 Materials

2.1 Media and Buffers

1. Lysogeny broth (LB): 10 g/l tryptone, 5 g/l yeast extract, 10 g/l NaCl (supplemented with 15 g/l agar to prepare solid medium) (*see* **Note 1**)
2. Super optimal broth with catabolite repression (SOC) medium: 20 g/l tryptone, 5 g/l yeast extract, 0.5 g/l NaCl, 10 ml/l 1 M $MgCl_2$, 10 ml/l 1 M $MgSO_4$, 20 ml/l 20% glucose
3. 10% sucrose; filter sterilized 0.22 μm
4. 1 M of 3-methylbenzoate (3mbz); filter sterilized 0.22 μm
5. 1 M rhamnose; filter sterilized 0.22 μm

2.2 Strains, Plasmids, and DNA

1. Chemically competent *E. coli* PIR2 cells (e.g., One Shot™ PIR2 Chemically Competent *E. coli*, Thermo Fisher Scientific)
2. Expression host of choice (see Sect. 3.1, e.g., *P. putida* P15, P15-L, or own host)
3. Template GOI DNA
4. Plasmid vectors (Table 1)

2.3 Molecular Biology

PCR reagents:

1. Phusion Polymerase and 5× HF Phusion buffer
2. DreamTaq polymerase and 10× Green DreamTaq buffer
3. dNTPs
4. Tile primers (100 μM) (see Sect. 3.2.1)
5. Plasmid primers (100 μM):
   a. pPUT_F: GCGTTCGGTCAAGGTTCTGGAC
   b. pPUT_R: GCGGCAACCGAGCGTTC
   c. glmS_UP: ACCATTCCGCTGCAGTTG
   d. Tn7TR: CACAGCATAACTGGACTGATTTC

   Gel electrophoresis:
6. Agarose 1%
7. 50 μg/ml ethidium bromide
8. DNA ladder (e.g., GeneRuler 1 kb DNA ladder, GeneRuler 100 bp ladder)

   BsaI restriction enzyme (20 U/μl)
   T4 DNA ligase (5 U/μl) and 10× T4 DNA ligation buffer

Table 1
Overview of plasmid vectors used in the pPUT expression method

Plasmid vectors	Characteristics	Source
pPUT-ST	pBG derivative with LUZ7T50-Pphi15BCD05-AATG-P14g-BCD2-msfGFP-TAAC-phi15T5T1 insertion; AATG-P14g-BCD2-msfGFP-TAAC is flanked with BsaI recognition sites and can be substituted for a desired insert; oriT; oriV(R6K); Tn7L and Tn7R extremes; Km^R/Gm^R	[7]
pTNS2	Helper plasmid; oriV(R6K); tnsABCD; Amp^R	[11]
pSTDesXa-phi15RNAP	pRO1600 ori, ColE1 ori, oriT, Kan^R, XylS-Pm expression; SacB cassette is substituted with phi15RNAP	[7]
pSTDesR-p14c-BCD22-phi15lys(G3RQ)	pSTDesR derivative in which the RhaRS/pRhaBAD inducible module was replaced by a weak constitutive promoter (p14c) and weak RBS (BCD22); oriT; oriV(RK2); p14c-BCD22→MCS; SmR/SpR; MCS is substituted with phi15lysozyme(G3RQ)	[7]
pSTDesR-phi15gp16	oriT; oriV(RK2); RhaRS/pRhaBAD→MCS; SmR/SpR; MCS is substituted with phi15gp16	[7]

2.4 Commercial Kits

1. PCR purification kit (e.g., GeneJet PCR purification kit, Thermo Fisher Scientific)
2. Miniprep kit for plasmid extraction (e.g., GeneJet Plasmid Miniprep kit, Thermo Fisher Scientific)

2.5 Equipment and Materials

1. Incubators at 37°C and 30°C (static and shaking)
2. PCR thermocycler
3. Gel electrophoresis tray and combs
4. UV-transilluminator
5. UltraBright blue light transilluminator
6. Electroporator
7. 0.2 cm electroporation cuvette
8. Eppendorf tubes (0.5 and 1.5 ml)
9. A tabletop centrifuge
10. Glass test tubes
11. Erlenmeyer (250 ml)

12. Falcon tubes (50 ml)
13. Syringe filters (0.22 µm)
14. Water bath
15. OD_{600} reader
16. Microvolume spectrophotometer (e.g., simplinano)
17. Freezers (−20°C and −80°C)
18. Fridge (4°C)
19. 10 mm petri dishes
20. Micropipettes (single and multichannel)
21. Sterile pipette tips

3 Methods

3.1 Selection of Strains and Modules

Due to the modularity of the pPUT platform, there are multiple options to approach expression (Fig. 1). Several aspects can be kept in mind to select the system most suited for your application.

3.1.1 Plasmid-Based vs Dedicated Expression Host

Expression can be done either with plasmid-based phi15 modules, or with a dedicated expression strain that has them readily integrated.

Plasmid Vector Modules

The plasmid system (Fig. 1a) is most suitable when expression is desired in a *Pseudomonas* host of your choice as it is readily transferrable to a new host. However, it comes with specific points-of-attention as it will require selection on (multiple) antibiotics, can have more variation in expression levels due to plasmid copy number variations, and will result in somewhat reduced yields compared with the dedicated strain, due to increased cellular burden and potentially suboptimal phi15 RNAP or lysozyme levels. Furthermore, in the current version of the system, it is not possible to combine both lysozyme and growth-decoupling modules together (see Sects. 3.1.2 and 3.1.3) with the plasmid vector system due to plasmid incompatibility.

Dedicated Expression Hosts

A set of dedicated expression hosts containing one or multiple modules was established in *Pseudomonas putida* SEM11 (Fig. 1b). In these strains, the modules have been integrated into the host genome at specific loci that have been selected to maximise expression levels. The base strain (P15) encodes the phi15-RNAP at locus PP0013, while strain P15-L has the additional phi15 lysozyme module at locus PP4305 [7]. The advantage of using a dedicated host is that it is fully optimised, has consistent expression results and does not require antibiotic selection, resulting in higher expression levels and assuages concerns for industrial fermentations that require the absence of antibiotics. Moreover, both hosts are compatible with the optional growth-decoupling module. However, it restricts the user to a fixed host that may not have all the desired features for the expression of interest (*see* **Note 2**).

3.1.2 Optional Lysozyme Module

The phi15 lysozyme module enables more stringent expression control as it represses base levels of phi15 RNAP. This is highly beneficial for the expression of toxic proteins or proteins placing a metabolic burden on the cell. However, while the system with this module still reaches high expression levels, the lysozyme will impact the maximal expression levels that can be achieved as some phi15 RNAP is repressed.

3.1.3 Optional Growth-Decoupling Module

For expressions where decoupling of the growth phase from the production phase is desired, the phi15 gp16 (a T7 gp2 homologue for *P. putida*) is available on plasmid pSTDesR (Fig. 1c). This module can increase yields per cell as the host no longer invests in growth but rather on recombinant protein production upon induction with rhamnose. However, it does require selection with an additional antibiotic and is not compatible with the plasmid-based phi15 lysozyme module.

3.2 Cloning Your Gene-of-Interest (GOI) in pPUT

Regardless of strains and modules that are selected, the pPUT-ST vector is at the heart of the expression system. It is required for integration of the GOI into the *glmS* locus and has all the necessary elements for optimal expression, initiated from the phi15 promoter. pPUT-ST is designed for cloning with SEVAtile, a Golden Gate-based cloning method compatible with SEVA plasmids [8, 9]. It enables scarless introduction of one or more DNA elements (SEVAtiles) simultaneously into the vector as each SEVAtile is flanked by a unique four-nucleotide position tag (PT) that guides their ligation towards a predefined sequence (Fig. 2). First, the method uses tail-PCR to introduce the position tag and BsaI restriction sites to the tile of interest. Next, the tile is assembled in the pPUT-ST backbone. Finally, the assembled constructs are transformed into *E. coli* for propagation and sequence verification.

3.2.1 Tile Creation

1. Design two primers to amplify the tile with your GOI. Each primer starts with three random bp to allow annealing of the BsaI restriction enzyme, followed by a BsaI restriction site to clone the tile in the pPUT-ST destination vector (*see* **Note 3**). The restriction site is then followed by a position tag (PT; in [square brackets]) to enable assembly of multiple parts in a one pot reaction. Template oligos (*see* **Notes 4** and **5**):
   a. Forward: 5′-NNN GCTCTTC N AGA GGTCTC [NAATG] sequence GOI-3′
   b. Reverse: 5′-NNN GCTCTTC N CTT GGTCTC [NGTTA] sequence GOI-3′

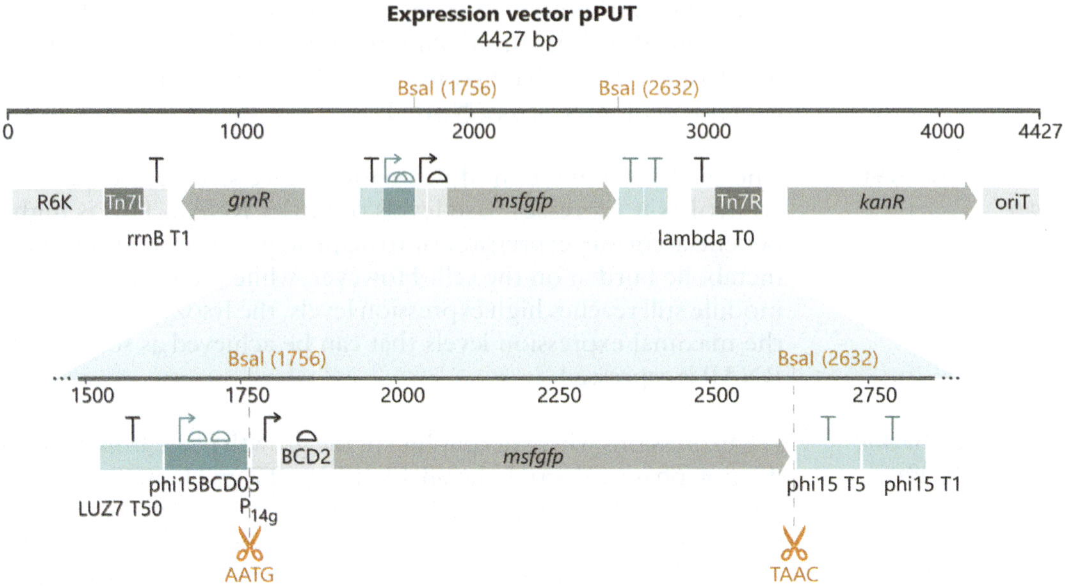

Fig. 2 Linear vector map of the pPUT-ST plasmid with R6K and oriT origins (R6K and oriT), kanamycin and gentamycin resistance markers (kanR and gmR), transposase Tn7 recognition sites (Tn7L and Tn7R), transcriptional terminators rrnB T1, LUZ7 T50, phi15 T5, phi15 T1 and lambda T0, phi15 promoter region phi15-BCD05, and a removable fluorescence cassette for cloning (P14g-BCD2-msfgfp) flanked with BsaI recognition sites. The pPUT-ST vector is compatible with SEVAtile shuffling (ST) and Golden Standard (GS) assembly (level 1). When cut with restriction enzyme BsaI, this results in specific overhangs (position tags) AATG (L) and TAAC (R). (Figure adapted from [7])

Amplify and purify the tiles as described in chapter 'Fluorescence-based assays to evaluate the functionality of phage-derived regulatory elements for applications in synthetic biology', step *Tile creation*.

3.2.2 Shuffling Tile(s) in the pPUT-ST Destination Vector and Vector Propagation

1. Shuffle the tile(s) in the pPUT-ST destination vector and propagate it as described in chapter 'Fluorescence-based assays to evaluate the functionality of phage-derived regulatory elements for applications in synthetic biology'.

3.3 Vector Assessment

1. Place the plate on a blue light transilluminator and indicate colonies that do not show green fluorescence (*see* **Note 6**).
2. Perform a colony PCR (cPCR) on eight indicated colonies to analyse the results (*see* **Note 7**).
   a. Reaction mixture: 1 colony, 0.5 µl of dNTPs, 0.5 µl of each plasmid primer (20 µM; pPUT_F and pPUT_R), 2.5 µl of Green Dream Taq buffer, and 0.25 µl of DreamTaq polymerase in a total volume of 25 µl.
   b. Cycling conditions: initial denaturation for 10 min at 95°C; 30 cycles of 30 s denaturation at 95°C, 30 s primer

annealing at 58.5°C, and 1 kb/min extension at 72°C; perform the final extension for 10 min at 72°C and hold at 12°C.

3. Verify the length of the PCR product by performing a gel electrophoresis using a 1% agarose gel. A DNA ladder can be used to assess the size of the PCR product (*see* **Note 8**).
4. Check your samples for correct, mutation-free assembly by Sanger Sequencing.
5. Inoculate a correct clone from the plate into 4 ml LB with Kan^{50} and incubate ON at 37°C.
6. Purify the plasmid with a plasmid purification kit (*see* **Note 9**).

3.4 Integration of the pPUT Expression Cassette in Host of Choice

To enable expression in *P. putida*, the expression cassette needs to be transferred from the suicide pPUT-ST plasmid to the *glmS* gene locus (*see* **Note 10**). This is achieved through co-electroporation with a secondary pTNS2 suicide plasmid that encodes all necessary transposase genes. Once introduced, transposase genes will be expressed and transpose the expression cassette encoded between Tn7-sites and the target locus. Insertion is selected for by a gentamycin resistance gene encoded on the transferred cassette and cassettes are checked for correct insertion by cPCR.

3.4.1 Transformation of P. putida by Electroporation

1. Make a 4 ml overnight culture of the strain of interest in LB.
2. Two times 1.5 ml of culture is transferred into 2 ml Eppendorf tubes.
3. Centrifuge 2 min (10,000 ×*g*) in tabletop centrifuge.
4. Remove the supernatants and dissolve in 1 ml of 10% sucrose solution.
5. Centrifuge 2 min (10,000 ×*g*).
6. Repeat Steps 4 and 5 two more times (three wash steps with 10% sucrose in total).
7. Remove the remnant supernatants with a pipette.
8. Resuspend the pellets in a single volume of 100 μl 10% sucrose (100 μl per two tubes).
9. Add 300 ng of pPUT-ST plasmid and 500 ng of pTNS2 (*see* **Note 11**).
10. Incubate for 5 min (room temperature) and transfer the mixture to a 0.2 cm cuvette.
11. Perform the electroporation (2.5 kV, 200 W, 25 μF, 1 pulse) (*see* **Note 12**).
12. Immediately add 450 μl SOC medium and incubate at 30°C (shaking).
13. Plate on LB containing Gm^{10} and incubate overnight at 30°C.

3.4.2 Verification of Correct Chromosomal Integration

Per clone two cPRCs are used to verify correct integration and presence of the integration cassette (*see* **Note 13**). The primers for the integration check bind on host genome on the one hand and on the integration cassette on the other hand, to ensure correct site and orientation of integration, while primers for checking the insert are used to check completeness of the integrated expression cassette.

1. Perform two different cPCRs on eight colonies as described in Sect. 3.3 Step 2 using following primer sets and settings (*see* **Notes 7** and **14**):
   - Integration check: glmS_up (on host genome) and Tn7TR (on integrated cassette); annealing temperatures 53°C; 30 s elongation → expected band of 235 bp.
   - Full insert check: same primers and settings as in Step 2 of Sect. 3.3.
2. Verify the length of the PCR product by performing a gel electrophoresis using a 1% agarose gel. A DNA ladder can be used to assess the size of the PCR product (*see* **Note 10**).

3.5 Equipping Your Host with the Desired (Stringent Expression or Growth Decoupling) Modules

When plasmid-based modules of phi15 RNAP or lysozyme are required (when not using P15 or P15-L expression strains), or when a growth-decoupler is required in your final expression strain, these modules can be added to your strain by electroporation after pPUT-ST integration. For each required modules, the reader can follow the same transformation procedure as described in Sect. 3.4.1. However, in Step 9 only 20–50 ng plasmid and no pTNS2 is required as these modules do not integrate and will need to be maintained by antibiotic selection (*see* **Note 15**). Furthermore, the antibiotic used for plasmid selection in Step 13 will differ depending on the module (i.e., kanamycin (50 μg/ml) for pSTDesXa-phi15RNAP, and streptomycin (200 μg/ml) for both pSTDesR-p14c-BCD22-phi15lys(G3RQ) and pSTDesR-phi15gp16).

3.6 phi15-Based Expression of GOI

The pPUT expression system is both robust and flexible with many variables that can be tweaked to optimise yields. The protocol outlined in this section is a guideline that has led to successful high-yield protein expressions at lab scale [7], though it can be adapted towards the readers GOI and application to further enhance yields. It involves growing cells until exponential phase, induction of the phi15 RNAP-based expression with 3mBz and finally, optional induction of a growth decoupler.

1. Make a 4 ml overnight culture of the strain of interest in LB with appropriate antibiotics (if any) (*see* **Note 16**)
2. Dilute in 50 ml fresh LB medium with appropriate antibiotics to an OD_{600} of 0.1 in a 250 ml shake flask.

3. Grow at 30°C with shaking at 180 rpm to an OD_{600} of 0.5.
4. Induce the phi15 RNAP with 0.3 mM 3 mBz and incubate for 20 h (*see* **Notes 17** and **18**).
5. Transfer the cells to a 50 ml falcon tube and centrifuge at 4500×*g* for 10 min at 4°C.
6. Cells can then be processed for purification of protein of interest.

4 Notes

1. To ensure plasmid selection, media were supplemented with the appropriate antibiotic: gentamycin (Gm; 10 μg/ml), kanamycin (Kan; 50 μg/ml), or ampicillin (Amp; 100 μg/ml), depending on the resistance marker encoded by the plasmid. The specific antibiotic associated with each plasmid is listed in Table 1.
2. For standard protein expressions in *P. putida*, we recommend the P15 strain without additional modules as a starting point.
3. When BsaI restriction sites are present, Gibson assembly can be used as an alternative [10].
4. The standard PT of pPUT-ST is NAATG and NTAAC (5' to 3'). The ATG and TAA in these PT are part of the coding sequence of interest to avoid scars and give an optimal distance of the start codon to the RBS.
5. It is possible to integrate multiple consecutive tiles in a single reaction (e.g., for multigene clusters). The reader is referred to chapter 'Fluorescence-based assays to evaluate the functionality of phage-derived regulatory elements for applications in synthetic biology' for an example on multitile assembly or to the original SEVAtile article for further details [9]. Alternative pPUT plasmids are available with different PT to enable integration of different elements [7].
6. This step is optional, but increases chance of selecting good clones due to selecting for clones that have msfGFP cut out of the cloning site.
7. To keep track of which colonies are screened using cPCR, the colonies that are picked can be plated to an agar plate grid where each square in the grid refers to the position in the PCR plate.
8. Note that unedited pPUT-ST vector has an msfGFP insert that can generate a band of a 'typical' gene generating a total expected band size of 1594 bp in unedited plasmids. Edited plasmids have a 719 bp size + the size of GOI.

9. pPUT-ST has a low copy number in PIR2 cells (15–20 copies). Generating larger or multiple smaller volumes of overnight cultures prior to plasmid purification, and loading them on the same silica column of your purification kit can increase yield.
10. While the *glmS* locus is quite conserved, some strains can have it at different locations or have multiple copies. When using your own strain, check if the Tn7 recognition site at the *glmS* locus is present.
11. Transformation efficiencies can vary between *P. putida* strains. If it fails at these plasmid concentrations, it can help to increase pTNS2 concentrations up to 1 μg. However, the volume of plasmid should be kept below 10 μl to avoid electrical arcing during electroporation.
12. Time constants between 4.5 and 5.5 are generally good indicators of successful electroporation.
13. While rarely occurring, we recommend whole genome sequencing of the strain to identify additional mutations.
14. As an alternative for picking the same colony twice—once for each cPCR—(certainly for small colonies), you can first dissolve the colony in 25 μl mQ and use 2.5 μl of that in each PCR reaction.
15. Co-electroporation of modules is possible but not recommended as it can have reduce transformation efficiency.
16. Alternative media can be used. pPUT is robust across different media, with documented success in LB, M9 minimal medium, and De Bont minimal medium.
17. When using the growth decoupler, this is a good time point to induce with 10 mM rhamnose as well, to combine protein expression and growth halt in one step. Tweaking the time of halting growth can be done to optimize yields further.
18. Incubation times can be modified to optimise yield for your specific application.

References

1. Studier FW (1991) Use of bacteriophage T7 lysozyme to improve an inducible T7 expression system. J Mol Biol 219:37–44
2. Rosenberg AH, Lade BN, Dao-shan C et al (1987) Vectors for selective expression of cloned DNAs by T7 RNA polymerase. Gene 56:125–135
3. Stargardt P, Feuchtenhofer L, Cserjan-Puschmann M et al (2020) Bacteriophage inspired growth-decoupled recombinant protein production in *Escherichia coli*. ACS Synth Biol 9:1336–1348
4. Bhatwa A, Wang W, Hassan YI et al (2021) Challenges associated with the formation

of recombinant protein inclusion bodies in *Escherichia coli* and strategies to address them for industrial applications. Front Bioeng Biotechnol 9:630551

5. Lammens E-M, Feyaerts N, Kerremans A et al (2023) Assessing the orthogonality of phage-encoded RNA polymerases for tailored synthetic biology applications in Pseudomonas species. Int J Mol Sci 24:7175
6. Weimer A, Kohlstedt M, Volke DC et al (2020) Industrial biotechnology of *Pseudomonas putida*: advances and prospects. Appl Microbiol Biotechnol 104:7745–7766
7. Lammens E-M, Volke DC, Kerremans A et al (2025) Engineering a phi15-based expression system for stringent gene expression in *Pseudomonas putida*. Commun Biol 8:171
8. Martínez-García E, Fraile S, Algar E et al (2023) SEVA 4.0: an update of the Standard European Vector Architecture database for advanced analysis and programming of bacterial phenotypes. Nucleic Acids Res 51:D1558–D1567
9. Lammens E-M, Boon M, Grimon D et al (2022) SEVAtile: a standardised DNA assembly method optimised for *Pseudomonas*. Microb Biotechnol 15:370–386
10. Gibson DG, Young L, Chuang R-Y et al (2009) Enzymatic assembly of DNA molecules up to several hundred kilobases. Nat Methods 6:343–345
11. Choi K-H, Schweizer HP (2006) mini-Tn7 insertion in bacteria with single attTn7 sites: example *Pseudomonas aeruginosa*. Nat Protoc 1:153–161

Check for updates

Chapter 2

New-to-Nature Photoenzymatic Catalysis

Chang-Liang Zhu, Jie Feng and Mao-Lin Li

Abstract

Photoenzymatic catalysis offers a novel methodology to control chemoselectivities and enantioselectivities in photochemistry and biocatalysis, is establishing its standing as an emerging frontier in synthetic chemistry. Herein, we describe a fundamental experimental protocol for preparing and purifying enzymes form *Escherichia coli* BL21(DE3), including cloning and expression, purification, concentration determination of enzymes, and an exemplar reaction for photoenzymatic catalysis. This methodology can offer constructive proposals and practical guidance for researchers in synthetic biology.

Keywords Photoenzymatic catalysis, Synthetic biology, New-to-nature

1 Introduction

The synergistic integration of the high reactivity from photochemistry and the exceptional selectivity from biocatalysis, offering highly promising strategies for addressing challenges in enantioselective radical reactions and exploring new-to-nature enzyme reactivity [1]. Photoenzymatic catalysis cooperativity strategy overcomes the limitations inherent in natural enzymatic catalysis and addresses the challenge of radical stereocontrol in photochemistry, particularly enabling enantioselective control at remote stereogenic centers. [2]. Current research primarily focuses on exploiting light-regenerated cofactors to maintain native enzymatic catalytic activity [3, 4]. Furthermore, enzymes can be engineered through directed evolution or protein engineering to precisely control reaction intermediates and chiral recognition processes, thereby achieving a highly chemoselectivity and enantioselectivity [5, 6].

More studies have revealed that various natural and engineered oxidoreductases, transferases, and cyclases can photochemically generate radicals from unnatural precursors or facilitate reactions between photochemically generated radicals and ground-state enzymatic intermediates [7]. These strategies have enabled numerous challenging enantioselective transformations, for example, hydroalkylation,

Huan Peng and S. Brian Kim (eds.), *Biological Macromolecule Engineering Technologies*,
Methods in Molecular Biology, vol. 3032, https://doi.org/10.1007/978-1-0716-5257-2_2,

hydroamination, hydrosulfonation, acylation, reductive, and oxidative couplings [8–15]. Although significant progress has been achieved in photoenzymatic catalysis, many of its reactivities remain undiscovered due to limitations in reaction types and enzyme diversity [7]. For researchers entering enzyme catalysis, it is essential not only to master multidisciplinary knowledge, including organic synthetic methodologies, photochemistry, and enzymology, but more importantly, to acquire fundamental experimental techniques.

In this chapter, we present an experimental methodology of how to preparate and purify enzymes, including photoenzymatic catalyzed reaction, with a fundamental specialized perspective on photoenzymatic systems. The resulting experimental protocol will enable researchers across disciplines to expedite the design and optimization of photoenzymatic systems and facilitate the discovery of new reactivities.

2 Materials

2.1 Solutions

1. Binding buffer: 50 mM NaH_2PO_4, 300 mM NaCl, and 10 mM imidazole in 1.0 L deionized H_2O, then adjust pH to 8.0 with NaOH.
2. Wash buffer: 50 mM NaH_2PO_4, 300 mM NaCl, and 10 mM imidazole in 1.0 L deionized H_2O, then adjust pH to 8.0 with NaOH.
3. Elution buffer: 50 mM NaH_2PO_4, 300 mM NaCl, and 250 mM imidazole in 1.0 L deionized H_2O, then adjust pH to 8.0 with NaOH.
4. Phosphate buffer (NaPi buffer): 5.35 g Na_2HPO_4 and 1.475 g NaH_2PO_4 dissolved in 1.0 L deionized H_2O, then adjust pH to 8.0 with NaOH.
5. 20 wt% SDS: 20.0 g sodium dodecyl sulfate dissolved in 180 g H_2O.
6. High-affinity Ni-charged resin FF (GenScript biotech corporation).
7. Ultrafiltration spin columns (15 mL, 30 kDa, PES) (Beyotime Biotechnology).
8. Acetate buffer: 12.0 mmol sodium acetate in 100 mL deionized H_2O, then adjust pH to 5.5 with acetic acid.
9. Citrate buffer: 12.0 mmol sodium citrate in 100 mL deionized H_2O, then adjust pH to 5.5 with citric acid.
10. GDH (L-glutamate, >60 U/mg): ~5 U in 20 μL 20% glycerine, and NaPi buffer (pH = 7.4).

11. NADP+ solution: 0.02 mmol NADP+ in 2 mL deionized H_2O.
12. Glucose solution: 1.0 mmol in 2 mL deionized H_2O.
13. α-methyl styrene solution: 0.2 mmol α-methyl styrene in 2 mL DMSO.
14. 2-bromomethyl pyridine solution: 0.4 mmol 2-bromomethyl pyridine hydrobromide in 2 mL deionized H_2O.
15. 1,1,1-trifluoro-2-iodoethane solution: 0.4 mmol 2-bromomethyl pyridine hydrobromide in 2 mL deionized H_2O.

2.2 Equipment

1. Hei-PLATE Mix'n'Heat Core+ Φ145.
2. Mikrouna Universal (2440/750/900) Glovebox (Fig. 1).
3. Parallel light reactor: SSSTECH-AF3 (Fig. 2).

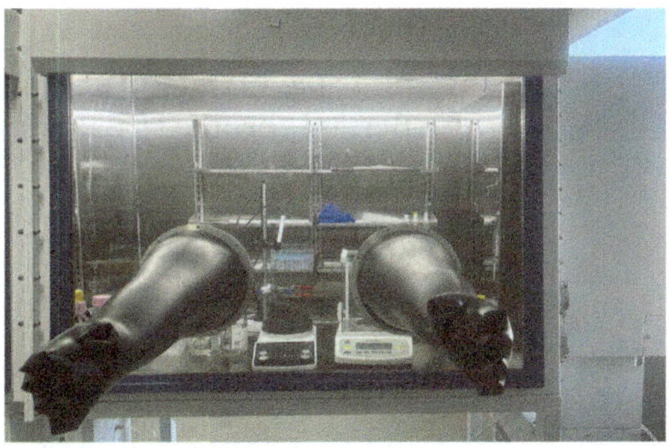

Fig. 1 Mikrouna universal glovebox

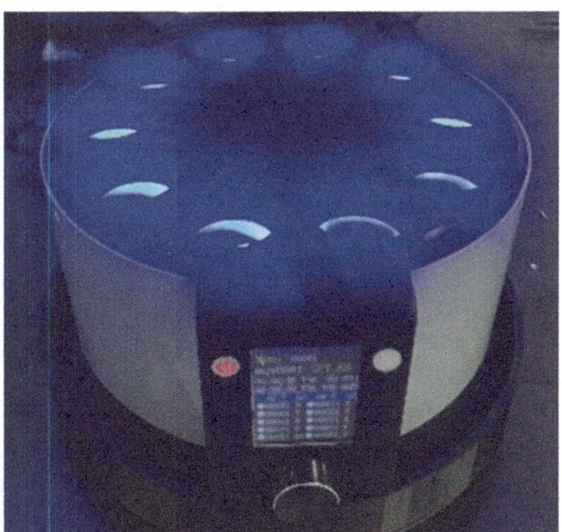

Fig. 2 Parallel light reactor AF-3

4. GC-MS: Agilent 8890 GC system and Agilent 5977C GC/MSD, GC: Agilent 8890 GC system.
5. HPLC: Agilent 1260 Infinity III.

3 Methods

3.1 Cell Culture (OYE1/OYE3)

1. Transfer 50 μL of competent *Escherichia coli* BL21(DE3) into a 1.5 mL tube, and then add 2 μL (2–5 ng) pET30a_OYE1/OYE3 plasmid using a micropipette. Incubate on ice for 30 min.
2. Perform heat shock at 42°C (water bath) for 30–45 s, then return to ice for 2 minimmediately, and add 1 mL Luria–Bertani broth (LB) to incubate at 37°C, 250 rpm for 1 h.
3. Centrifuge at 10,000 rpm for 1 min in a benchtop centrifuge. Discard supernatant, retaining ~50 μL cell suspension. Spread plate and incubate at 37°C for 12 h.
4. pET30a_OYE1/OYE3 was inoculated into 5 mL of Luria–Bertani broth (LB) containing 50 μg/mL kanamycin and grown overnight at 37°C, 250 rpm.
5. The seed cultures were then inoculated into 500 mL of fresh TB medium (1%, v/v) containing 50 μg/mL kanamycin, which were grown at 37°C, 250 rpm until OD600 up to ~0.6.
6. Protein expression was subsequently induced with the addition of 0.1 mM isopropyl β-D-thiogalactoside (IPTG). The induced cultures were placed at 25°C, 250 rpm, for 16 h for protein production.

Cells and media:
Escherichia coli BL21 (DE3) (Ranjeck Technology Co., Ltd.)
Tryptone (Shanghai Sangon Biotech Co., Ltd. CAS number: 73049-73-7)
Yeast Extract (Shanghai Sangon Biotech Co., Ltd. CAS number: 8013-01-2)
Agarose (Bide Pharmatech Co., Ltd. CAS number: 9012-36-6)
Agar (Shanghai Biyuntian Biological Co., Ltd.)
Riboflavin 5′-monophosphate sodium salt (CAS number: 130-40-5)
Kanamycin (CAS number: 25389-94-0)
IPTG (isopropyl β-D-thiogalactoside, CAS number: 367-93-1)
Luria–Bertani broth medium: 10.0 g tryptone, 5.0 g yeast extract, and 10.0 g NaCl in 950 mL deionized H_2O.
Terrific broth (TB) medium: 20.0 g tryptone, 24.0 g yeast extract, 4.0 mL glycerol in 900 mL deionized H_2O, and 100 mL 0.017 M KH_2PO_4 and 0.072 M K_2HPO_4.

50 µg/mL kanamycin (1000×): 4.0 g kanamycin dissolved in 80 mL H_2O.

100 mg/mL ampicillim (1000×): 8.0 g ampicillim dissolved in 80 mL H_2O.

0.1 M IPTG: 1.906 g IPTG dissolved in 80 mL H_2O

3.2 Purification Protocol for His-Tagged Protein

1. Following expression, *E. coli* BL21 (DE3) cells were collected by centrifugation (6000 rpm for 10 min) and resuspended with NaPi buffer, pH = 7.4.
2. For the purifications of protein, FMN-Na (1 mg/mL) was added to the cell suspension. The mixture was then lysed by sonication and clarified by centrifugation (4°C, 13,000 rpm for 60 min).
3. 2 mL (50 mg/mL in 20% ethanol) Ni-NTA resin add to 20 mL gravity column, then using ice cold Binding buffer to remove ethanol. Subsequently, Ni-NTA resin add to the protein with binding buffer, and incubating resin with protein in binding buffer for ≤10 min on ice.
4. The Ni–enzyme solution pour into gravity column rapidly. After completing drainage, washing with 5 mL binding buffer first, then wash buffer (20 bed volumes).
5. Elute into ultrafiltration spin columns (15 mL, 30 kDa) with elution buffer (10 bed volumes) until yellow color disappears.
6. Exchange with phosphate buffer (NaPi buffer) to remove imidazole at 4°C, 4000 rpm for 30 min, and repeat four times.
7. The resulted proteins were stored in 20% glycerol at −20°C. The purity of the proteins was determined by denaturing sodium dodecyl sulfate (SDS)-PAGE.

3.3 Concentration Determination of Enzymes

A 10 µL enzyme solution is added to an aqueous solution containing 980 µL water and 10 µL 10%wt% sodium dodecyl sulfate solution. The mixture is vortexed for 1 min and the absorbance at 446 nm ($\varepsilon_{446\,nm} = 12{,}200\ M^{-1}\cdot cm^{-1}$) is taken to calculate the concentration of the enzyme using the equation below:

$$c = \frac{Abs}{\varepsilon l}$$

3.4 Exemplar Photoenzymatic Reaction

All the photoenzymatic catalytic reactions introduced in this chapter must be performed in a strictly oxygen-free environment to ensure the smooth progress of the reactions or to avoid interference from oxygen, and the detail information as follows:

3.4.1 Analysis Methods for Photoenzymatic Reaction 1 [12]

As shown in Fig. 3, (1) in a nitrogen-filled glovebox, the acetate buffer (120 mM, 800 μL), GDH (~5U in 20 μL), NADP+ solution (10 mM, 20 μL), glucose solution (500 mM, 40 μL), 2-bromomethyl pyridine solution (200 mM, 40 μL), and α-methyl styrene DMSO solution (100 mM, 40 μL) are introduced into a 4 mL glass. (2) OYE1 (1 mol%, 40 μL) is added into the vial, the mixture was stirred at 25°C. (3) The reaction mixture is subsequently loaded on the following photoreactor and illuminated with 15 W 440 nm blue LED at 25°C for 12 h. (4) Afterward, dodecane and 1 mL saturated Na_2CO_3 are added to the reaction mixture, which was extracted with 2 × 1 mL EtOAc and characterization.

Analysis Methods

GC–MS: The target compound was verified by GC–MS based on its m/z.

The GC yield was calculated according to the calibration curve constructed with pure racemates of the products.

HPLC condition: Chiralcel OJ-H column (25 cm × 0.46 cm ID), hexane/2-propanol = 90:10, flow rate = 1.0 mL/min, 210 nm UV detector, t_R = 7.12 min (minor), and t_R = 9.00 min (major).

3.4.2 Analysis Methods for Photoenzymatic Reaction 2 [13]

As shown in Fig. 4, (1) in a nitrogen-filled glovebox, the citrate buffer (120 mM, 800 μL), GDH-105 (approx. 5 U, 20 μL), NADP+ (10 mM, 20 μL), glucose (500 mM, 40 μL), 1,1,1-trifluoro-2-iodoethane (800 mM, 40 μL), and α-methyl styrene in DMSO (100 mM, 40 μL) are sequentially introduced to a 4 mL glass vial. (2) To this mixture, OYE3 (1 mol%, 40 μL) is added, and the contents are stirred at 25°C. (3) The mixture is then subjected to illumination using a photoreactor equipped with a 540 mW, 440 nm blue LED, maintained at 25°C for 12 h. (4) Post-illumination, $PhCF_3$ (100 mM, 40 μL) is introduced, and the solution is extracted using 2 × 0.6 mL $CDCl_3$ for 19F NMR analysis on an Agilent 600 MHz spectrometer to determine yield. For optional GC-MS characterization, dodecane and 1 mL of saturated Na_2CO_3 are added, and the solution is extracted with 2 × 0.8 mL EtOAc before GC-MS analysis.

Analysis Methods

GC-MS: The target compound was verified by GC-MS based on its m/z.

The GC yield was calculated according to the calibration curve establish with pure racemates of the products.

GC condition: Chirasil-Dex CB, inject temperature = 250°C, detector temperature = 300°C, inlet pressure = 17.7 psi, flow

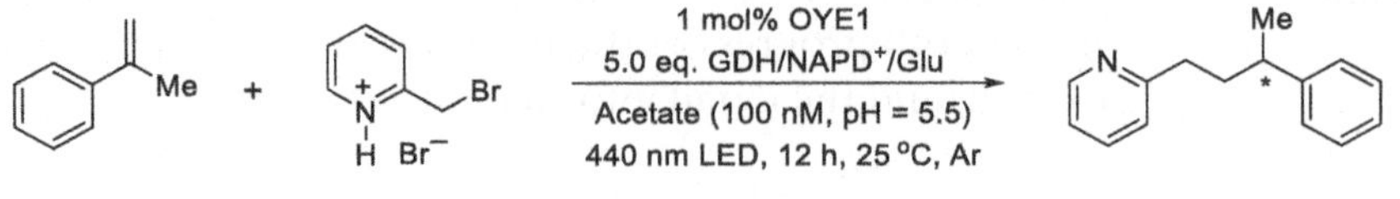

Fig. 3 Enantioselective radical hydroalkylation reaction with an azaarene

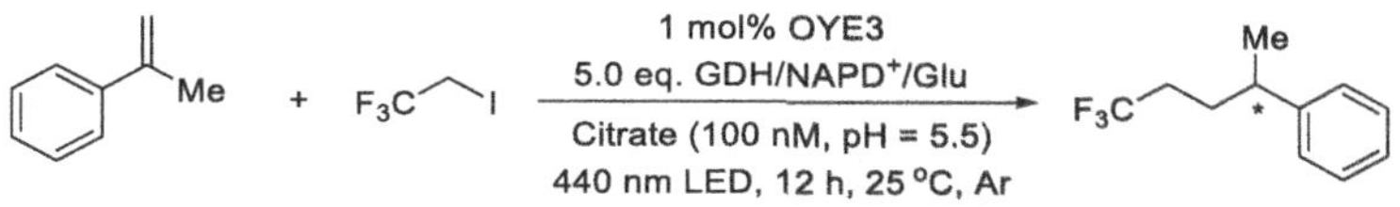

Fig. 4 Asymmetric photoenzymatic catalysis using fluorinated motifs

rate = 2.0 mL/min, split ratio = 10/1, temperature program = 50°C (0 min)–1°C/min–150°C.

4 Notes

Buffer for purification of polyhistidine-tagged proteins under native conditions. The photoenzymatic catalysis reaction described in this chapter needs to be carried out in a strictly oxygen-free conditions, and the reaction conduct well at a temperature with 25–28°C.

Acknowledgments

We thank the Fundamental Research Funds for the Central Universities, and the China Postdoctoral Science Foundation under Grant Number 2025M780963 for financial support.

References

1. Wang Y, Xue P, Cao M, Yu T, Lane ST, Zhao H (2021) Directed evolution: methodologies and applications. Chem Rev 121:12384–12444
2. Buller R, Lutz S, Kazlauskas RJ, Snajdrova R, Moore JC, Bornscheuer UT (2023) From nature to industry: harnessing enzymes for biocatalysis. Science 382:eadh8615
3. Harrison W, Huang X, Zhao H (2022) Photobiocatalysis for abiological transformations. Acc Chem Res 55:1087–1096
4. Litman ZC, Wang Y, Zhao H, Hartwig JF (2018) Cooperative asymmetric reactions combining photocatalysis and enzymatic catalysis. Nature 560:355–359
5. Miller DC, Athavale SV, Arnold FH (2022) Combining chemistry and protein engineering for new-to-nature biocatalysis. Nat Synth 1:18–23
6. Emmanuel MA, Bender SG, Bilodeau C, Carceller JM, DeHovitz JS, Fu H, Liu Y, Nicholls BT, Ouyang Y, Page CG, Qiao T, Raps FC, Sorigué DR, Sun SZ, Turek-Herman J, Ye Y, Rivas-Souchet A, Cao J, Hyster TK (2023) Photobiocatalytic strategies for organic synthesis. Chem Rev 123:5459–5520
7. Zhang Z, Li ML, Zhao H (2025) Mechanistic investigation of repurposed photoenzymes with new-to-nature reactivity. Curr Opin Green Sustain Chem 52:101009–101019
8. Fu H, Hyster TK (2024) From ground-state to excited-state activation modes: flavin-dependent "ene"-reductases catalyzed non-natural radical reactions. Acc Chem Res 57:1446–1457
9. Biegasiewicz KF, Cooper SJ, Gao X, Oblinsky DG, Kim JH, Garfinkle SE, Joyce LA, Sandoval BA, Scholes GD, Hyster TK (2019) Photoexcitation of flavoenzymes enables a stereoselective radical cyclization. Science 364:1166–1169
10. Huang X, Wang B, Wang Y, Jiang G, Feng J, Zhao H (2020) Photoenzymatic enantioselective intermolecular radical hydroalkylation. Nature 584:69–74
11. Chen X, Zheng D, Jiang L, Wang Z, Duan X, Cui D, Liu S, Zhang Y, Yu X, Ge J, Xu J (2023) Photoenzymatic hydrosulfonylation for the stereoselective synthesis of chiral sulfones. Angew Chem Int Ed 62:e20221814
12. Li ML, Harrison W, Zhang Z, Yuan Y, Zhao H (2024) Remote stereocontrol with azaarenes

via enzymatic hydrogen atom transfer. Nat Chem 16:277–284
13. Li ML, Yuan Y, Harrison W, Zhang Z, Zhao H (2024) Asymmetric photoenzymatic incorporation of fluorinated motifs into olefins. Science 385:416–421
14. Xu Y, Chen H, Yu L, Peng X, Zhang J, Xing Z, Bao Y, Liu A, Zhao Y, Tian C, Liang Y, Huang X (2024) A light-driven enzymatic enantioselective radical acylation. Nature 625:74–78
15. Xing Z, Liu F, Feng J, Lu Y, Wu Z, Zhao B, Chen B, Heng P, Xu Y, Liu A, Zhao Y, Wang C, Wang B, Huang X (2025) Synergistic photobiocatalysis for enantioselective triple-radical sorting. Nature 637:1118–1123

Chapter 3

CRISPR-Hybrid: Intracellular Selection of CRISPR-Associated Aptamers in Bacteria

Qiwen Su-Tobon and Jia Niu

Abstract

CRISPR technologies have evolved from nuclease-based genome editing to programmable systems for transcriptional and epigenetic regulations. Emerging CRISPR systems expand editing versatility by incorporating CRISPR-associated aptamers (CAPs) into single-guide RNAs (sgRNAs), enabling recruitment of RNA-binding proteins (RBPs) fused to diverse effectors. However, the limited availability of orthogonal aptamer–RBP pairs has hindered broad application, as conventional SELEX-based aptamer discovery is time-intensive and often fails to yield aptamers functional in cells. We developed the CRISPR-Hybrid platform, an intracellular selection method that directly evolves CAPs within bacterial cells. This system links aptamer–RBP interactions to a fluorescent reporter readout, allowing fluorescence-activated cell sorting (FACS) to enrich functional variants from libraries exceeding 10^7 sequences. Each selection round can be completed in 2 days, enabling rapid enrichment of aptamers that retain activity in both bacterial and mammalian contexts. This protocol details construction of randomized DNA libraries, preparation of host cells, execution of intracellular selection and FACS enrichment, and recovery of aptamer sequences for downstream analysis. By providing a fast, in-cell, and physiologically relevant approach to aptamer discovery, CRISPR-Hybrid expands the repertoire of CAPs available for modular and multiplexed CRISPR editing.

Keywords CRISPR-associated aptamers (CAPs), Intracellular aptamer selection, Fluorescence-activated cell sorting (FACS), sgRNA engineering, Multiplexed CRISPR system

1 Introduction

The CRISPR–Cas system, first identified as an adaptive immune defense in bacteria and archaea, has rapidly transformed into one of the most versatile platforms for genome engineering. Early implementations focused on Cas9 nuclease activity, enabling targeted gene knockouts and precise edits through double-strand DNA breaks and repair [1–4]. The subsequent development of catalytically inactive Cas9 (dCas9) expanded the scope of CRISPR technology beyond editing by creating a programmable DNA-binding scaffold that could be fused to transcriptional or epigenetic regulators [5–16]. While

Huan Peng and S. Brian Kim (eds.), *Biological Macromolecule Engineering Technologies*,
Methods in Molecular Biology, vol. 3032, https://doi.org/10.1007/978-1-0716-5257-2_3,

these non-nuclease CRISPR tools provided powerful means for gene regulation, their reliance on direct fusion of exogenous effectors to dCas9 limited the flexibility and multiplexing potential of the system, and raised immunogenicity and cytotoxicity risks.

To overcome these constraints, new CRISPR systems were developed in which single-guide RNAs (sgRNAs) are engineered to carry structured RNA motifs that function as affinity reagents, also known as aptamers [17]. The sgRNA-integrated RNA aptamers, hereafter referred to as CRISPR-associated aptamers (CAPs), can recruit intracellular RNA-binding proteins (RBPs) fused to transcriptional effectors or endogenous effectors. By incorporating multiple orthogonal aptamer–RBP pairs, CRISPR systems can be programmed to simultaneously direct different effectors to distinct genomic loci, enabling multiplexed and modular control of gene expression, epigenetic modifications, and beyond. This strategy expands the CRISPR toolbox, opening avenues for increasingly complex genetic circuits and synthetic biology applications.

Despite their promise, the broader adoption of CAPs has been constrained by the scarcity of orthogonal aptamer–RBP pairs that retain robust function inside living cells. Traditional aptamer discovery through Systematic Evolution of Ligands by Exponential Enrichment (SELEX) [18, 19] is time-consuming, requires purified targets, and often yields aptamers that fail to fold properly or maintain binding activity in vivo [20]. These limitations have created a critical bottleneck for advancing new CRISPR technologies, underscoring the need for methods that can directly select aptamers in cells under physiological conditions.

To address this challenge, we developed the CRISPR-Hybrid platform, an intracellular selection strategy that enables rapid identification of functional CAPs in bacteria cells (Fig. 1) [21]. In this system, sgRNA scaffolds are engineered to contain randomized RNA regions, and aptamer activity is coupled to a fluorescent reporter whose expression is activated when a cognate RBP–effector fusion

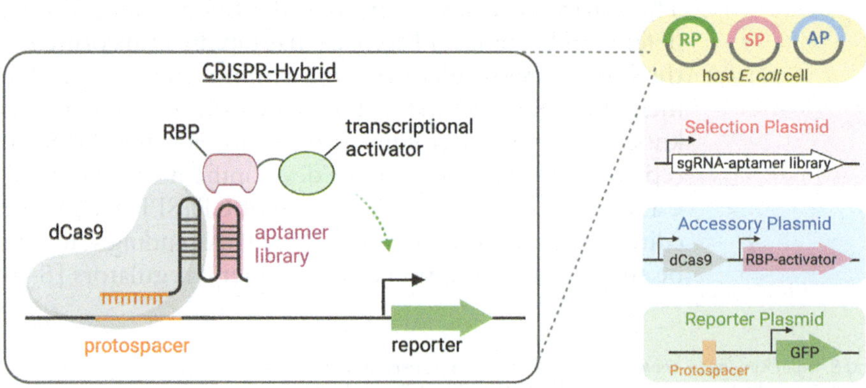

Fig. 1 Design of CRISPR-Hybrid

is successfully recruited. This design links aptamer function to a selectable phenotype, allowing fluorescence-activated cell sorting (FACS) to enrich functional variants from libraries containing more than ten million sequences. Crucially, each selection round, including library transformation, growth, sorting, and plasmid recovery, can be completed in just 2 days, greatly accelerating the discovery process compared with SELEX. Aptamers identified using CRISPR-Hybrid not only exhibit high specificity and orthogonality in bacterial cells but also retain robust activity in mammalian cells, underscoring their versatility and translational potential.

This chapter presents a comprehensive protocol for applying the CRISPR-Hybrid platform to discover new CRISPR-associated aptamers. We describe the preparation of randomized DNA libraries and host cells, the execution of intracellular selection and FACS enrichment, and the recovery and analysis of aptamer sequences using high-throughput sequencing. By enabling rapid, scalable, and intracellularly relevant selection of CAPs, this method represents a significant advance for expanding the modularity and multiplexing capacity of CRISPR systems, thereby unlocking new opportunities for genetic regulation and synthetic biology.

2 Materials

2.1 Equipment

- 0.1–2.5, 1–10 (optional), 2–20, 20–200, and 100–1000 μL pipettes
- PAGE gel electrophoresis system (BioRad)
- Analytical balance
- Hand-held UV lamp
- Clean razor blade
- PCR thermocycler (Thermo Scientific)
- Agarose gel electrophoresis system (BioRad)
- Microcentrifuge
- NanoDrop™ One/OneC Microvolume UV-Vis Spectrophotometer (Thermo Scientific)
- SpeedVac
- Freezers −80°C, −20°C, and 4°C
- Electroporation system
- Incubator shaker
- 0.2 mL PCR tubes
- Eppendorf tubes 2.0, 1.5, and 0.5 mL
- Electroporation cuvettes 0.1 cm (USA Scientific 9104-1050)
- Erlenmeyer flasks 2 and 0.5 L

- Culture tube (USA Scientific 5618-7262)
- JA10 tubes
- Petri dish
- Falcon test tube with cell strainer snap cap, 5 mL (Fisher 08-771-23)
- Spectrophotometer to measure optical density at 600 nm
- Becton Dickinson FACSAria Sorter

2.2 Materials

- Tris–EDTA buffer (Fisher BP2473100)
- 10% TBEU–urea gel electrophoresis (Invitrogen EC6875BOX)
- 2× TBE–UREA loading buffer (Invitrogen LC6876)
- Ultra Low Range DNA Ladder (ThermoFisher Scientific 10597012)
- ZR small-RNA PAGE Recovery Kit (Zymo Research R1070)
- Q5 HiFi DNA polymerase (NEB 0491)
- dNTP mix (NEB: N0447L)
- Amplification primers listed in Tables 1 and 7
- pCD061 (Addgene # 113315)
- DpnI (NEB R0176S)
- Agarose (Fisher BP1356-500)

Table 1
Amplification primers used to generate the aptamer libraries

Oligo	Sequence (5′ to 3′)	Description
QSL2	GTTAAAATAAGGCTAGTCCGTTATCAACTTGAAAAA GTGCGCACATNNNNNNNNNNNATGTGCTTTTTTG AAGCTTGGGCCCGAACAAA	Forward primer for piece A of loop library
QSL5	TAAAATAAGGCTAGTCCGTTATCAACTTGAAAAAGT GCGCNNNNTAGCTAAGCTANNNNGCTTTTTTGAA GCTTGGGCCCGAACAAAA	Forward primer for piece A of stem library
QS94	ACTCGAGTAAGGATCCAGTTCACCGAC	Reverse primer for piece A
QS91	GAAAAGTGCCACCTGACGTCTAAGAAACCATTATT	Forward primer for piece B
QS92	TTTTCAAGTTGATAACGGACTAGCCTTATT	Reverse primer for piece B
QS95	GTCGGTGAACTGGATCCTTACTCGAGT	Forward primer for piece C
QS96	AATAATGGTTTCTTAGACGTCAGGTGGCACTTTTC	Reverse primer for piece C

- 1× TAE electrophoresis buffer: 40 mM Tris base, 20 mM acetate, 1 mM EDTA (Fisher BP1332500)
- Ethidium bromide (Invitrogen 15585011)
- 6× gel loading dye purple (NEB: B7024A)
- DNA ladder 1-kb (NEB N3232)
- QIAquick gel extraction kit (Qiagen 28704)
- NEBuilder HiFi DNA Assembly Master Mix (NEB E2621L)
- Yeast-tRNA (Invitrogen AM7119)
- 3 M sodium acetate pH 5.2 (Fisher Scientific: AAA131840B)
- Ethanol, absolute (200 Proof) (Sigma: BP2818500)
- SOC media (100 mL): dissolve 0.5 g yeast extract (Fisher BP9727-500) and 2 g tryptone (Fisher BP1421-2) into 80 mL of MilliQ H_2O and add stock solutions to these final concentrations: 10 mM NaCl, 2.5 mM KCl, 10 mM $MgCl_2$, 10 mM $MgSO_4$, and 20 mM glucose (add after autoclave). Sterilize by autoclaving at 121°C for 15 min
- Sterile Falcon tube 50 mL (Fisher 14-432-22)
- LB media: dissolve 12.5 g LB Broth, Miller (Fisher BP97235) into 0.5 L of MilliQ H_2O in a 1 L flask and sterilize by autoclaving at 121°C for 15 min
- Electrocompetent Top10 cells (ThermoFisher Scientific C404050)
- Carbenicillin (Fisher BP26485)
- ZymoPURE II Plasmid Midiprep Kit (Zymo Research D4200)
- LB agar, Miller (Fisher BP1524-500)
- Glycerol (Fisher BP229-1)
- GeneJET Plasmid Miniprep Kit (Thermo Scientific K0503)
- pCD185 (Addgene #113317)
- pJF076Sa (Addgene #11332)
- Chloramphenicol (Fisher BP904-100)
- Kanamycin (Fisher BP906-5)
- SOB media 500 mL: dissolve 10 g tryptone and 2.5 g yeast extract in 500 mL MilliQ H_2O and autoclave at 121°C for 15 min. After the media has cooled down, add 855.8 μL of filter-sterilized 5 M NaCl
- Phosphate-buffered saline (PBS) solution, pH 7.4 (Fisher BP3991)
- Recovery-PCR buffer: 1 mM Triton X-100, 20 mM Tris–HCl pH 8.0, 2 mM EDTA

3 Methods

Before Selection

3.1 Construction of the DNA Library Encoding CAPs

3.1.1 DNA Library Purification, PCR, and Gel Extraction

The following protocol is to purify synthesized ssDNA libraries to remove residual truncated products, synthesis impurities, and other undesired oligonucleotides by using PAGE gel. This is highly recommended to ensure high purity and full-length sequence integrity for downstream selection.

1. Synthesize single-stranded DNA (ssDNA) libraries from IDT, using a customized recipe (A: 25%, C: 25%, G: 25%, T: 25%) for random regions. (see oligonucleotides sequences in Table 1). Resuspend the ssDNA to 100 μM in Tris–EDTA buffer.
2. Purify 2 pmol of ssDNA libraries on Invitrogen 10% TBEU–Urea PAGE gel by mixing 20 μL of 100 μM ssDNA with 20 μL 2× TBE–UREA loading buffer. Also, prepare a diluted sample by mixing 2 μL of ssDNA with 20 μL 2× TBE–UREA loading buffer. Heat denature the mixtures on a thermocycler at 95°C for 5 min and load onto the PAGE gel before the mixture cooled down. Load the undiluted and diluted samples on separate ends of the gel. Load 0.5 μg of Ultra Low Range DNA Ladder next to the diluted sample. Set gel electrophoresis conditions to 200 V for 45 min.
3. Visualize the DNA on a clean UV light table and excise the DNA band using a clean razor blade. Make sure to use a UV face shield and work as fast as possible to avoid exposure to UV light.
4. Place the excised gel into a pre-weighted Zymo-Spin IV column from the ZR small-RNA PAGE Recovery Kit. Weigh the gel and tube to determine the mass of the gel.
5. Purify the gel extract using ZR small-RNA PAGE Recovery Kit. Crush the gel with a Squisher-Single against the side of the column. Add 400 μL RNA recovery buffer and incubate at 65°C for 15 min. Quick freeze the sample on dry ice or in a −80°C freezer for 5 min, followed by thawing at 65°C for 5 min. Snap off the tip of the column and place into a Collection tube and centrifuge at 1500 × *g* for 30 s. Transfer the flow-through into a Zymo-Spin IIICG column into another collection tube and centrifuge at 1500 × *g* for 30 s. Separate the flow-through into a collection tube and add two volumes of RNA max buffer. Mix well by pipetting up and down and transfer into a Zymo-Spin IC column in a collection tube and centrifuge again. Add 400 μL RNA prep buffer and centrifuge. Discard the flow-through. Add 700 μL

RNA wash buffer and centrifuge. Discard the flow-through. Add 400 μL RNA wash buffer and centrifuge for 1 min. Transfer the column into a 1.5 mL Eppendorf tube and elute DNA using 20 μL nuclease-free water. Measure DNA concentration and quality using NanoDrop.

Plasmid pQS14.3, modified from pCD061 (Addgene # 113315), was used as the cloning vector. The MS2 aptamer region was deleted to generate a blank plasmid lacking any aptamer sequences. The ssDNA library served as a primer to PCR amplify insert A. To avoid sequence bias near the randomized region during assembly, insert A was extended by primer-less overlap PCR with insert B. This is designed such that the assembly overlapping region lie outside the randomized library sequences. The resulting extended fragment (Insert C) is then assembled into pQS14.3 vector using Gibson assembly.

6. Set up two 50 μL reactions for each desired PCR product (see Table 2) following these thermocycler conditions: denaturation at 98°C, followed by 30 cycles of 98°C for 15 s, primer-specific annealing temperature for 30 s, and 72°C for extension (use 20 s per kilobase of product and add an additional 20 s at the end). Perform a final extension at 72°C for 2 min and hold at 4°C.
7. Follow Table 3 for the corresponding primers for each insert and vector, along with their specific annealing temperatures.
8. Treat PCR products with *DpnI* at 37°C for 2 h to digest template DNA.
9. Gel purify the PCR products on a 1.5% agarose gel by dissolving 1.5 g agarose in 100 mL of 1× TAE buffer and adding 2 μL of ethidium bromide. Let the gel solidify at room temperature. Place the gel into an electrophoresis system and cover it with enough 1× TAE buffer. Load one well with 2 μL of DNA ladder. For each 50 μL PCR reaction, add 10 μL of

Table 2
PCR with NEB Q5 HiFi DNA polymerase

Order added	Component	Volume (μL)
1	H_2O	32.5
2	5× Q5 reaction buffer	10
3	10 mM dNTPs	1.0
4	10 μM forward primer	2.5
5	10 μM reverse primer	2.5
6	10 ng/μL DNA template	1.0
7	Q5 HiFi DNA polymerase	0.5

Table 3
PCR primers details

	Insert A	Insert B	Insert C	Vector
Forward primer	QSL2 or QSL5	QS91	/	QS95
Reverse primer	QS94	QS92	/	QS96
Annealing temperature	72°C	67°C	72°C	72°C
Product length (bp)	456	251	670	1985

10× DNA loading dye and load into individual wells. Run the gel at ~120 V for 35 min.

10. Visualize the DNA on a clean UV light table and excise the DNA band using a clean razor blade. Make sure to use a UV face shield and work as fast as possible to avoid exposure to UV light.
11. Pool gel slices of the same PCR product type into a 1.5 mL Eppendorf tube. Weigh the gel and tube to determine the mass of the gel.
12. Purify each PCR product type using one column of the QIAquick gel extraction kit. Add five volumes of buffer PB to one volume of the PCR sample and mix by pipetting up and down. Add 10 μL of 3 M sodium acetate, pH 5.0, and mix. Transfer the PCR sample into a QIAquick spin column in a 2 mL collection tube, and centrifuge at 17,900 × *g* for 1 min. Discard the flow-through and add 750 μL buffer PE to the QIAquick column and centrifuge again. Discard the flow-through and centrifuge the QIAquick column again without adding any buffers to remove any residual buffer PE. Place the QIAquick column into a 1.5 mL Eppendorf tube and elute the DNA with 20 μL of nuclease-free water. Measure DNA concentration and quality using a NanoDrop.
13. After inserts A and B are gel purified, perform primer-less overlap PCR to generate insert C. Use 100 ng of each insert as template and adjust the water volume accordingly. Add DMSO to a final concentration of 4%. Perform only 20 PCR cycles for this reaction. Follow Tables 2 and 3 for specific conditions.
14. Gel purify insert C by repeating Sect. 3.1.1, Steps 10–12. As shown in Fig. 2, the PCR amplification of insert C results in multiple side products, but the desired 670 bp band is prominent.

3.1.2 Gibson Assembly and Ethanol Precipitation of DNA Library

To achieve desired transformation efficiency for maximum library coverage, a total of ten assembly reactions are performed. The assembly products are pooled and ethanol precipitated to remove residual

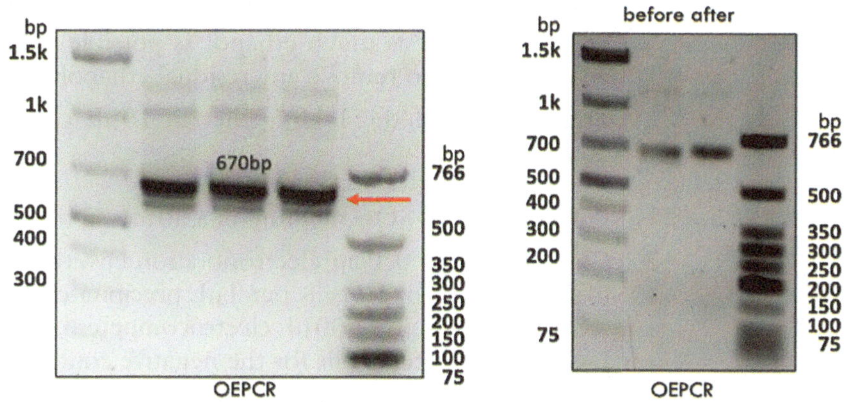

Fig. 2 The left agarose gel shows PCR products of insert C prior to gel purification. The desired 670 bp fragment is indicated by the red arrow. The right agarose gel shows insert C before and after gel purification

Table 4
Ethanol precipitation for 1 μg of DNA

Reagent	Volume (μL)	Proportion
3 M NaOAc, pH 5.2	10	0.1
10 mg/mL yeast tRNA	2	0.02
100% ethanol	300	3
Total volume	412	

buffers and enzymes and to concentrate the assembled DNA library for subsequent transformation in Sect. 3.1.3.

1. Set up ten reactions of 10 μL each to assemble insert C into the vector using a 4:1 molar ratio with NEBuilder HiFi Gibson Assembly Master Mix; for each reaction use 5 μL of Master Mix and add water up to 10 μL total volume. Incubate at 50°C for 30 min, then cool to 4°C.
2. Combine all assembly products into a single 1.5 mL Eppendorf tube and add the components listed in Table 4 for ethanol precipitation.
3. Incubate at −80°C for 1 h.
4. Spin down the DNA pellet by centrifuging at 20,000 × *g* in the cold room for 20 min. Carefully remove the supernatant with a pipette.
5. Wash the pellet by adding 500 μL of cold 70% ethanol and gently mixing (keep the tube on ice). Add another 500 μL of cold 70% ethanol and mix gently.

6. Centrifuge again at 20,000 × *g* in the cold room for 10 min. Carefully remove as much ethanol as possible with a pipette. Use a SpeedVac to remove any residual ethanol and water.
7. Gently resuspend the DNA pellet in 10 μL of nuclease-free water.

3.1.3 Transformation of the DNA Library into E. coli

1. Prewarm 10 mL SOC medium in a 50 mL sterile Falcon tube.
2. Chill on ice the 0.1 cm electroporation cuvettes, 50 μL electrocompetent Top10 cells per 1 μL precipitated DNA library (one library requires 500 μL electrocompetent cells), and 50 μ L electrocompetent cells for the negative control pQS14.3.
3. Add 1 μL of precipitated DNA library to the thawed electrocompetent cells and mix by flicking; avoid excessive agitation of cells, and do not pipette to mix. Transfer the mixture to a cuvette, avoid creating bubbles, and ensure the mixture covers the electrodes completely. Wipe the sides of the cuvette with a Kimwipe to remove moisture. Electroporate with an exponential pulse of 1.6 kV, 200 Ω, and 25 μF. If a pop is heard or the time constant is below 4.2, repeat the electroporation due to arcing.
4. Add 1 mL of prewarmed SOC medium to the cuvette, transfer cells and media to another sterile 50 mL Falcon tube, and repeat until the entire library is completely transformed. For negative control: transfer cells and media to a sterile 10 mL culture tube.
5. Incubate at 37°C, 250 rpm for 1 h.
6. Separate 10 μL of cells transformed with DNA library into a sterile 1.5 mL Eppendorf tube to determine transformation efficiency (see Steps 11–16). Put aside the cells transformed with the negative control and proceed to Step 11.
7. Transfer the recovered transformed cells to a 500 mL sterile flask with 150 mL LB supplemented with 100 μg/mL carbenicillin.
8. Incubate at 37°C, 250 rpm for 17 h.
9. Separate 1 mL bacterial culture for glycerol stock (see Sect. 3.1.3, Steps 16–18).
10. Isolate DNA library with the ZymoPURE II Plasmid Midiprep Kit, and this library is referred as aptamer plasmid hereafter. Separate the overnight bacterial culture into three 50 mL Falcon tube and centrifuge at 3400 × *g* for 10 min. Discard the supernatant and resuspend the cell pellet with 8 mL of ZymoPURE P2 by pipetting up and down. Add 8 mL of ZympPURE P3 and immediately mix by gently inverting the tube ten times. Must avoid vortexing. Let the mixture sit at room temperature for 5 min to lyse the cells

and the solution will appear clear, purple, and viscous. Add 8 mL of ZymoPURE P3 and mix by inverting until the sample turns completely yellow. Centrifuge the lysate at 3400 × *g* for 10 min and load the supernatant in the ZymoPURE Syringe Filter-X. Remove the Luer Lock plug from the Filter and place it into a clean 50 mL conical tube. Use the plunger to push the solution through the filter until approximately 20 mL of cleared lysate is recovered. Add 8 mL ZymoPURE binding buffer to the cleared lysate and invert eight times. Connect the 15 mL Reservoir-X and Zymo-Spin V-PS Column, and place the assembly into a 50 mL conical tube. Add 10 mL lysate mixture into the assembly and centrifuge at 500 × *g* for 2 min. Empty the flow-through and repeat step until all the lysate has passed through the column. Add 5 mL ZymoPURE Wash 1 and centrifuge at 500 × *g* for 2 min. Wash twice by adding 5 mL ZymoPURE Wash 2 and centrifuge at 500 × g for 2 min. Remove the Reservoir-X and place the Zymo-Spin VP-S column in a collection tube and centrifuge at 16,000 × *g* for 1 min to remove residual wash buffer. Transfer the column into a 1.5 mL Eppendorf tube and elute DNA with 200 μL ZymoPURE elution buffer by centrifuging at 16,000 × *g* for 1 min. Measure DNA concentration and quality with NanoDrop.

To determine the transformation efficiency (TE):

11. Prepare serial dilutions (1:10^6, 1:10^5, 1:10^4, 1:10^3, 1:100) of the 10 μL recovered cells (from Step 6) in sterile LB medium.
12. Plate 100 μL of dilutions :10^6, 1:10^5, 1:10^4 onto LB agar plates containing appropriate antibiotics (100 μg/mL carbenicillin). Spread evenly using sterile a spreader.
13. Plate 100 μL of recovered cells with negative control onto a LB agar plate containing appropriate antibiotics (100 μg/mL carbenicillin). Spread evenly using sterile a spreader.
14. Incubate plates overnight at 37°C.
15. Count colonies on plates containing between 30 and 300 colonies. Ensure more than tenfold transformants relative to library size is obtained to achieve full library coverage (*see* **Note 1** for negative control). Calculate transformation efficiency as colony-forming units per μg of DNA using the formula:

$$\mathrm{TE} = \frac{\text{Number of colonies} \times \text{dilution factor} \times \text{total recovery volume}}{\text{volume plated} \times \text{amount of DNA transformed}}.$$

To prepare glycerol stocks for long-term storage:

16. Transfer 1 mL of bacterial culture from Step 9 into a sterile 1.5-mL Eppendorf tube and centrifuge at 12,000 rpm for 2 min at room temperature.
17. Discard the supernatant and resuspend the cell pellet in LB medium containing 25% (v/v) sterile glycerol by pipetting up and down gently.
18. Transfer the cells into a sterile cryovial and store at –80°C.

To assess library diversity in selected transformants:

19. Pick 20 individual colonies randomly on TE plates from Step 16.
20. Inoculate each colony into 3 mL LB medium containing 100 μg/mL carbenicillin and grow overnight at 37°C with shaking at 250 rpm.
21. Isolate plasmid DNA from each culture using GeneJET Plasmid Miniprep Kit. Harvest 2 mL of overnight culture in a 2 mL Eppendorf tube by centrifuging at 6800 × *g* for 2 min. Discard the media and resuspend the cell pellet in 250 μL resuspension solution by pipetting up and down until no cell clumps remain. Add 250 μL lysis solution and invert six times until the solution becomes viscous and slightly clear. Do not incubate for more than 5 min. Add 350 μL neutralization solution and invert six times until the solution becomes cloudy. Centrifuge at 12,000 × *g* for 5 min and transfer the supernatant to a GeneJet spin column by pipetting to avoid transferring the white precipitate. Centrifuge at 12,000 × g for 1 min, discard the flow-through. Wash DNA by adding 500 μL wash solution, centrifuge and discard the flow-through. Repeat wash step two times. Centrifuge empty column to remove any residual buffer. Elute DNA with 30 μL elution buffer, and measure DNA concentration and quality with NanoDrop.
22. Submit plasmid DNA for Sanger sequencing using primer L4440.
23. Analyze sequencing results by aligning the nucleotide sequences from 20 colonies and examining variation at each position within the randomized region to assess library diversity.

3.2 Preparation of Electrocompetent Cells with Accessory and Reporter Plasmids

The CRISP-Hybrid system consists of three components: an aptamer plasmid (SP), an accessory plasmid (AP), and a reporter plasmid (RP). This section describes the preparation of electrocompetent *E. coli* MG1655 K12 cells carrying both the AP and RP plasmids (referred to as AP + RP electrocompetent cells). Preparing cells with AP and RP allows for efficient transformation of the SP and ensures high transformation efficiency and full coverage of the aptamer library during selection. AP is modified from pCD185, Addgene #113317, by replacing MCP with desired RNA-binding protein for intracellular selection, and RP is modified from pJF076Sa, Addgene #11332, by replacing mRFP1 with sfGFP.

To transform AP and RP in MG1655 K12 cells:

1. Dilute AP and RP to concentration of 10 ng/μL, and add 6 ng of each plasmid into a 0.5 mL Eppendorf tube.
2. Prewarm 1 mL SOC medium in a 10 mL sterile culture tube.
3. Chill on ice a 0.1 cm electroporation cuvette, 50 μL electrocompetent *E. coli* MG1655 K12 cells, and plasmid mixture for transformation.
4. Once electrocompetent cells are thawed, add plasmids and mix by flicking; avoid excessive agitation of cells, and do not pipette to mix. Transfer the mixture to a cuvette, avoid creating bubbles, and ensure the mixture covers the electrodes completely. Wipe the sides of the cuvette with a Kimwipe to remove moisture. Electroporate with an exponential pulse of 1.6 kV, 200 Ω, and 25 μF. If a pop is heard or the time constant is below 4.2, repeat the electroporation due to arcing.
5. Add 1 mL of prewarmed SOC medium to the cuvette, transfer cells and media to the sterile culture tube, and incubate at 37°C, 250 rpm for 1 h.
6. Separate 10 μL of transformed cells into a sterile 1.5 mL Eppendorf tube, and calculate transformation efficiency by following Sect. 3.1.3, Steps 11–15 and supplementing appropriate antibiotics 25 μg/mL chloramphenicol and 50 μg/mL kanamycin.
7. Inoculate a single colony in 3 mL LB media, supplemented with appropriate antibiotics, and incubate overnight at 37°C, 250 rpm.
8. Separate 1 mL bacterial culture for glycerol stock (see Sect. 3.1.3, Steps 16–18).

To prepare AP + RP electrocompetent cells:

Day 1

9. Inoculate a single colony of MG1655 cells with AP and RP in two sterile culture tubes each with 5 mL LB supplemented with 25 μg/mL chloramphenicol and 50 μg/mL kanamycin.
10. Incubate at 37°C, 250 rpm for 16 h.
11. Autoclave 1 L Super Optimal Broth (SOB) media (500 mL each in 2 L Erlenmeyer flasks that have never been washed with detergent as it will drastically decrease cell competency), 1 L glycerol (15%v/v), 250 mL LB agar, 50 Eppendorf tubes (0.5 mL), and three JA-10 centrifuge bottles.

Day 2

12. Separate 1 mL SOB media in a 1.5 mL Eppendorf tube to use in Step 17 as blank.
13. Inoculate 5 mL of overnight culture into each 500 mL SOB flask (1:100 dilution) and incubate at 37°C, 250 rpm for 2–3 h.
14. Set the centrifuge rotor to 4°C and chill three sterile JA-10 centrifuge bottles and 15% (v/v) glycerol on ice.
15. Transfer 1 mL of bacteria culture to a cuvette to measure OD_{600} (use SOB media from Step 14 as blank). Stop incubation when cells reach mid-log phase (OD_{600} = 0.5–0.6). If cells in one flask reaches $OD_{600} \approx 0.55$ before the other, place that flask on ice and continue incubating the other flask.
16. Once both cultures reach ~0.55 OD_{600}, divide the culture evenly into the three precooled sterile JA-10 bottles under aseptic conditions. Let bottles sit on ice for 10 min, and chill the sterile 0.5 mL Eppendorf tubes.
17. Centrifuge at 5000 × g for 15 min at 4°C.
18. Immediately decant the supernatant and gently wipe the interior walls of each tube with Kimwipes to remove residual media and salts and maximize cell competency
19. Resuspend each cell pellet in 15 mL of prechilled 15% (v/v) glycerol using a serological pipette, and combine the resuspended cells into one tube containing ~300 mL of 15% glycerol.
20. Repeat Steps 19 and 21 twice for a total of three washes.
21. After the third wash, resuspend the final pellet in 1.5 mL of 15% glycerol and aliquot into 55 μL volumes in prechilled 0.5 mL Eppendorf tubes. Perform this step rapidly to minimize cell death.
22. Flash-freeze aliquots by dropping directly into liquid nitrogen.
23. Store these AR+RP electrocompetent cells at −80°C.
24. To measure transformation efficiency, transform 10 ng of pQS14.3 lacking aptamer or SP carrying DNA library and follow Sect. 3.2, Steps 2–6 supplying appropriate antibiotics.

3.2.1 CRISPR-Hybrid Intracellular Selection

Each round of CRISPR-Hybrid intracellular selection consists of three parts with a 2-day turnaround per round: (1) library introduction where SP are transformed into AP+RP electrocompetent cells, (2) fluorescence-activated cell sorting (FACS) where the most active aptamer binders enabling maximum reporter expression are collected, and (3) sequence recovery where aptamer sequences are

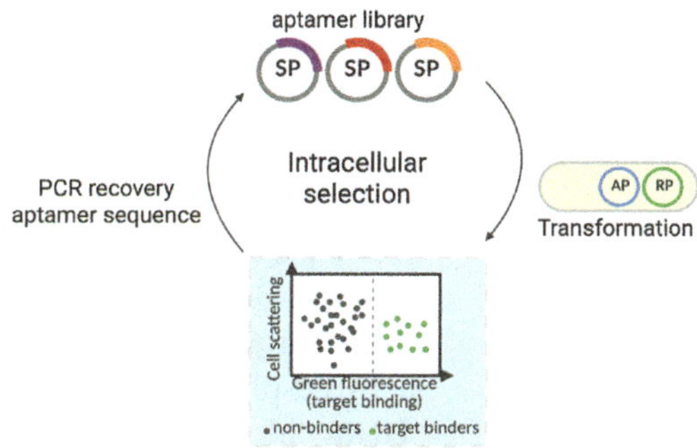

Fig. 3 One round of CRISPR-Hybrid intracellular selection

extracted and assembled into a new library to be reintroduced in the next round (Fig. 3).

3.3 Aptamer Library Introduction

1. Dilute SP to concentration of 10 ng/μL.
2. Prewarm 1 mL SOC medium in a 10 mL sterile culture tube.
3. Chill on ice a 0.1 cm electroporation cuvette, 50 μL electrocompetent AP + RP electrocompetent cells, and plasmid for transformation.
4. Once electrocompetent cells are thawed, add 1 μL or 10 ng of plasmid and mix by flicking; avoid excessive agitation of cells, and do not pipette to mix. Transfer the mixture to a cuvette, avoid creating bubbles, and ensure the mixture covers the electrodes completely. Wipe the sides of the cuvette with a Kimwipe to remove moisture. Electroporate with an exponential pulse of 1.6 kV, 200 Ω, and 25 μF. If a pop is heard or the time constant is below 4.2, repeat the electroporation due to arcing.
5. Add 1 mL of prewarmed SOC medium to the cuvette, transfer cells and media to the sterile culture tube, and incubate at 37°C, 250 rpm for 1 h.
6. Separate 10 μL of transformed cells into a sterile 1.5 mL Eppendorf tube, and calculate transformation efficiency by following Sect. 3.1.3, Steps 11–15 and supplementing appropriate antibiotics (100 μg/mL carbenicillin, 25 μg/mL chloramphenicol, and 50 μg/mL kanamycin).
7. Transfer cells to a 50 mL sterile Falcon tube containing 15 mL LB supplemented with 100 μg/mL carbenicillin, 25 μg/mL chloramphenicol, and 50 μg/mL kanamycin.
8. Incubate at 37°C, 250 rpm for 17 h to maximize GFP expression for FACS.
9. Separate 2 mL bacterial culture to prepare two glycerol stocks (see Sect. 3.1.3, Steps 16–18).

10. Repeat Steps 1–8 to prepare negative control for FACS by transforming pQS14.3 lacking aptamer (minimum GFP expression).
11. Repeat Steps 2–8 to prepare positive control for FACS by transforming pQS1.3 containing a MS2 aptamer (modified from pQS14.3), AP containing MS2-Coat Protein (MCP), and RP. Because the prepared AP + RP electrocompetent cells carry target RNA-binding protein in place of MCP, perform a three-plasmid transformation by diluting each plasmid to 10 ng/μL and adding 0.6 μL each into a 0.5 mL Eppendorf tube. Then repeat Steps 2–8 with *E.coli* MG1655 K12 electrocompetent cells.

3.4 FACS

1. Separate 1 mL of each overnight culture from Sect. 3.3 into sterile 1.5 mL Eppendorf tubes and pellet cells.
2. Centrifuge at 17,900 × *g* for 5 min at 4°C.
3. Remove supernatant and wash cells with cold phosphate-buffered saline (PBS, pH 7.4).
4. Repeat Steps 13 and 14 for a total of two washes.
5. Dilute cells 1:50 in 2 mL cold PBS and filter through a Falcon test tube with a cell strainer snap cap, then keep on ice.
6. Add 1 mL SOC to another Falcon test tube with a cell strainer snap cap and keep on ice.
7. Analyze negative control cells on the Becton Dickinson FACSAria Sorter until the cell count reaches 10,000.
8. Apply a side scatter height (SSC-H) threshold of 10^3 to enrich for single bacterial cells.
9. Set up gate 1 by isolating single cells centered on the FSC-A vs SSC-A plot.
10. Display only gate 1 events on the FSC-H vs FSC-A plot and define gate 2 along the diagonal to further select single cells.
11. Exclude events at the edges of the fluorescence histogram to remove outliers.
12. Define sorting gate 3 on FITC-A vs PE-A plot by excluding all events from the negative control and including only cells with GFP fluorescence above the background.
13. Analyze the positive control sample and verify that gates 1–3 correctly capture this population.
14. Adjust FITC-A vs PE-A compensation accordingly using the compensation calculator corresponding to the FACS sorter.
15. Perform a SIP Clean on the FACSAria Sorter by flushing PBS for 2 min to remove residual positive control cells.

16. Analyze the library sample and save fluorescence data for comparison with additional rounds of selection.
17. Set gating stringency to sort 7–14 cells per 100,000 events.
18. Adjust sorting gate 3 on FITC-A vs PE-A plot to isolate the top 0.007% of the fluorescent library population.
19. Sort cells until total coverage exceeds 100-fold of the library size (e.g., for a 10^7 library, sort at least 10^9 cells); typically, ~1500 cells are collected from a 5×10^6 variant library.
20. Collect sorted cells into 1 mL SOC for recovery.

3.5 Aptamer Sequence Recovery

1. Transfer 1 mL SOC media with sorted cells to a 1.5 mL Eppendorf tube and centrifuge at 17,900 × g for 5 min at 4°C. Use a marker to label the location of the expected cell pellet.
2. Carefully pipet out the supernatant without disturbing the cell pellet.
3. Resuspend the cell pellet in 12 μL of recovery-PCR buffer and transfer to a 0.2 mL PCR tube.
4. Lyse cells by incubating at 95°C for 3 min, then immediately cool to 4°C.
5. Use the lysate as DNA template for three 50 μL PCR reactions to amplify the region containing the aptamer library (see Tables 5 and 6 for PCR conditions).
6. Follow Sect. 3.1.1, Step 10 to gel purify the PCR products on a 1% TAE-agarose gel and to elute with 25 μL nuclease-free water.
7. Subclone the amplified aptamer library back into the plasmid backbone using NEBuilder HiFi Gibson Assembly Master Mix: add insert and vector in a 2:1 molar ratio, 5 μL of Master

Table 5
Recovery-PCR with NEB Q5 HiFi DNA polymerase

Order added	Component	Volume (μL)
1	H_2O	28.5
2	5× Q5 reaction buffer	10
3	10 mM dNTPs	1.0
4	10 μM forward primer	2.5
5	10 μM reverse primer	2.5
6	DNA template (sorted cells)	4.0
7	Q5 HiFi DNA polymerase	0.5

Table 6
Recovery-PCR primers details

	Insert A	Vector
Forward primer	QS91	QS95
Reverse primer	QS94	QS96
Annealing temperature	72°C	72°C
Product length (bp)	670	1985

Mix, and add water up to 10 μL total volume. Incubate at 50°C for 30 min, then cool to 4°C.

8. Prewarm 1 mL SOC medium in a 10 mL sterile culture tube.
9. Chill on ice a 0.1 cm electroporation cuvette, 50 μL electrocompetent Top10 cells, and plasmid for transformation.
10. Once electrocompetent cells are thawed, add 1 μL of the assembly reaction and mix by flicking; avoid excessive agitation of cells, and do not pipette to mix. Transfer the mixture to a cuvette, avoid creating bubbles, and ensure the mixture covers the electrodes completely. Wipe the sides of the cuvette with a Kimwipe to remove moisture. Electroporate with an exponential pulse of 1.6 kV, 200 Ω, and 25 μF. If a pop is heard or the time constant is below 4.2, repeat the electroporation due to arcing
11. Add 1 mL of prewarmed SOC medium to the cuvette, transfer cells and media to the sterile culture tube, and incubate at 37°C, 250 rpm for 1 h.
    a. Optional to calculate transformation efficiency by isolating 10 μL of transformed cells into a sterile 1.5 mL Eppendorf tube, and following Sect. 3.1.3, Steps 11–15, supplementing with 100 μg/mL carbenicillin.
    b. Optional to assess sequence diversity of enriched aptamer sequences after each round by following Sect. 3.1.3, Steps 19–23.
12. Transfer cells to a 50 mL sterile Falcon tube containing 15 mL LB supplemented with 100 μg/mL carbenicillin.
13. Incubate at 37°C, 250 rpm for 17 h.
14. Isolate 2 mL overnight bacterial culture to prepare two glycerol stocks (see Sect. 3.1.3, Steps 16–18).
15. Isolate plasmid using GeneJET Plasmid Miniprep Kit and conclude this round of selection.
16. Return to Part 1- Library Introduction in Sect. 3.3 for next round.

3.6 Post-Selection Analysis with High-Throughput Sequencing

3.6.1 Sample Preparation

1. Set up two 50 μL PCR reactions using NEB Q5 HiFi DNA polymerase (see conditions on Tables 2 and 7) and 10 ng of enriched aptamer library plasmid (assembled and isolated in Sect. 3.5) as PCR DNA template.
2. Design PCR amplification primers to include unique Illumina-compatible adapters for Amplicon sequencing (Table 7), and limit PCR product length to be less than 250 base pairs.
3. Gel purify the PCR products on a 2% TAE-agarose gel by following Sect. 3.1.1, Step 10 with modifications of washing twice with PE buffer and elute from one column with 20 μL nuclease-free water.
4. Confirm size and purity by analyzing the gel extracted product on another 2% TAE-agarose gel.
5. Measure product concentration and purity using a NanoDrop.
6. Dilute product to 20 ng/μL with nuclease-free water and submit minimum of 500 ng (less than 25 μL total volume) for Amplicon sequencing.

3.6.2 Enrichment Analysis

1. Download raw FASTQ files and process using Python scripts available on GitHub (https://github.com/JNiulab/CRISPR-hybrid.git).
2. Apply a quality score (Q-score) filter to discard any reads below the specific threshold (Q-scores below 10).
3. Use a mismatch filter to align each read to expected sequence in the fixed regions of the library plasmid, skipping the aptamer pool region containing the randomized sequences.
4. Extract the aptamer pool region (enriched sequences) from each valid read and count read occurrences.

Table 7
PCR for NGS. Sequencing adapters are highlighted in bold

Forward primer (sequence 5' to 3')	QS_seq18 **ACACTCTTTCCCTACACGACGCTCTTCCGATCTT**TGACAGCT AGCTCAGTC
Reverse primer (sequence 5' to 3')	**GACTGGAGTTCAGACGTGTGCTCTTCCGATCT**GTTTTTGTTC GGGCCCAAG
Annealing temperature	68°C
Product length (bp)	235

5. Export these enriched aptamer sequences and their read counts into a comma-separated values (CSV) file.
6. Identify the most abundant aptamer sequences and characterize them individually using additional assays (cell-based reporters, ELISA, SPR, etc.).
7. Predict the secondary structures of the enriched aptamers using the Mfold web server (https://www.unafold.org).

4 Notes

1. Transformation with the negative control plasmid should yield less than ten colonies. If more than ten colonies are observed, increasing *DpnI* digestion time and reducing DNA template amount during PCR are recommended.

Acknowledgments

This work was supported by NIH Director's New Innovator Award 1DP2HG011027-01 (J.N.) and National Science Foundation Graduate Research Fellowship Program (Q.S.-T.).

References

1. Cong L et al (2013) Multiplex genome engineering using CRISPR/Cas systems. Science 339:1088–1093. https://doi.org/10.1126/science.1231143
2. Hsu PD, Lande ES, Zhang F (2014) Development and applications of CRISPR-Cas9 for genome engineering. Cell 157:1262–1278. https://doi.org/10.1016/j.cell.2014.05.010
3. Jinek M et al (2012) A Programmable dual-RNA – guided DNA endonuclease in adaptive bacterial immunity. Science 337:816–822. https://doi.org/10.1126/science.1225829
4. Mali P et al (2013) RNA-guided human genome engineering via Cas9. Science 823:823–827. https://doi.org/10.1126/science.1232033
5. Billon P et al (2017) CRISPR-mediated base editing enables efficient disruption of eukaryotic genes through induction of STOP codons. Mol. Cell 67:1068–1079.e1064. https://doi.org/10.1016/j.molcel.2017.08.008
6. Cheng AW et al (2016) Casilio: a versatile CRISPR-Cas9-Pumilio hybrid for gene regulation and genomic labeling. Cell Res 26:254–257. https://doi.org/10.1038/cr.2016.3
7. Choudhury SR, Cui Y, Lubecka K, Stefanska B, Irudayaraj J (2016) CRISPR-dCas9 mediated TET1 targeting for selective DNA demethylation at BRCA1 promoter. Oncotarget 7:46545–46556. https://doi.org/10.18632/oncotarget.10234
8. Dominguez AA, Lim WA, Qi LS (2015) Beyond editing: repurposing CRISPR–Cas9 for precision genome regulation and interrogation. Nat Rev Mol Cell Biol 17:5–15. https://doi.org/10.1038/nrm.2015.2
9. Dong C, Fontana J, Patel A, Carothers JM, Zalatan JG (2018) Synthetic CRISPR-Cas gene activators for transcriptional reprogramming in bacteria. Nat Commun 7:2489. https://doi.org/10.1038/s41467-018-04901-6
10. Halperin SO et al (2018) CRISPR-guided DNA polymerases enable diversification of all nucleotides in a tunable window. Nature 560:248–252. https://doi.org/10.1038/s41586-018-0384-8
11. Hilton IB et al (2015) Epigenome editing by a CRISPR-Cas9-based acetyltransferase activates genes from promoters and enhancers. Nat Biotechnol 33:510–517. https://doi.org/10.1038/nbt.3199

12. Jia Y et al (2018) Next-generation CRISPR/Cas9 transcriptional activation in Drosophila using fly-SAM. Proc Natl Acad Sci U S A 115:4719–4724. https://doi.org/10.1073/pnas.1800677115
13. Katayama S, Shiraishi K, Gorai N, Andou M (2021) A CRISPR/Cas9-based method for targeted DNA methylation enables cancer initiation in B lymphocytes. Adv Genet 2:1–9. https://doi.org/10.1002/ggn2.10040
14. Komor AC, Badran AH, Liu DR (2017) CRISPR-based technologies for the manipulation of eukaryotic genomes. Cell 168:20–36. https://doi.org/10.1016/j.cell.2016.10.044
15. Maeder ML et al (2013) CRISPR RNA-guided activation of endogenous human genes. Nat Methods 10:977–979. https://doi.org/10.1038/nmeth.2598
16. Martella A et al (2019) Systematic evaluation of CRISPRa and CRISPRi modalities enables development of a multiplexed, orthogonal gene activation and repression system. ACS Synth Biol 8:1998–2006. https://doi.org/10.1021/acssynbio.8b00527
17. Shechner DM, Hacisuleyman E, Younger ST, Rinn JL (2015) Multiplexable, locus-specific targeting of long RNAs with CRISPR-display. Nat Methods 12:664–670. https://doi.org/10.1038/nmeth.3433
18. Tuerk C, Gold L (1990) Systematic evolution of ligands by exponential enrichment: RNA ligands to bacteriophage T4 DNA polymerase. Science 249:505–510. https://doi.org/10.1126/science.2200121
19. Ellington AD, Szostak JW (1990) In vitro selection of RNA molecules that bind specific ligands. Nature 346:818–822
20. Sola M et al (2020) Aptamers against live targets: is in vivo SELEX finally coming to the edge?. Mol Ther Nucleic Acids 21:192–204. https://doi.org/10.1016/j.omtn.2020.05.025
21. Su-Tobon Q et al (2025) CRISPR-hybrid: a CRISPR-mediated intracellular directed evolution platform for RNA aptamers. Nat Commun 16: 595. https://doi.org/10.1038/s41467-025-55957-0

Check for updates

Chapter 4

Designing and Engineering Enzymes for New-to-Nature Photobiocatalysis

Jie Feng, Chang-Liang Zhu and Mao-Lin Li

Abstract

Photobiocatalysis, an interdisciplinary frontier integrating chemistry, biology, and catalysis, drives enzymatic catalysis from natural reactions to new-to-nature transformations via protein engineering and photochemistry. This review centers on enzyme classes, covering nine key families: fatty acid photodecarboxylases (FAPs), nicotinamide-dependent reductases, flavin-dependent "ene"-reductases (EREDs), and others like artificial photoenzymes. We detail photoenzymatic development, mechanisms, and protein engineering for reactivity expansion. Despite enzymes' superior stereocontrol, their reaction scope is narrower than small-molecule catalysts. Future exploration of universal activation mechanisms and AI-assisted engineering will expand photobiocatalysis' potential for sustainable chiral synthesis and green chemistry.

Keywords Photobiocatalysis, New-to-nature, Nature cofactor-dependent enzymes, Metalloenzymatic, Artificial photoenzymes

1 Introduction

As a frontier in the interdisciplinary field of chemistry, biology, and catalysis, photobiocatalysis is driving a leapfrog development in enzymatic catalysis from natural reactions to nonnatural, high-value transformations through the deep integration of protein engineering and photochemical techniques [1]. Combining the dual advantages of visible-light catalysis and enzyme, this field has emerged as a powerful tool for sustainable and selective chemical synthesis [2]. In recent years, through the synergistic innovation of photoredox catalysis and biocatalysis, researchers have successfully expanded the catalytic boundaries of enzymes, achieving efficient and stereoselective synthesis of various new-to-nature reactions [3]. According to the type of light excitation, it can be divided into direct visible-light excited reaction and indirect visible-light excited reaction, according to the direction of electron transfer, it can be classified as reduction-initiated reaction (substrate gains electrons), oxidation-initiated reaction

Huan Peng and S. Brian Kim (eds.), *Biological Macromolecule Engineering Technologies*,
Methods in Molecular Biology, vol. 3032, https://doi.org/10.1007/978-1-0716-5257-2_4,

(substrate loses electrons), and triplet-triplet energy transfer reaction (EnT) [4].

Unlike many existing reviews that primarily summarize the general advances in photobiocatalysis [1–21] this article is organized around enzyme classes to provide a mechanistically grounded perspective. Specifically, this review covers fatty acid photodecarboxylase, nicotinamide-dependent reductases, flavin-dependent "ene"-reductases, flavin-dependent Baeyer–Villiger monooxygenases, pyridoxal-5′-phosphate (PLP)-dependent oxidoreductases, thiamine-dependent oxidoreductases, Pyrroloquinoline Quinone (PQQ)-dependent enzyme, metalloenzymes, and artificial photoenzymes. For each class, we systematically discuss the historical development of photoenzymatic catalysis, dissect the underlying catalytic mechanisms, and highlight how protein engineering has been leveraged to expand their reactivity. By structuring the review in this way, we aim to provide readers with a comprehensive and comparative framework for understanding how diverse enzymes have been adapted and engineered to perform new-to-nature photochemical transformations.

2 Fatty Acid Photodecarboxylases (FAPs)

Fatty acid photodecarboxylase (FAP) is the third natural photoenzyme discovered after DNA photolyase [22] and protochlorophyllide oxidoreductase [23]. *Cv*FAP, which is derived from *Chlorella variabilis* NC64A, is present in unicellular photosynthetic green microalgae and can catalyze the decarboxylation of fatty acid substrates driven by blue light, generating the corresponding *n*-alkane or alkene products. This enzyme contains a flavin adenine dinucleotide (FAD) cofactor and has a preference for long-chain fatty acid substrates [24].

As shown in Fig. 1, mechanistic studies have shown that upon visible light excitation, the cofactor FAD reaches an excited state (FAD*), which promotes the single-electron oxidation of the fatty acid anion, forming $FAD^{\bullet-}$ and triggering a radical decarboxylation (intramolecular cyclization)/dehalogenation reaction to generate an alkyl radical. Subsequently, a back electron transfer (BET) occurs from $FAD^{\bullet-}$ to the alkyl radical. Then, the arginine residue in the enzyme's active site protonates the alkyl anion, thus forming the hydrocarbon product [25].

A number of recent studies have shown that FAPs has the potential to be used in the field of industrial biotechnology [26]. The substrate scope of *Cv*FAP can be altered through protein engineering methods [27] and the use of decoy molecules [28]. In addition, the understanding of the photochemical reaction mechanism of *Cv*FAP provides important references for the design of flavin-based photoenzymes [29]. In 2021, Wu and coworkers developed a light-driven

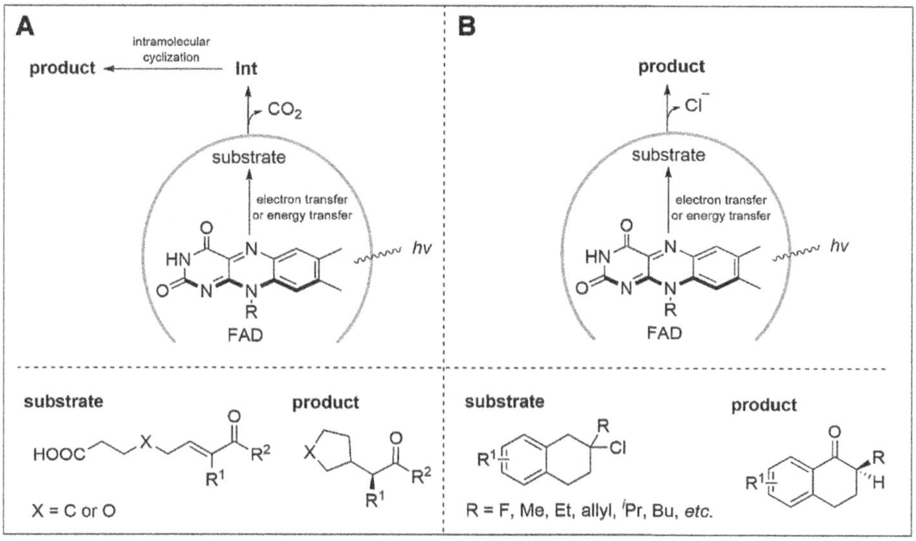

Fig. 1 Catalytic process of acid photodecarboxylase (**A**) decarboxylative cyclization, (**B**) dehalogenation reaction

decarboxylative deuteration reaction using D_2O as the deuterium source with *Cv*FAP [30] and engineered *Cv*FAP based on the FRISM (focused rational iterative site-specific mutagenesis) strategy, aiming to enable it to selectively catalyze the decarboxylation of trans-fatty acids. They identified the key mutation sites through molecular docking, introduced specific amino acids for mutation screening, and obtained the mutant V453E. This mutant has showed 1000-fold higher trans-over-cis selectivity for the model substrates than wild type [31]. In 2024, they engineered the *Cv*FAP and obtained two highly stereocomplementary variants with excellent selectivity (both giving products with an enantiomeric excess value of up to 99%) [32].

In 2019, Wu and coworkers reported a new strategy for the direct photoinduced kinetic resolution of α-functionalized carboxylic acids by engineered *Cv*FAP for the first time [33]. In 2020, Zheng and coworkers applied the engineered *Cv*FAP to perform kinetic resolution on racemic phosphinothricin at the gram scale, obtaining unreacted L-phosphinothricin with a high conversion (49%) and excellent stereoselectivity (ee up to 96%) [34]. In 2024, Yang and coworkers engineered the natural *Cv*FAP and obtained a set of variants defined as radical photocyclases (RAPs), which were capable of synthesizing a series of chiral cyclization products [35]. In the same year, Wu and coworkers reported that *Cv*FAP catalyzed unnatural reductive dehalogenation (deacetoxylation) reactions for the synthesis of a series of chiral α-substituted tetralones [36]. This is a novel mechanism involving the FAD_{ox}/FAD_{sq} redox pair and an electron transfer/proton transfer (ET/PT) process for radical termination.

The new-to-nature reactions catalyzed by *Cv*FAP mainly focus on the control of stereocenters, asymmetric cyclization, asymmetric dehalogenation, or deacetoxylation. Although the mechanism of *Cv*FAP is now well understood, their applications in chemical synthesis remain under-explored. It is of great significance to expand the catalytic boundaries of *Cv*FAP through strategies, such as genetic engineering, protein engineering, and computational chemistry.

3 Nicotinamide-Dependent Reductases

Nicotinamide-dependent reductases catalyzing new-to-nature chemical reactions under light excitation have been made some progress in recent years. As shown in Fig. 2, mechanistic studies revealed the formation of EDA complex between the substrate and reduced cofactor nicotinamide adenine dinucleotide phosphate (NADPH) in the enzyme active site. Upon visible-light irradiation, the SET process followed by the cleavage of the C–Br bond or radical addition generate $NADP^{\bullet+}$ and the prochiral radical. Subsequent hydrogen atom transfer (HAT) from $NADP^{\bullet+}$ to the prochiral radical formed enantioselective products. $NADP^{+}$ was then reduced by cofactor regeneration to complete the catalytic cycle.

In 2016, Hyster and coworkers reported the highly enantioselective radical dehalogenation of lactones catalyzed by

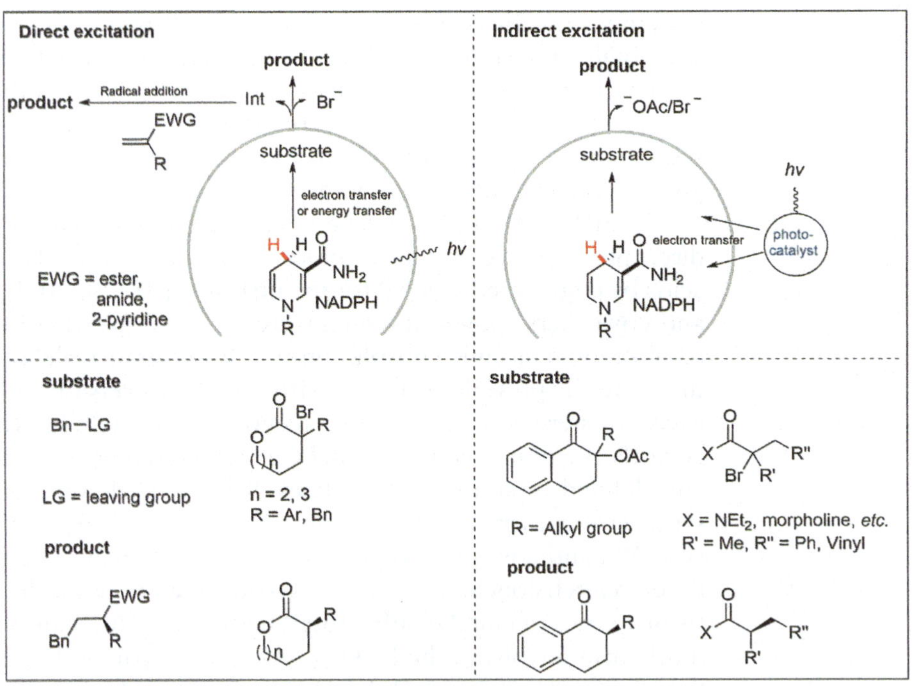

Fig. 2 Catalytic process of nicotinamide-dependent reductases

nicotinamide-dependent ketoreductases, this is the first example of photo-induced enzyme promiscuity [37]. In 2022, Zhao group found that nicotinamide-dependent ketoreductases can catalyze the intermolecular coupling of *N*-(acyloxy)phthalimide with Michael acceptors [38]. In 2024, Huang and coworkers reported that nicotinamide-dependent imine reductases (IRED) achieve enantioselective radical hydroalkylation of enamides [39].

In 2018, Hyster and coworkers developed photoredox catalysts to endow nicotinamide-dependent enzymes with new catalytic functions. Under visible-light irradiation, xanthene-based photocatalysts enable nicotinamide-dependent oxidoreductases to catalyze enantioselective deacetoxylation reactions [40].

Nicotinamide-dependent oxidoreductases can currently catalyze reactions, such as radical dehalogenation, radical deacetoxylation, and radical addition.

4 Flavin-Dependent "Ene"-Reductases (EREDs)

Flavin-dependent "ene"-reductases (EREDs) are biocatalysts with high-substrate promiscuity in catalyzing nonnatural reactions [11]. According to the reaction mechanism, it can be divided into direct visible-light-excited reduction-initiated reaction, indirect visible-light-excited reduction-initiated reaction, and direct visible-light-excited oxidation-initiated reaction.

4.1 Direct Visible-Light-Excited Reduction-Initiated Reaction

As shown in Fig. 3A, reactions in which a cofactor within the protein active site is photoexcited, promoting electron transfer from the cofactor to the substrate to convert the starting material into products. In 2019, Hyster and coworkers discovered that photoexcitation of flavin-dependent "ene"-reductases can alter their catalytic functions, promoting asymmetric radical cyclization to construct various lactams, with the stereochemical preference endowed by the enzyme's active site [41]. Next year, they achieved the asymmetric reductive cyclization reaction by generating unstable alkyl radicals through photoenzymatic catalysis, providing a new method for the synthesis of related compounds [42]. In 2020, Zhao and coworkers achieved a photoenzymatic enantioselective intermolecular radical hydroalkylation reaction, which holds significant synthetic application value [43]. In 2020, Hyster's research showed that quaternary charge-transfer complexes could enable photoenzymatic intermolecular hydroalkylation of olefins, expanding the reaction types of photoenzymatic catalysis [44]. In 2020, they discovered that flavin-dependent "ene"-reductases could achieve photoenzymatic reduction reactions, further enriching the catalytic applications of this enzyme [45]. In 2021, Hyster and coworkers discovered that ground-state electron transfer could serve as the initiating mechanism for biocatalytic C–C

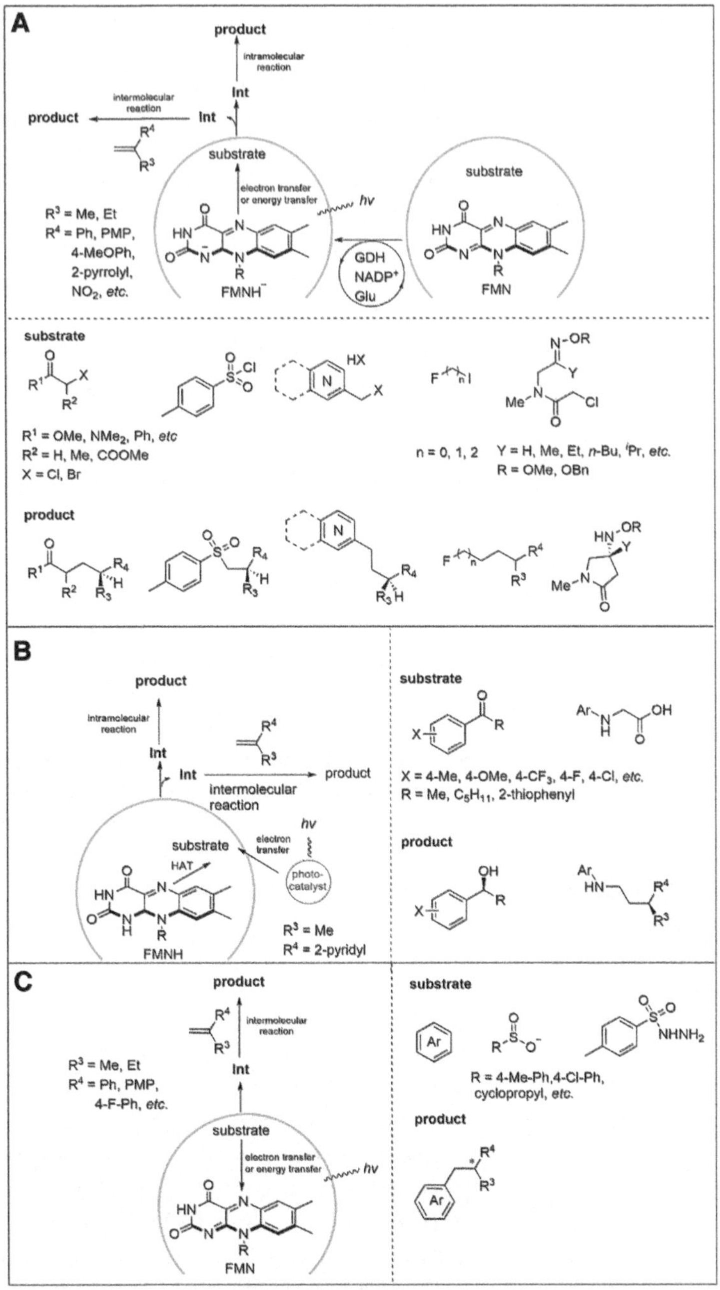

Fig. 3 Catalytic process of flavin-dependent "ene"-reductases (**A**) direct visible-light-excited reduction-initiated; (**B**) indirect visible-light-excited reduction-initiated; (**C**) direct visible-light-excited oxidation-initiated

bond formation reactions, providing a new perspective for understanding the mechanisms of related reactions [46]. In 2021, Hyster and coworkers reported the engineering of flavin-dependent "ene"-reductases (EREDs) via directed evolution to enable catalytic radical addition reactions of oximes, achieving highly stereoselective synthesis of α-tertiary amines. Mechanistic studies revealed that oximes participate in forming an enzyme-templated charge-transfer complex, and the products can be further derivatized, providing a new strategy for the asymmetric synthesis of chiral α-tertiary amines [47]. In the same year, they developed a protein engineering platform for high-throughput optimization of photoenzymes. Through one round of engineering optimization, the catalytic function and photon efficiency of photoenzymes in synthesizing specific compounds were improved [48]. In 2022, they compared four methods for preparing γ, δ, ε, and ζ-lactams and found that the photoenzymatic method using flavin-dependent "ene"-reductases had the highest product selectivity [49]. Inspired by photosynthesis, Hyster and coworkers designed a conjugate composed of covalently linked photoenzyme and light antenna, achieving the separation of light-harvesting and catalytic functions. Through energy transfer, the "ene"-reductase catalyzes the hydroalkylation of olefins under low-energy green light, with the yield increased by approximately four times, and intermolecular reactions are realized, providing a new strategy for nonnatural photoenzymatic catalysis [50]. In 2022, a highly chemoselective and enantioselective Csp^3–Csp^3 cross-electrophilic coupling reaction between alkyl halides and nitroalkanes catalyzed by flavin-dependent "ene"-reductases was reported, providing a new method for constructing such chemical bonds [51]. This year, Hyster and coworkers reported the photoenzymatic allylic alkylation of α-chloroamides with allylsilanes using flavin-dependent EREDs via a β-fragmentation radical-termination mechanism, enabling the highly enantioselective preparation of 5, 6, and 7-membered lactams [52]. They achieved control over nitrogen-centered radicals through directed evolution, successfully catalyzing various types of enantioselective hydroamination reactions, such as 5-exo and 6-endo, providing new strategies for enzymatic radical reactions and C–N bond formation [53]. In 2023, Hyster and coworkers reported the photoenzymatic reductive coupling reaction of electron-deficient heterocyclic compounds (such as chloromethylpyridine) and olefins catalyzed by flavin-dependent EREDs [54]. In 2023, a photoenzymatic stereoselective hydrosulfonylation reaction for the synthesis of chiral sulfones was reported by Xu and coworkers, providing an effective method for the preparation of chiral sulfones [55]. In 2023, highly chemoselective and stereoselective reactions between nitroalkanes and alkyl halides were achieved using engineered flavin-dependent EREDs [56]. In 2023, Zhao and coworkers achieved remote stereocontrol of azaarenes through a photoinduced chiral hydrogen atom transfer mechanism, successfully synthesizing diverse azaarene compounds with remote γ-stereocenters [57]. In 2023, Hyster and coworkers reported that the evolution of flavin-dependent "ene"-reductases enabled the enantioselective carbohydroxylation of alkenes by enzymatic catalysis, allowing the preparation of enantiomer-enriched tertiary alcohols [58]. In 2023, Hyster and coworkers developed variants of flavin-dependent EREDs such as PagER through directed evolution, enabling these enzymes to catalyze regioselective radical alkylation of electron-rich/electron-deficient arenes [59]. In 2023, Huang and coworkers enhanced the reducing ability of flavin-hydroquinone cofactors of EREDs by directly exciting them with visible light, achieving enantioselective single-electron reduction of enamides [60]. In 2023, Zhao and coworkers reported a visible-light-induced intermolecular radical hydroalkylation reaction catalyzed by "ene"-reductases. Using α-halocarbonyl compounds and terminal alkenes as substrates, the enantioselective radical reaction is triggered by forming an electron donor–acceptor complex [61]. In 2024, Zhao and coworkers utilized flavin-dependent "ene"-reductases to catalyze the visible-light-induced generation of amino-radical cations from hydroxylamines, which then underwent asymmetric intermolecular hydroamination reactions with vinylarenes [62]. Hyster and coworkers reported the photoenzymatic hydroalkylation of α,α-dichloroamides with alkenes catalyzed by flavin-dependent "ene"-reductases engineered via protein engineering [63]. In 2024, a photoenzymatic catalysis method was developed Zhao and coworkers to asymmetrically introduce fluorine-containing groups into alkenes, providing a new approach for the synthesis of fluorinated compounds [64]. In 2024, Hyster and coworkers reported two engineered flavin-dependent EREDs that can catalyze the regioselective hydroalkylation of α-bromoketones with cyclic and linear dienes. By regulating the selectivity of HAT in the radical termination step, γ, δ- and δ, ε-unsaturated ketones, and other constitutional isomers were produced in high yields, respectively [65]. In 2025, Xu and coworkers developed a ground-state flavin-dependent enzyme-catalyzed enantioselective radical trifluoromethylation strategy. They utilized engineered enzymes to catalyze hydrotrifluoromethylation and trifluoromethyl-alkyl cross-electrophilic coupling reactions. The reactions were initiated by single-electron transfer from the ground-state flavin hydroquinone and quenched by hydrogen atom transfer from the flavin semiquinone, achieving high yields and high enantioselectivity [66]. In the same year, Zhao and coworkers develops a photoenzymatic strategy using repurposed flavin-dependent "ene"-reductases for the stereoablative enantioconvergence of γ-chiral oximes via hydrogen atom transfer, enabling the conversion of racemic γ-chiral oximes into enantioenriched γ-chiral ketones with high enantioselectivity through a sequence involving photoinduced single-electron reduction, stereoablative 1,5-HAT, enzymatic enantioconvergent HAT, and spontaneous imine hydrolysis [67]. Recently, Huang and colleagues developed a photoenzymatic kinetic resolution strategy by repurposing “ene”-reductases as unnatural photohydrolases, which enables the synthesis of enantioenriched arylsulfinamides (up to 99.5:0.5 er) through an excited “ene”-reductase-triggered nitrogen-centered radical pathway under visible light excitation, providing an alternative biocatalytic route to chiral sulfinamides [68].

4.2 Indirect Visible-Light-Excited Reduction-Initiated Reaction

As shown in Fig. 3B, the reaction involves the excitation of an exogenous cofactor, with electrons transferring from the exogenous cofactor to the substrate to enable chemical transformations within the protein active site. In 2019, Hyster and coworkers discovered that in the presence of a photoredox catalyst, flavin-dependent EREDs can achieve the reduction of aromatic ketones to chiral alcohols via the generation of ketyl radical intermediates, without the need to modify the enzyme or cofactor. This demonstrates the potential of synergizing photocatalysis with biocatalysis to expand new pathways for enzymatic reactions [69]. In 2020, it was reported that in the presence of a photoredox catalyst, flavin-dependent EREDs achieve the asymmetric hydrogenation of heteroaromatic olefins (such as vinyl pyridines) through a radical mechanism. A neutral benzylic radical intermediate is generated, and then a chiral product is obtained via stereoselective hydrogen atom transfer at the enzyme active site. This demonstrates a new pathway for the synergistic expansion of the asymmetric reduction of heterocyclic compounds by enzymes and photocatalysis [70]. In 2023, Hyster and coworkers achieved enantioselective decarboxylative alkylation of amino acids with vinylpyridines through a synergistic photoenzymatic catalysis strategy, which combined flavin-dependent EREDs with the exogenous $Ru(bpy)_3^{2+}$ cofactor [71]. In 2025, Huang and colleagues developed a synergistic photo/biocatalytic system consisting of an ene-reductase and green-light-excited Rhodamine B, which enables enantioselective hydroamination of enamides via nitrogen-centered radicals to synthesize diverse enantioenriched vicinal diamines with excellent enantioselectivities (up to >99% enantiomeric excess) under green light excitation [72].

4.3 Direct Visible-Light-Excited Oxidation-Initiated Reaction

As shown in Fig. 3C, reactions in which a cofactor within the protein active site is photoexcited, promoting electron transfer from the substrate to the cofactor to convert the starting material into products. In 2023, Huang and coworkers achieved redox-neutral asymmetric radical hydroarylation of alkenes with electron-rich arenes using flavin-dependent ene-reductases, initiating the reaction via single-electron oxidation [73]. In 2024, Ye and coworkers reported an oxidation-initiated photoenzymatic enantioselective hydrothiolation of alkenes, which utilized an engineered Gluconobacter “ene”-reductase mutant (GluER-W100F-W342F). Without the need for electron donor–acceptor complexes or coenzyme regeneration systems, sulfonyl radicals are generated via single-electron oxidation of sulfites, and high enantioselectivity is achieved through tyrosine-mediated hydrogen atom transfer [74]. In 2024, Xu and coworkers developed a photoenzymatic-catalyzed redox-neutral radical hydrothiolation reaction initiated by oxidized flavin (FMN). Using sulfites or sulfonyl hydrazides as radical precursors, the excited-state FMN abstracts an electron from the substrate to generate a sulfonyl radical, which is then trapped by an alkene. The reaction is quenched by the stereoselective hydrogen-atom transfer of flavin semiquinone [75]. In 2024, the Ye group catalyzed the hydrosulfonylation reaction of olefins with sulfonates using a GluER mutant [76].

The research on flavin-dependent "ene"-reductases is continuously deepening, the applications of photoenzymatic catalysis are gradually expanding, and protein engineering modifications have provided a powerful means for improving their performance and developing new functions. In the future, they are expected to be applied in more fields and achieve more research achievements.

5 Flavin-Dependent Baeyer–Villiger Monooxygenases

Mechanistic studies revealed that reactions in which a cofactor within the protein active site is photoexcited, promoting electron transfer to convert the starting material into products (Fig. 4).

In 2022, Wu and coworkers engineered cyclohexanone monooxygenase (CHMO) via structure-guided protein engineering, achieving

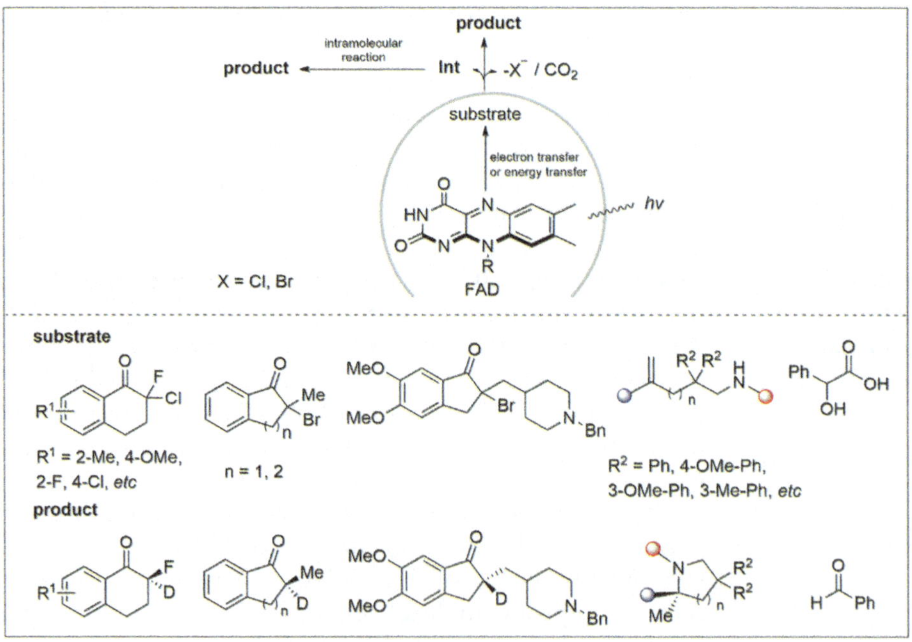

Fig. 4 Catalytic process of Baeyer–Villiger monooxygenases

enantioselective reductive dehalogenation of α,α-halofluoroketones through a photoinduced electron transfer/proton transfer (ET/PT) mechanism to efficiently synthesize chiral α-fluoroketones [77]. In 2023, Scholes and coworkers investigated through femtosecond spectroscopy, site-directed mutagenesis, and quantum-classical calculations, revealing the light-induced electron transfer mechanism from histidine to flavin in the photodecarboxylation reaction catalyzed by lactate monooxygenase (LMO), and developed a new reaction for the catalytic oxidative decarboxylation of mandelic acid to produce benzaldehyde [78]. In 2024, Wu and coworkers have demonstrated that engineered CHMO enables the highly selective synthesis of α -deuterated carbonyl compounds via photoinduced reductive dehalogenation in D_2O, achieving up to 92% yield, 98% enantiomeric excess (e.e.), and 96% deuterium incorporation rate. This study also validated the application potential of this method in the synthesis of deuterated drugs [79]. Recently, Scholes and coworkers reported that engineered cyclohexanone monooxygenase (CHMO) successfully synthesized chiral α-deuterated carbonyl compounds via photoinduced enantioselective reductive dehalogenation [80].

These four publications collectively focus on photoenzymatic catalysis and protein engineering, with a specific emphasis on the light-induced nonnatural catalytic activities of monooxygenase. Through protein engineering and mechanistic investigations (such as electron transfer, proton transfer, and radical intermediate regulation), these studies have achieved enantioselective synthesis of chiral fluorinated/deuterated carbonyl compounds, C–N bond formation, and decarboxylation reactions. They demonstrate the potential of photoenzymatic catalysis to construct complex chiral molecules and isotope-labeled compounds under mild conditions, providing new strategies for drug synthesis and green chemistry.

6 Pyridoxal-5′-Phosphate (PLP)-Dependent Enzymes

Mechanistic studies revealed that the excited-state photocatalyst promotes single-electron transfer (SET) of the radical precursor to generate an alkyl radical intermediate. Simultaneously, within the enzyme active site, a serine derivative is converted into an electrophilic intermediate Int via a pyridoxal 5′-phosphate (PLP)-catalyzed pathway. The alkyl radical subsequently enters the active site and adds to Int. Through electron transfer/proton transfer (ET/PT) or proton-coupled electron transfer (PCET) processes, an external aldimine is generated, which releases the target product upon hydrolysis (Fig. 5).

In 2020, Hyster and coworkers reported a method for dynamic kinetic resolution (DKR) by combining pyridoxal 5′-phosphate (PLP)-catalyzed racemization of α-amino ketones with ketoreductases

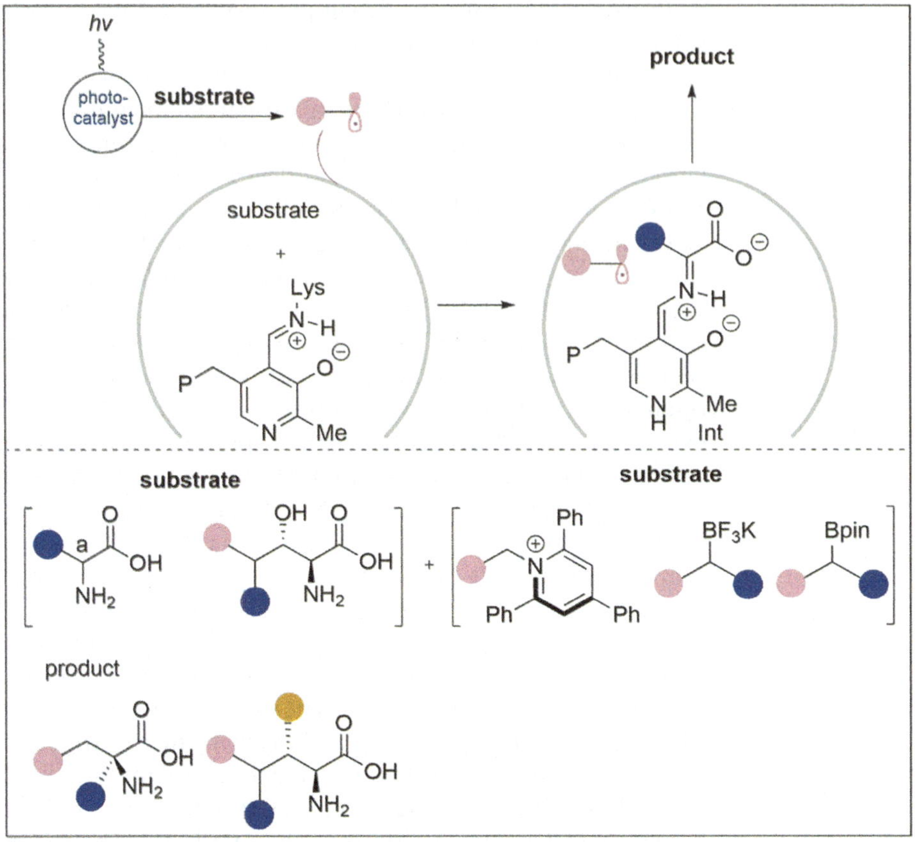

Fig. 5 Catalytic process of pyridoxal-5′-phosphate (PLP)-dependent enzymes

(KREDs), enabling efficient synthesis of highly diastereoselective and enantioselective 1,2-amino alcohols. Four stereoisomers can be obtained using commercially available KREDs [81]. In 2023, Yang and coworkers developed a synergistic photoredox–pyridoxal radical biocatalysis strategy. By using engineered PLP-dependent enzymes, they achieved stereoselective radical coupling between amino acids and alkyl boron reagents, successfully synthesizing α-trisubstituted and tetrasubstituted noncanonical amino acids (ncAAs) including products with three consecutive chiral centers. This study also revealed the stereocontrol mechanism of the enzyme active site over radical addition [82]. In 2024, researchers developed a photobiocatalytic asymmetric sp^3–sp^3 oxidative cross-coupling reaction. Through the synergistic action of engineered pyridoxal 5′-phosphate (PLP)-dependent threonine aldolase and a photoredox catalyst, they achieved α-C–H functionalization of glycine and α-branched amino acids, synthesizing noncanonical amino acids (ncAAs) containing consecutive chiral centers [83]. In 2024, Yang and coworkers reported a light-induced radical α-alkylation reaction catalyzed by threonine aldolase. Using Katritzky pyridinium salts as radical precursors, the reaction initiates radicals through PLP-derived aldimine intermediates. Without the need for traditional photocatalysts, this approach achieves enantioselective alkylation of amino acids and synthesizes α-trisubstituted/tetrasubstituted products [84]. In 2025, Yang and colleagues developed second-generation PLP radical enzymes through directed evolution for the enantiodivergent photobiocatalytic synthesis of noncanonical amino acids [85]. Recently, they developed a stereoselective three-component radical-mediated C–C coupling via cooperative photobiocatalysis, using directed evolution of repurposed pyridoxal decarboxylases to achieve full fragment variability [86]. They also developed a photobiocatalytic oxidative coupling method for the enantioselective synthesis of α -tetrasubstituted noncanonical amino acids [87].

Photobiocatalysis by pyridoxal 5′-phosphate (PLP)-dependent enzymes or dynamic kinetic resolution (DKR) for the stereoselective synthesis of noncanonical amino acids (ncAAs) and chiral compounds demonstrates the potential of photoenzyme synergistic catalysis and PLP enzyme engineering in constructing complex chiral molecules.

7 Thiamine Pyrophosphate (TPP)-Dependent Oxidoreductases

Mechanistic studies show that the excited-state photocatalyst initiates single-electron oxidation of the enzyme-bound Breslow intermediate, leading to radical generation within the enzyme active site. The photocatalyst-derived radical anion reduces the radical precursor, generating a prochiral radical (initially located outside the active site). Finally, the prochiral radical undergoes radical cross-coupling with the radical in the active site to enantioselectively form a $C(sp^2)–C(sp^3)$ bond, with the final product released after cleavage of the catalytically active cofactor (Fig. 6).

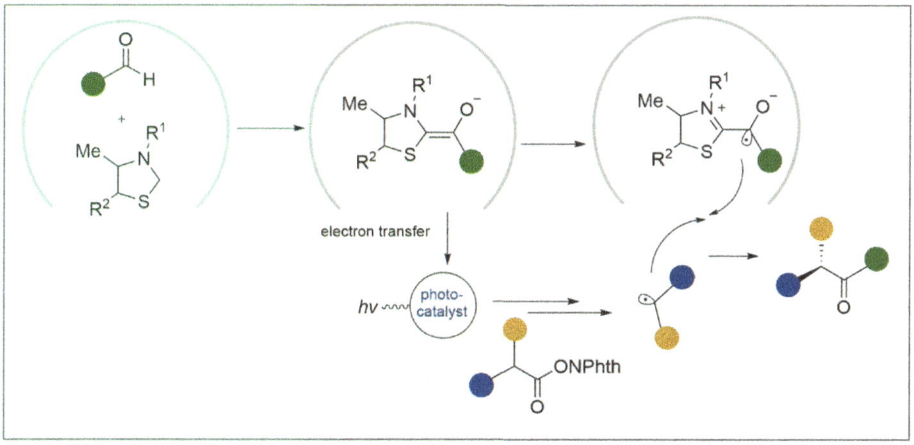

Fig. 6 Catalytic process of thiamine pyrophosphate (TPP)-dependent oxidoreductases

Huang and coworkers utilized the cross-coupling of enzyme-bound TPP-derived ketyl radicals and alkyl radicals to efficiently synthesize various chiral ketones [88]. In 2024, Huang and coworkers utilized the cross-coupling of enzyme-bound TPP-derived ketyl radicals and alkyl radicals to efficiently synthesize various chiral ketones [89]. Yang and coworkers developed an asymmetric decarboxylative alkylation reaction of benzaldehydes and α-keto acids by synergistically using a fluorescein photocatalyst and engineered TPP-dependent enzymes (such as BFD and BAL), generating highly enantioselective α-branched ketones. The study revealed the mechanism involving single-electron oxidation of enzyme-bound Breslow intermediates and cross-coupling with photo-generated alkyl radicals [90].

Through the synergy of protein engineering and photoredox catalysis, TPP-dependent enzymes have broken through the limitations of traditional two-electron pathways in enzymatic catalysis, enabling radical-mediated enantioselective carbon–carbon bond formation.

8 Pyrroloquinoline Quinone (PQQ)-Dependent Enzyme

PQQ-dependent alcohol dehydrogenases are a key tool that helps bacteria "break down alcohol," functioning with the aid of the cofactor "PQQ" and metal ions [91].

As shown in Fig. 7, Zeymer group reported that natural redox cofactor pyrroloquinoline quinone (PQQ) enables photocatalytic radical cyclizations under visible-light irradiation, with its trimethyl ester ($PQQMe_3$) functioning as a standalone photoredox catalyst in organic solvents and PQQ-dependent dehydrogenases achieving stereoselective photoenzymatic catalysis in aqueous buffers (e.g., 69% yield and 82:18 enantiomeric ratio for oxindole formation), thereby expanding the toolbox of stereoselective photobiocatalysis [92]. This was the first report on the PQQ enzyme-catalyzed new-to-nature reactions.

9 Metalloenzymatic

As shown in Fig. 8A, in the first cycle, the excited-state photocatalyst reduces the NHPI ester via a SET process, generating a $C(sp^3)$ radical through sequential fragmentation and decarboxylation steps. The oxidized photocatalyst then oxidizes the Fe(II) center to form an Fe(III)-N_3 intermediate for radical capture. In 2024, Huang and coworkers developed a synergistic photoredox-non-heme iron enzyme (SavHPPD) platform, enabling enantioselective decarboxylative

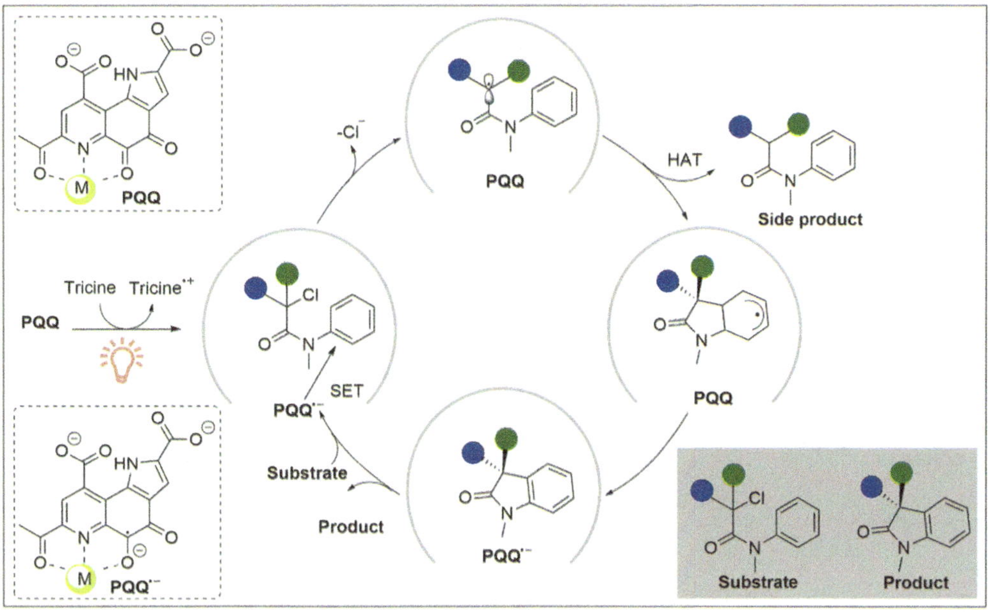

Fig. 7 Catalytic process of pyrroloquinoline quinone (PQQ)-dependent enzyme

azidation and thiocyanation of *N*-hydroxyphthalimide esters using Eosin Y as the photocatalyst [93].

As shown in Fig. 8B, the *N*,*O*-bidentate coordinated substrate undergoes deprotonation to form the enzymatic enolate intermediate **Int. B**. Meanwhile, the excited-state photosensitizer triggers the generation of a ketyl radical **Int. C** from the α-halo ketone. **Int. B** adds to the ketyl radical to form **Int. D**, and this enolate intermediate **Int. D** initiates the reductive quenching of excited $[Ru]^{2+*}$, generating product-bound **Int. E**. Finally, product/substrate exchange initiates the next catalytic cycle. In 2025, Huang and coworkers developed a synergistic Lewis acid enzyme-photoredox catalysis platform, achieving enantiodivergent radical alkylation of 2-acyl imidazoles through directed evolution of copper-binding cupin variants, enabling efficient synthesis of α-chiral ketones with high enantioselectivity [94].

By integrating the radical-generating capability of photoredox catalysis with the enantioselective control of metalloenzymes and Lewis acid enzymes, new asymmetric synthesis methods have been developed.

10 Artificial Photoenzymes

As shown in Fig. 9, Wu and coworkers harnessed genetic code expansion technology to incorporate unnatural photosensitizers into the protein LmrR. Through multiple rounds of directed evolution

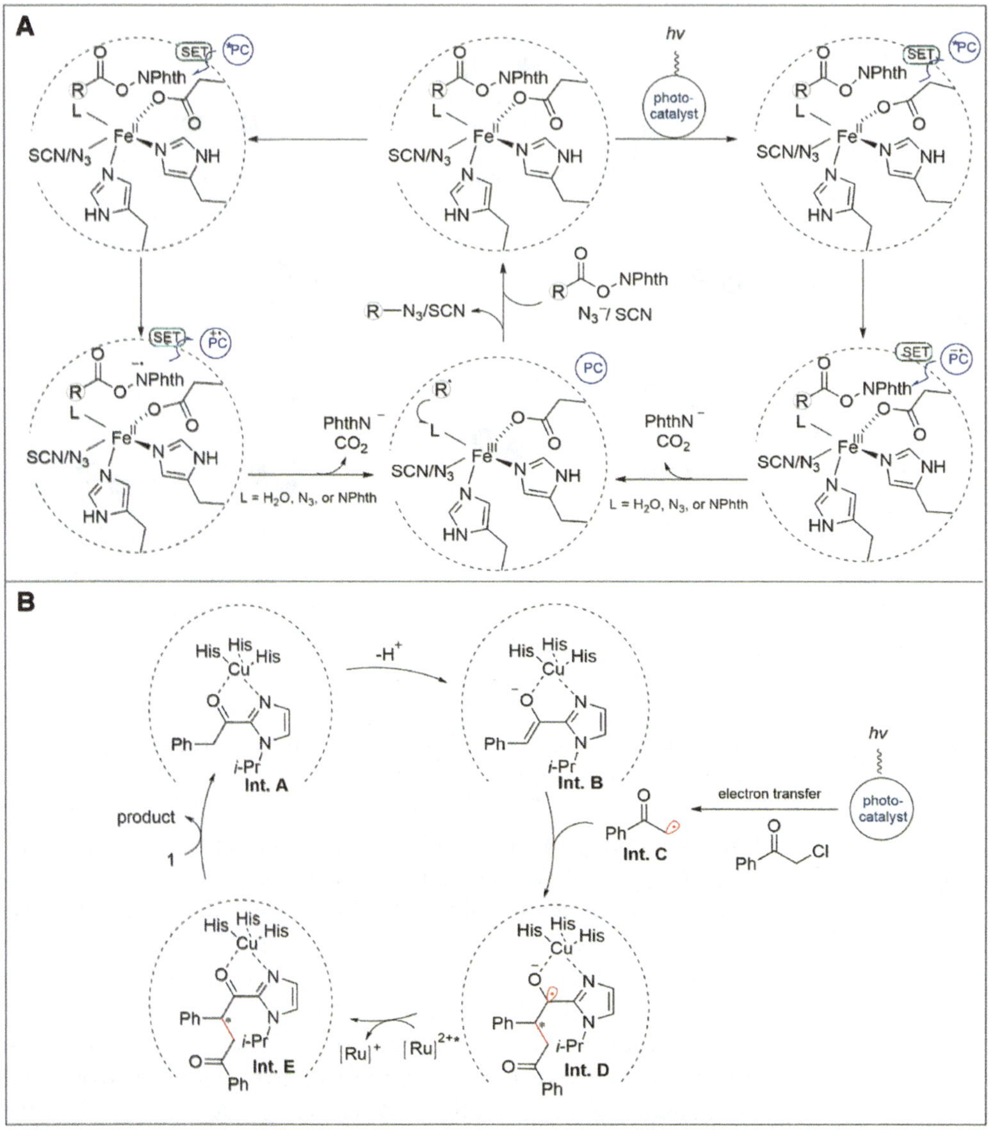

Fig. 8 Catalytic process of metalloenzymatic (**A**) iron enzyme catalysis, (**B**) copper enzyme catalysis [93, 94]

and crystal structure-guided optimization, they developed the triplet-state photoenzyme TPe4.0_FBpA, which efficiently catalyzes the enantioselective [2+2] cycloaddition of indole derivatives (yields 80%–97%, enantiomeric excess ≥90%) and enables whole-cell catalytic applications [95]. Green and coworkers designed a photoenzyme, EnT1.3, using genetic code expansion technology. Following directed evolution, this photoenzyme can catalyze enantioselective [2+2] cycloaddition reactions via triplet energy transfer. It operates efficiently under aerobic conditions, and its catalytic mechanism has been structurally elucidated [96]. In 2024, Wu and coworkers

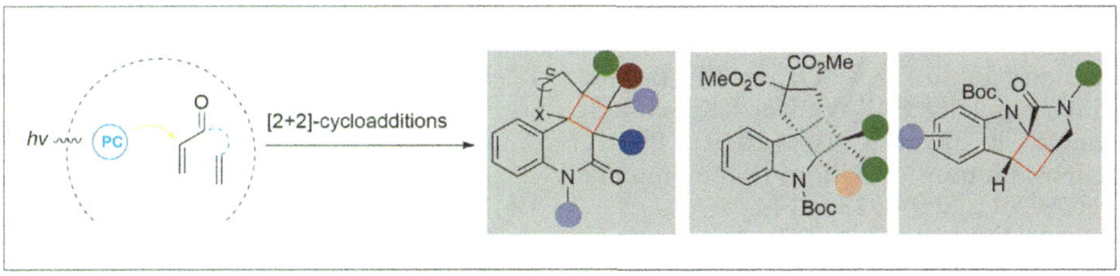

Fig. 9 Catalytic process of artificial photoenzymes

developed a chemogenetic modification strategy to introduce artificial photosensitizers into the protein scaffold LmrR via cysteine bio-conjugation chemistry. Through directed evolution, they optimized the photoenzyme CMP4.0, enabling enantioselective [2+2] cycloaddition of 2-carboxamide indoles under visible light and whole-cell catalysis [97].

The development of artificial photoenzymes has broken through the limitations of traditional organic catalysis and can be applied to enantioselective [2+2] cycloaddition reactions, providing a new biocatalytic paradigm for light-driven asymmetric synthesis and demonstrating the potential of combining protein engineering with photochemistry.

11 Conclusion Remarks

The field of biocatalysis is undergoing vigorous and exciting development, with enzymes holding the potential to expand the known chemical space and guide chemists to explore underexplored transformation reactions. Unlike small-molecule catalysts, engineered enzymes can exert unique and precise control over individual steps in catalytic cycles, including conformational control of substrates and reaction intermediates, such as carbocations and radicals.

Although enzymes are well-suited for stereoselective reactions, the range of reaction types currently accessible to biocatalysts remains narrower compared with small-molecule catalysts. Therefore, exploring the reactivity of general activation mechanisms in engineered enzymes represents an extremely valuable opportunity for chemists.

Acknowledgments

We thank the Fundamental Research Funds for the Central Universities for financial support.

References

1. Schmermund L, Jurkaš V, Özgen FF, Barone GD, Büchsenschütz HC, Winkler CK, Schmidt S, Kourist R, Kroutil W (2019) Photo-biocatalysis: biotransformations in the presence of light. ACS Catal 9:4115–4144
2. Emmanuel MA, Bender SG, Bilodeau C, Carceller JM, DeHovitz JS, Fu H, Liu Y, Nicholls BT, Ouyang Y, Page CG (2023) Photobiocatalytic strategies for organic synthesis. Chem Rev 123:5459–5520
3. Jain S, Ospina F, Hammer SC (2024) A new age of biocatalysis enabled by generic activation modes. J Am Chem Soc 4:2068–2080
4. Yu J, Chen B, Huang X (2024) Single-electron oxidation triggered by visible-light-excited enzymes for asymmetric biocatalysis. Angew Chem Int Ed 64:e202419262
5. Hyster TK (2019) Radical biocatalysis: using non-natural single electron transfer mechanisms to access new enzymatic functions. Synlett 31:248–254
6. Seel CJ, Gulder T (2019) Biocatalysis fueled by light: on the versatile combination of photocatalysis and enzymes. ChemBioChem 20:1871–1897
7. Huang X, Cao M, Zhao H (2020) Integrating biocatalysis with chemocatalysis for selective transformations. Curr Opin Chem Biol 55:161–170
8. Chen K, Arnold FH (2020) Engineering new catalytic activities in enzymes. Nat Catal 3:203–213
9. Pagar D et al (2021) Recent advances in biocatalysis with chemical modification and expanded amino acid alphabet. Chem Rev 121:6173–6245
10. Peng Y et al (2022) Recent advances in photobiocatalysis for selective organic synthesis. Org Process Res Dev 26:1900–1913
11. Harrison W, Huang X, Zhao H (2022) Photobiocatalysis for abiological transformations. Acc Chem Res 55:1087–1096
12. Özgen FF, Runda ME, Schmidt S (2022) Photo-biocatalytic cascades: combining chemical and enzymatic transformations fueled by light. ChemBioChem 22:790–806
13. Alphand V et al (2023) Exciting enzymes: current state and future perspective of photobiocatalysis. ChemPhotoChem 7:e202200325
14. Wang JP, Zong MH, Li N (2024) Photobiocatalysis: a promising tool for sustainable synthesis. Chem Catal 4:100933. https://doi.org/10.1016/j.checat.2024.100933
15. Klein AS et al (2024) A de novo metalloenzyme for cerium photoredox catalysis. J Am Chem Soc 146:25976–25985
16. Runda ME, Schmidt S (2024) Light-driven bioprocesses. Phys Sci Rev 9:2287–2320
17. Fu H, Hyster TK (2024) From ground-state to excited-state activation modes: flavin-dependent 'ene'-reductases catalyzed non-natural radical reactions. Acc Chem Res 57:1446–1457
18. Birch-Price Z et al (2024) Noncanonical amino acids in biocatalysis. Chem Rev 124:8740–8786
19. Xu Y, Liu F, Zhao B, Huang X (2024) Repurposing naturally occurring enzymes using visible light. Chin J Chem 42:3553–3558
20. Liu F, Peng X, Yu J, Huang X (2025) Recent advances in repurposing natural enzymes for new-to-nature asymmetric photobiotransformations. Org Chem Front 12:4608–4634
21. Zhang Z, Li M, Zhao H (2025) Mechanistic investigation of repurposed photoenzymes with new-to-nature reactivity. Curr Opin Green Sustain Chem 52:101009. https://doi.org/10.1016/j.cogsc.2025.101009
22. Sancar A (2003) Structure and function of DNA photolyase and cryptochrome blue-light photoreceptors. Chem Rev 103:2203–2237
23. Begle TP (1994) Photoenzymes: a novel class of biological catalysts. Acc Chem Res 27:394–401
24. Sorigué D, Légeret B, Cuiné S, Blangy S, Moulin S, Billon E, Richaud P, Brugière S, Couté Y, Nurizzo D, Müller P, Brettel K, Pignol D, Arnoux P, Li-Beisson Y, Peltier G, Beisson F (2017) An algal photoenzyme converts fatty acids to hydrocarbons. Science 357:903–907
25. Sorigué D, Hadjidemetriou K, Blangy S, Gotthard G, Bonvalet A, Coquelle N, Samire P, Aleksandrov A, Antonucci L, Benachir A (2021) Mechanism and dynamics of fatty acid photodecarboxylase. Science 372:eabd5687
26. Huijbers MME, Zhang W, Tonin F, Hollmann F (2018) Light-driven enzymatic decarboxylation of fatty acids. Angew Chem Int Ed 57:13648–13651
27. Zhang W, Lee JH, Younes SHH, Tonin F, Hagedoorn PL, Pichler H, Baeg Y, Park JB, Kourist R, Hollmann F (2020) Photobiocatalytic synthesis of chiral secondary fatty alcohols from renewable unsaturated fatty acids. Nat Commun 11. https://doi.org/10.1038/s41467-020-16099-7
28. Zhang WY, Ma M, Huijbers M, Filonenko GA, Pidko EA, Schie M, Boer S, Burek BO, Bloh JZ, Berkel WJH, Smith WA, Hollmann F (2019) Hydrocarbon synthesis via photoenzymatic decarboxylation of carboxylic acids. J Am Chem Soc 141:3116–3120
29. Heyes DJ, Lakavath B, Hardman SJO, Sakuma M, Hedison TM, Scrutton NS (2020) Photochemical mechanism of light-driven fatty acid photodecarboxylase. ACS Catal 10:6691–6696
30. Xu J, Fan J, Lou Y, Xu W, Wang Z, Li D, Zhou H, Lin X, Wu Q (2021) Light-driven

decarboxylative deuteration enabled by a divergently engineered photodecarboxylase. Nat Commun 12:3983. https://doi.org/10.1038/s41467-021-24259-6
31. Li D, Han T, Xue J, Xu W, Xu J, Wu Q (2021) Engineering fatty acid photodecarboxylase to enable highly selective decarboxylation oftransfatty acids. Angew Chem Int Ed 60:20695–20699
32. Mou K, Guo Y, Xu W, Li D, Wang Z, Wu Q (2024) Stereodivergent protein engineering of fatty acid photodecarboxylase for light-driven kinetic resolution of sec-alcohol oxalates. Angew Chem Int Ed 63:e202318374
33. Xu J, Hu Y, Fan J, Arkin M, Li D, Peng Y, Xu W, Lin X, Wu Q (2019) Light-driven kinetic resolution of α-functionalized carboxylic acids enabled by an engineered fatty acid photodecarboxylase. Angew Chem Int Ed 58:8474–8478
34. Cheng F, Li H, Wu DY, Li JM, Fan Y, Xue YP, Zheng YG (2020) Light-driven deracemization of phosphinothricin by engineered fatty acid photodecarboxylase on a gram scale. Green Chem 22:6815–6818
35. Ju S, Li D, Mai BK, Liu X, Vallota-Eastman A, Wu J, Valentine DL, Liu P, Yang Y (2024) Stereodivergent photobiocatalytic radical cyclization through the repurposing and directed evolution of fatty acid photodecarboxylases. Nat Chem 16:1339–1347
36. Xu W, Mou K, Lu Z, Kang X, Guo Y, Ding B, Chen Z, Wang Z, Wu Q (2024) Catalytic promiscuity of fatty acid photodecarboxylase enables stereoselective synthesis of chiral α-tetralones. Angew Chem Int Ed 63:e202412862
37. Emmanuel MA, Greenberg NR, Oblinsky DG, Hyster TK (2016) Accessing non-natural reactivity by irradiating nicotinamide-dependent enzymes with light. Nature 540:414–417
38. Huang X, Feng J, Cui J, Jiang G, Harrison W, Zang X, Zhou J, Wang B, Zhao H (2022) Photoinduced chemomimetic biocatalysis for enantioselective intermolecular radical conjugate addition. Nat Catal 5:586–593
39. Chen B, Li R, Feng J, Zhao B, Zhang J, Yu J, Xu Y, Xing Z, Zhao Y, Wang B et al (2024) Modular access to chiral amines via imine reductase-based photoenzymatic catalysis. J Am Chem Soc 146:14278–14286
40. Biegasiewicz KF, Cooper SJ, Emmanuel MA, Miller DC, Hyster TK (2018) Catalytic promiscuity enabled by photoredox catalysis in nicotinamide-dependent oxidoreductases. Nat Chem 10:770–775
41. Biegasiewicz KF, Cooper SJ, Gao X, Oblinsky DG, Kim JH, Garfinkle SE, Joyce LA, Sandoval BA, Scholes GD, Hyster TK (2019) Photoexcitation of flavoenzymes enables a stereoselective radical cyclization. Science 364:1166–1169
42. Clayman PD, Hyster TK (2020) Photoenzymatic generation of unstabilized alkyl radicals: an asymmetric reductive cyclization. J Am Chem Soc 142:15673–15677
43. Huang X, Wang B, Wang Y, Jiang G, Feng J, Zhao H (2020) Photoenzymatic enantioselective intermolecular radical hydroalkylation. Nature 584:69–74
44. Page CG, Cooper SJ, DeHovitz JS, Oblinsky DG, Biegasiewicz KF, Antropow AH, Armbrust KW, Ellis JM, Hamann LG, Horn EJ et al (2020) Quaternary charge-transfer complex enables photoenzymatic intermolecular hydroalkylation of olefins. J Am Chem Soc 143:97–102
45. Sandoval BA, Clayman PD, Oblinsky DG, Oh S, Nakano Y, Bird M, Scholes GD, Hyster TK (2021) Photoenzymatic reductions enabled by direct excitation of flavin-dependent 'ene'-reductases. J Am Chem Soc 143:1735–1739
46. Fu H, Lam H, Emmanuel MA, Kim JH, Sandoval BA, Hyster TK (2021) Ground-state electron transfer as an initiation mechanism for biocatalytic C–C bond forming reactions. J Am Chem Soc 143:9622–9629
47. Gao X, Turek-Herman JR, Choi YJ, Cohen RD, Hyster TK (2021) Photoenzymatic synthesis of α-tertiary amines by engineered flavin-dependent 'ene'-reductases. J Am Chem Soc 143:19643–19647
48. Nicholls BT, Oblinsky DG, Kurtoic SI, Grosheva D, Ye Y, Scholes GD, Hyster TK (2022) Engineering a non-natural photoenzyme for improved photon efficiency. Angew Chem Int Ed 61:e202113842
49. Nicholls BT, Qiao T, Hyster TK (2022) A photoenzyme for challenging lactam radical cyclizations. Synlett 33:1204–1208
50. Cesana PT, Page CG, Harris D, Emmanuel MA, Hyster TK, Schlau-Cohen GS (2022) Photoenzymatic catalysis in a new light: gluconobacter 'ene'-reductase conjugates possessing high-energy reactivity with tunable low-energy excitation. J Am Chem Soc 144:17516–17521
51. Fu H, Cao J, Qiao T, Qi Y, Charnock SJ, Garfinkle S, Hyster TK (2022) An asymmetric sp^3–sp^3 cross-electrophile coupling using 'ene'-reductases. Nature 610:302–307
52. Laguerre N, Riehl PS, Oblinsky DG, Emmanuel MA, Black MJ, Scholes GD, Hyster TK (2022) Radical termination via β-scission enables photoenzymatic allylic alkylation using 'ene'-reductases. ACS Catal 12:9801–9805
53. Ye Y, Cao J, Oblinsky DG, Verma D, Prier CK, Scholes GD, Hyster TK (2022) Using enzymes to tame nitrogen-centred radicals for enantioselective hydroamination. Nat Chem 15:206–212
54. Bender SG, Hyster TK (2023) Pyridylmethyl radicals for enantioselective alkene

hydroalkylation using 'ene'-reductases. ACS Catal 13:14680–14684
55. Chen X, Zheng D, Jiang L, Wang Z, Duan X, Cui D, Liu S, Zhang Y, Yu X, Ge J et al (2023) Photoenzymatic hydrosulfonylation for the stereoselective synthesis of chiral sulfones. Angew Chem Int Ed 62:e202218140
56. Fu H, Qiao T, Carceller JM, MacMillan SN, Hyster TK (2023) Asymmetric C-alkylation of nitroalkanes via enzymatic photoredox catalysis. J Am Chem Soc 145:787–793
57. Li M, Harrison W, Zhang Z, Yuan Y, Zhao H (2023) Remote stereocontrol with azaarenes via enzymatic hydrogen atom transfer. Nat Chem 16:277–284
58. Ouyang Y, Turek-Herman J, Qiao T, Hyster TK (2023) Asymmetric carbohydroxylation of alkenes using photoenzymatic catalysis. J Am Chem Soc 145:17018–17022
59. Page CG, Cao J, Oblinsky DG, MacMillan SN, Dahagam S, Lloyd RM, Charnock SJ, Scholes GD, Hyster TK (2023) Regioselective radical alkylation of arenes using evolved photoenzymes. J Am Chem Soc 145:11866–11874
60. Zhang J, Zhang Q, Chen B, Yu J, Wang B, Huang X (2023) Photoenzymatic conversion of enamides to enantioenriched benzylic amines enabled by visible-light-induced single-electron reduction. ACS Catal 13:15682–15690
61. Zhang Z, Feng J, Yang C, Cui H, Harrison W, Zhong D, Wang B, Zhao H (2023) Photoenzymatic enantioselective intermolecular radical hydroamination. Nat Catal 6:687–694
62. Harrison W, Jiang G, Zhang Z, Li M, Chen H, Zhao H (2024) Photoenzymatic asymmetric hydroamination for chiral alkyl amine synthesis. J Am Chem Soc 146:10716–10722
63. Liu Y, Bender SG, Sorigue D, Diaz DJ, Ellington AD, Mann G, Allmendinger S, Hyster TK (2024) Asymmetric synthesis of α-chloroamides via photoenzymatic hydroalkylation of olefins. J Am Chem Soc 146:7191–7197
64. Li M, Yuan Y, Harrison W, Zhang Z, Zhao H (2024) Asymmetric photoenzymatic incorporation of fluorinated motifs into olefins. Science 385:416–421
65. Petchey MR, Ye Y, Spelling V, Finnigan JD, Gittings S, Johansson MJ, Hayes MA, Hyster TK (2024) Regiodivergent radical termination for intermolecular biocatalytic C–C bond formation. J Am Chem Soc 146:5005–5010
66. Duan X, Cui D, Wang M, Jin C, Cai X, Wang Z, Xu J (2025) Ground-state flavin-dependent enzymes catalyzed enantioselective radical trifluoromethylation. Nat Commun 16:1225. https://doi.org/10.1038/s41467-025-56437-1
67. Zhang Z, Li M, Harrison W, Lu J, Zhao Z, Yuan Y, Zhao H (2025) Photoenzymatic stereoablative enantioconvergence of γ-chiral oximes via hydrogen atom transfer. Nat Catal 8:548–555
68. Zhang Z, Zhang Q, Wang T, Zhao B, Chen B, Wang X, Chun J, Zhang T, Wang B, Huang X (2025) Kinetic hydrolysis of sulfinamides toward S(IV) chirality enabled by visible-light-excited 'ene'-reductases. ACS Catal 15:12684–12690
69. Sandoval BA, Kurtoic SI, Chung MM, Biegasiewicz KF, Hyster TK (2019) Photoenzymatic catalysis enables radical-mediated ketone reduction in ene-reductases. Angew Chem Int Ed 58:8714–8718
70. Nakano Y, Black MJ, Meichan AJ, Sandoval BA, Chung MM, Biegasiewicz KF, Zhu T, Hyster TK (2020) Photoenzymatic hydrogenation of heteroaromatic olefins using 'ene'-reductases with photoredox catalysts. Angew Chem Int Ed 59:10484–10488
71. Sun SZ, Nicholls BT, Bain D, Qiao T, Page CG, Musser AJ, Hyster TK (2023) Enantioselective decarboxylative alkylation using synergistic photoenzymatic catalysis. Nat Catal 7:35–42
72. Shi F, Chen B, Yu J, Zhu R, Zheng Y, Huang X (2025) Enantioselective biosynthesis of vicinal diamines enabled by synergistic photo/biocatalysis consisting of an 'ene'-reductase and a green-light-excited organic dye. Chin J Catal 68:223–229
73. Zhao B, Feng J, Yu L, Xing Z, Chen B, Liu A, Liu F, Shi F, Zhao Y, Tian C et al (2023) Direct visible-light-excited flavoproteins for redox-neutral asymmetric radical hydroarylation. Nat Catal 6:996–1004
74. Shi Q, Kang XW, Liu Z, Sakthivel P, Aman H, Chang R, Yan X, Pang Y, Dai S, Ding B et al (2024) Single-electron oxidation-initiated enantioselective hydrosulfonylation of olefins enabled by photoenzymatic catalysis. J Am Chem Soc 146:2748–2756
75. Jiang L, Zheng D, Chen X, Cui D, Duan X, Wang Z, Ge J, Xu J (2024) Photoenzymatic redox-neutral radical hydrosulfonylation initiated by FMN. ACS Catal 14:6710–6716
76. Ye J, Sakthivel P, Shi Q (2024) Chiral sulfones via single-electron oxidation-initiated photoenzymatic catalysis. Synlett 36:8–14
77. Peng Y, Wang Z, Chen Y, Xu W, Hu Y, Chen Z, Xu J, Wu Q (2022) Photoinduced promiscuity of cyclohexanone monooxygenase for the enantioselective synthesis of α-fluoroketones. Angew Chem Int Ed 61:e202211199
78. Li X, Page CG, Zanetti-Polzi L, Kalra AP, Oblinsky DG, Daidone I, Hyster TK, Scholes GD (2023) Mechanism and dynamics of photodecarboxylation catalyzed by lactate monooxygenase. J Am Chem Soc 145:13232–13240
79. Peng Y, Xia K, Wu Q (2024) Photoinduced promiscuity of cyclohexanone monooxygenase

for the enantioselective synthesis of α-deuterated carbonyl compounds. Mol Catal 553:113709. https://doi.org/10.1016/j.mcat.2023.113709

80. Raps FC, Rivas-Souchet A, Jones CM, Hyster TK (2024) Emergence of a distinct mechanism of C–N bond formation in photoenzymes. Nature 637:362–368
81. Cao J, Hyster TK (2020) Pyridoxal-catalyzed racemization of α-aminoketones enables the stereodivergent synthesis of 1,2-amino alcohols using ketoreductases. ACS Catal 10:6171–6175
82. Cheng L, Li D, Mai BK, Bo Z, Cheng L, Liu P, Yang Y (2023) Stereoselective amino acid synthesis by synergistic photoredox-pyridoxal radical biocatalysis. Science 381:444–451
83. Wang TC, Mai BK, Zhang Z, Bo Z, Li J, Liu P, Yang Y (2024) Stereoselective amino acid synthesis by photobiocatalytic oxidative coupling. Nature 629:98–104
84. Wang TC, Zhang Z, Rao G, Li J, Shirah J, Britt RD, Zhu Q, Yang Y (2024) Threonine aldolase-catalyzed enantioselective alpha-alkylation of amino acids through unconventional photoinduced radical initiation. J Am Chem Soc 146:22476–22484
85. Cheng L, Bo Z, Krohn-Hansen B, Yang Y (2025) Directed evolution and unusual protonation mechanism of pyridoxal radical C–C coupling enzymes for the enantiodivergent photobiocatalytic synthesis of noncanonical amino acids. J Am Chem Soc 147:4602–4612
86. Zhang C, Zhou J, Xie PP, Rivera SM, Alturaifi TM, Finnigan J, Charnock S, Liu P, Yang Y (2025) Diversity-oriented photobiocatalytic synthesis via stereoselective three-component radical coupling. Science 389:1312–1323. https://doi.org/10.1126/science.adx2935
87. Wang TC, Zhang X, Zhang Z, Rao G, Bo Z, Xie PP, Li J, Yang Q, Li X, Britt RD et al (2025) Directed evolution of threonine aldolases as radical C–C bond forming enzymes for photobiocatalytic synthesis of β-branched α-tetrasubstituted non-canonical amino acids. ChemRxiv. https://doi.org/10.26434/chemrxiv-2025-gcljn
88. Xu Y, Chen H, Yu L, Peng X, Zhang J, Xing Z, Bao Y, Liu A, Zhao Y, Tian C et al (2023) A light-driven enzymatic enantioselective radical acylation. Nature 625:74–78
89. Xing Z, Liu F, Feng J, Yu L, Wu Z, Zhao B, Chen B, Ping H, Xu Y, Liu A et al (2024) Synergistic photobiocatalysis for enantioselective triple-radical sorting. Nature 637:1118–1123
90. Liu X, Xu S, Chen H, Yang Y (2024) Unnatural thiamine radical enzymes for photobiocatalytic asymmetric alkylation of benzaldehydes and α-ketoacids. ACS Catal 14:9144–9150
91. Sarmiento-Pavía PD, Sosa-Torres ME (2021) Bioinorganic insights of the PQQ-dependent alcohol dehydrogenases. J Biol Inorg Chem 26:177–203
92. Bahukhandi SB, Klein AS, Mustafa G, Weyh M, Walter A, Thyrhaug E, Hauer J, Storch G, Zeymer C (2025) The natural redox cofactor pyrroloquinoline quinone (PQQ) enables photocatalytic radical cyclizations. Angew Chem Int Ed 64:e202505431
93. Rui J, Mu X, Soler J, Paris JC, Guo Y, Garcia-Borràs M, Huang X (2024) Merging photoredox with metalloenzymatic catalysis for enantioselective decarboxylative $C(sp^3)$-N_3 and $C(sp^3)$-SCN bond formation. Nat Catal 7:1394–1403
94. Zhang J, Zhang Q, Ge R, Liu A, Chen B, Zhang Z, Zhao B, Yu J, Zhao Y, Yu L et al (2025) Enantiodivergent radical alkylation by synergistic Lewis-acid-enzyme and photoredox catalysis. Angew Chem Int Ed 64:e202500338
95. Sun N, Huang J, Qian J, Zhou TP, Guo J, Tang L, Zhang W, Deng Y, Zhao W, Wu G et al (2022) Enantioselective [2+2]-cycloadditions with triplet photoenzymes. Nature 611:715–720
96. Trimble JS, Crawshaw R, Hardy FJ, Levy CW, Brown MJB, Fuerst DE, Heyes DJ, Obexer R, Green AP (2022) A designed photoenzyme for enantioselective [2 + 2] cycloadditions. Nature 611:709–714
97. Guo J, Qian J, Cai D, Huang J, Yang X, Sun N, Zhang J, Pang T, Zhao W, Wu G et al (2024) Chemogenetic evolution of diversified photoenzymes for enantioselective [2 + 2] cycloadditions in whole cells. J Am Chem Soc 146:19030–19041

Chapter 5

A Yeast Two-Hybrid Assay to Develop Optogenetic Tools for Controlling Protein–Protein Interactions

Ryan M. Woloschuk and G. Andrew Woolley

Abstract

Yeast two hybrid (Y2H) assays serve as powerful platforms for the detection and engineering of protein–protein interactions. Here, we describe a strategy for using Y2H-based screening to assess and develop optogenetic tools based on light-controlled protein–protein interactions.

Keywords Yeast two hybrid, Optogenetics, Protein engineering, Photoswitchable proteins, Library screening, High-throughput screening

1 Introduction

Since its development in 1989 by Henry and Fields, yeast two hybrid (Y2H)-based screening has been a cornerstone in the detection of protein–protein interactions [1]. Y2H involves the use of a split transcription factor, where a DNA-binding domain is fused to one protein of interest and a transcriptional-activation domain to a second protein. If there is an interaction between the two proteins, the transcription factor is reconstituted, and one or more reporter genes are expressed (Fig. 1). Y2H screening efforts have mapped the protein-–protein interactome and have been used in protein engineering efforts to identify novel protein–protein interactions [2–4]. The wide use of Y2H stems from its ability to enable high-throughput screening in the context of a eukaryotic system.

Optogenetic tools offer superb spatial and temporal control of biological activity. For example, the use of microbial channelrhodopsins has revolutionized neuroscience by allowing patterned cell-type selective control of neuronal firing [5, 6]. Optogenetic tools are often created through rational design, or by transplanting naturally occurring photoswitchable proteins from existing systems into a system of interest [7, 8]. However, initial designs can require extensive optimization to function effectively in their desired application.

Huan Peng and S. Brian Kim (eds.), *Biological Macromolecule Engineering Technologies*,
Methods in Molecular Biology, vol. 3032, https://doi.org/10.1007/978-1-0716-5257-2_5,

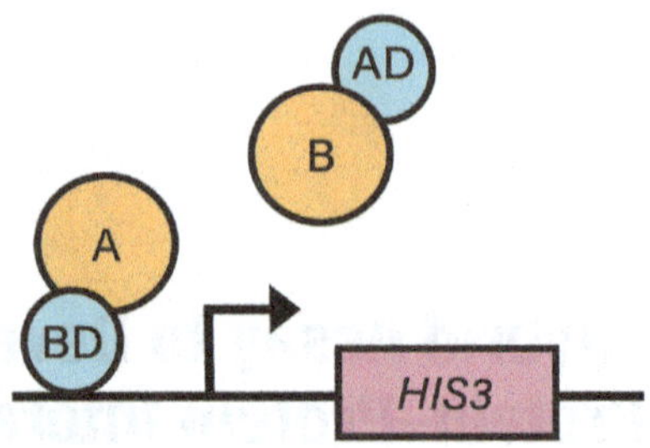

Fig. 1 Y2H schematic. In Y2H, a transcription factor is split between its DNA-binding domain (BD) and its transcriptional-activation domain (AD). The two halves of the transcription factor are then fused to two proteins (A and B). If the two proteins interact, they activate the transcription of a *HIS3* reporter gene, which allows for survival on media lacking histidine

As the field has progressed, there has been a need for high-throughput screening methods for tool optimization. The method employed must be amenable to screening both the light and dark states of the tool to maximize its performance. These requirements can be met using Y2H and, as a result, our group and others have been able to successfully improve existing optogenetic tools using Y2H-based screening [9, 10].

In this article, we present a simple drop test assay to examine the activity of a photoswitchable protein (PSP) interacting with a binding partner (BP) protein using Y2H. We use the Y2HGold yeast strain from Takara Bio, which contains several reporter genes useful for Y2H analysis of protein–protein interactions. Importantly, the strain possesses a *HIS3* reporter gene which enables histidine biosynthesis. When a protein–protein interaction is present between the PSP and its BP, *HIS3* expression allows for yeast growth on media lacking histidine. This protocol is based on our experience with light oxygen voltage (LOV) and photoactive yellow protein (PYP)-based tools [9, 11, 12]. The protocol would also be useful for other photoswitchable proteins but parameters like illumination conditions and chromophore delivery will likely need to be optimized for any given protein. One additional consideration is the strength of protein–protein interaction being assayed by Y2H. The *HIS3* reporter results in reduced or no growth when histidine synthesis is insufficient. However, once adequate *HIS3* expression is achieved, further improvements in expression do not result in an observable change in yeast growth. This is potentially a problem with optogenetic systems if both the light and dark states are sufficient for yeast growth as no light-dependent changes in growth will be observed. However, 3-amino-1,2,4-triazole (3-AT), a competitive inhibitor of the *HIS3* gene product, can be used to change the dynamic range of the assay and to select for variants with higher affinities [13].

In the second half of the chapter, we describe how the assay can be scaled up to screen libraries of optogenetic tools for improved or altered properties (e.g., altered kinetics or affinities). This section is broken up into three workflows. First, an initial selection is performed to remove nonfunctional protein variants within the library. This is followed by screening to look for clones that show light-dependent changes in growth. Finally, the activity of clones that successfully pass the selection and screening processes is validated.

2 Materials

2.1 Plasmid Construction

1. Matchmaker Gold Yeast Two-Hybrid System (Takara).
2. Genes corresponding to a PSP of interest and its BP.
3. Oligonucleotides to amplify genes of interest and add overhangs upstream and downstream to the pGBKT7 and pGADT7 vectors from the Matchmaker Gold Yeast Two-Hybrid kit. Additionally, oligonucleotides to amplify the corresponding insertion sites in the pGBKT7 and pGADT7 vectors (*see* **Note 1**).
4. Thermocyler.
5. Q5 DNA polymerase (NEB).
6. 10 mM deoxynucleotide (dNTP) solution mix.
7. *Dpn*I restriction enzyme.
8. Agarose and 1× TAE solution for DNA gel electrophoresis.
9. GelRed DNA stain (Biotium).
10. Qiagen PCR cleanup kit.
11. NEBuilder® HiFi DNA Assembly Master Mix (NEB).
12. Competent DH10β *E. coli* cells.
13. SOC broth (20 mM glucose, 10 mM magnesium chloride, 10 mM magnesium sulfate, 2.5 mM potassium chloride, 10 mM sodium chloride, 20 g/L tryptone, and 5 g/L yeast extract).
14. Ampicillin and kanamycin.
15. LB agar plates.
16. Qiagen mini prep kit.

2.2 Screening for Light-Dependent Protein–Protein Interactions

1. Matchmaker Gold Yeast Two-Hybrid System (Takara).
2. Fast Yeast Transformation™ kit (G-Biosciences).
3. Sterile water.

4. SD -W-L broth: 2 g/L drop out media, 100 mg/L adenine, 20 mg/L uracil, 50 mg/L histidine, and 2% dextrose (*see* **Note 2**).
5. SD -W-L agar: 15 g/L agar, 2 g/L drop out media, 100 mg/L adenine, 20 mg/L uracil, 50 mg/L histidine, and 2% dextrose (*see* **Notes 2 and 3**).
6. SD -W-L-H broth: 2 g/L drop out media, 100 mg/L adenine, 20 mg/L uracil, and 2% dextrose (*see* **Note 2**).
7. SD -W-L-H agar: 15 g/L agar, 2 g/L drop out media, 100 mg/L adenine, 20 mg/L uracil, and 2% dextrose (*see* **Notes 2 and 3**).
8. Blue LED (e.g., Thorlabs M455L4).
9. Tinfoil.
10. Optical power meter (e.g., Thorlabs PM160T or PM100D).
11. Camera or plate imager.

2.3 Library Screening

1. Matchmaker Gold Yeast Two-Hybrid System (Takara).
2. Fast Yeast Transformation™ kit (G-Biosciences).
3. Sterile water.
4. SD -W broth: 2 g/L drop out media, 100 mg/L adenine, 20 mg/L uracil, 100 mg/L leucine, 50 mg/mL histidine, and 2% dextrose (*see* **Note 2**).
5. SD -W-L-H agar: 15 g/L agar, 2 g/L drop out media, 100 mg/L adenine, 20 mg/L uracil, and 2% dextrose (*see* **Notes 2 and 3**).
6. Electroporation wash solution: 1 mM calcium chloride ($CaCl_2$) and 1 M sorbitol.
7. Competence solution: 0.1 M lithium acetate (LiAc) and 10 mM dithiothreitol (DTT).
8. Electroporator (e.g., BTX ECM 630).
9. 0.2-cm gene pulser cuvette (BioRad).
10. YPDA broth: 10 g/L Bacto Yeast extract, 20 g/L Bacto peptone, 20 g/L dextrose, and 40 mg/L adenine hemisulfate (*see* **Note 2**).
11. Blue LED (e.g., Thorlabs M455L4).
12. Aluminum foil.
13. Breathe-Easy® sealing membrane (Sigma-Aldrich).
14. Optical power meter (e.g., Thorlabs PM160T or PM100D).
15. Glass beads.
16. HT mini homogenizer (OPS Diagnostics).
17. Qiagen mini prep kit.

3 Methods

This section is composed of three Sects.: 3.1 Plasmid Construction, 3.2 Screening for Light-Dependent Protein–Protein Interactions, and 3.3 Library Screening. Section 3.1 describes the preparation of plasmids necessary to test for light-dependent Y2H interactions. Section 3.2 details a basic procedure for screening light-dependent protein–protein interactions between a PSP and its BP. Section 3.3 describes how to scale up the protocol to a library format that can be used to evaluate up to 10^5 different protein variants.

3.1 Plasmid Construction

This section describes the preparation of the plasmids necessary to perform Y2H testing between a PSP and its BP. For this protocol, the PSP and BD proteins are expressed using the pGADT7 and pGBKT7 plasmids using *pADH1* promoters. Alternatively, both components can be inserted into the yeast genome for expression which has worked well for some optogenetic Y2H protein pairs [14]. Cloning is carried out using Gibson assembly but other methods, such as restriction enzyme-based cloning, can also be used. One important consideration is what yeast plasmid to clone the PSP and its BP into. Previous library screening efforts have shown that library generation in proteins with an attached DNA-binding domain often results in greater numbers of false positives due to spurious activation of transcription [15]. The pGADT7 plasmid adds a N-terminal Gal4 transcriptional-activation domain while the pGBKT7 adds a N-terminal Gal4 DNA-binding domain (Fig. 2). For this reason, we recommend that the protein selected for library generation be cloned into the pGADT7 plasmid. This protocol is written around making libraries of PSPs, and as such the PSP will be cloned into

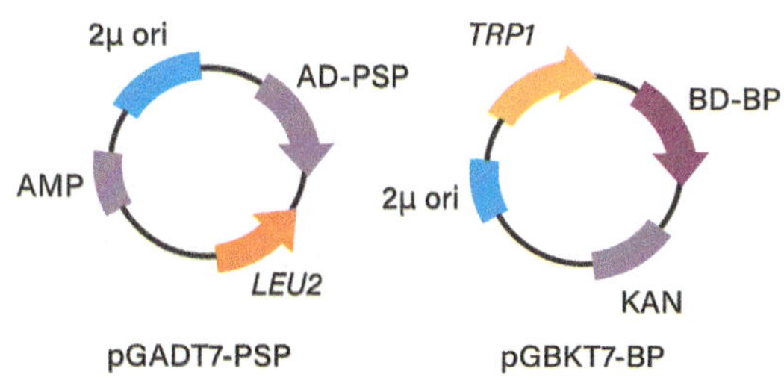

Fig. 2 Plasmid map of pGBKT7 and pGADT7 vectors with binding partner (BP) and photoswitchable protein (PSP) inserts. The pGADT7 vector appends a N-terminal Gal4 transcriptional-activation domain under the control of *pADH1* promoter. Additionally, the plasmid contains *TRP1* and ampicillin selection markers and a 2µ origin of replication. The pGBKT7 plasmid appends a N-terminal Gal4 DNA-binding domain under the control of a pADH1 promoter. The pGBKT7 vector also possess *LEU2* and kanamycin selection markers and a 2µ origin of replication

the pGADT7 plasmid (pGADT7-PSP) and its BP into the pGBKT7 plasmid (pGBKT7-BP).

1. pGADT7 and pGBKT7 yeast two hybrid vectors were obtained from the Matchmaker Gold Yeast Two-Hybrid System kit (Takara-Bio).
2. Using Q5 polymerase (NEB), carry out a PCR reaction to amplify the PSP and its BP with overhangs to the pGADT7 or pGBKT7 vectors respectively, according to the manufacturer's instructions.
3. Using Q5 polymerase, carry out a 25 μL scale PCR reaction to amplify the pGADT7 and pGBKT7 vectors according to the manufacturer's instructions.
4. Verify PCR reaction success by running 2 μL of PCR reaction product on a 1% agarose gel with an appropriate DNA stain.
5. Digest residual template DNA by adding 1 μL *Dpn*I (NEB) restriction enzyme to each PCR reaction along with 3 μL NEB CutSmart buffer and 3 μL DNAse free water. Reaction mixture incubated at 37°C for 1 h and 80°C for 20 min.
6. Purify the PCR reaction using a Qiagen PCR cleanup kit according to manufacturer's instructions.
7. Assemble constructs using NEB 2× HiFi Gibson assembly mix according to the manufacturer's instructions.
8. Combine 4 μL of the Gibson reaction with 50 μL chemically competent DH10β *E. coli*.
9. Leave cells on ice for 20 min before heat shocking at 42°C for 30 s.
10. Allow cells to recover for 2 min on ice before adding 200 μL SOC.
11. Incubate cells at 37°C for 1 h before plating on LB (100 μg/mL) ampicillin (for pGAD) or (50 μg/mL) kanamycin plates (for pGBK).
12. Incubate plates at 37°C overnight.
13. The next day, pick three to five colonies from each plate into separate 2 mL LB aliquots with (100 μg/mL) ampicillin or (50 μg/mL) kanamycin.
14. Grow cells at 37°C overnight with shaking at 200 rpm.
15. Mini-prep overnight cultures using a Qiagen mini prep kit according to the manufacturer's instructions.
16. Confirm successful plasmid assembly by Sanger or whole plasmid sequencing.

3.2 Screening for Light-Dependent Protein–Protein Interactions

This section details how to perform a Y2H drop test to test for interactions between a PSP and a potential BP. In addition to the pGADT7-PSP and pGBKT7-BP plasmids prepared in Sect. 3.1, you will also require the empty pGADT7, pGBKT7, pGBKT7-53, and pGADT7-T control plasmids that come with the Matchmaker Gold Yeast Two-Hybrid System kit. To verify that any light-dependent changes in growth are genuine, several controls are necessary. First to ensure that growth is not due to self-activation, both the pGADT7-PSP and pGBKT7-BP plasmids need to be co-transformed with empty pGBKT7 and pGADT7 vectors, respectively (Table 1). To ensure that growth differences are not due to light toxicity, a positive control in the form of the p53 and T-antigen interaction is used (Table 1).

A diagram of the drop test workflow is shown in Fig. 3. When testing a new photoswitchable interaction for the first time, it maybe necessary to test several light intensities. Additionally, for particularly strong protein–protein interactions (sub micromolar), it may be necessary to add additives, such as 3-AT, to observe light-dependent changes in growth [9, 12]. For more information on optimizing the Y2H workflow with different PSP and BP pairs (*see* **Note 4**).

1. Prepare competent Y2H Gold yeast using the G-bioscience EZ yeast transformation kit according to the manufacturer's instructions (*see* **Note 5**).
2. Combine 1 μg of each plasmid according to Table 1.
3. Combine DNA solutions with competent Y2HGold yeast according to the manufacturer's instructions and plate yeast on SD -W-L agar plates.
4. Grow yeast for 72 h until colonies appear.

Table 1
Drop test assay samples

Sample name	Plasmid 1	Plasmid 2	Purpose
Positive control	pGADT7-T	pGBKT7-53	Confirm media quality and the absence of any light effects on yeast growth.
Experimental	pGADT7-PSP	pGBKT7-BP	Test for a light-dependent interaction between the two proteins of interest
Negative control + PSP	pGADT7-PSP	pGBKT7	Confirm that the PSP does not affect yeast growth on its own.
Negative control + BP	pGADT7	pGBKT7-BP	Confirm that the BP does not affect yeast growth on its own.

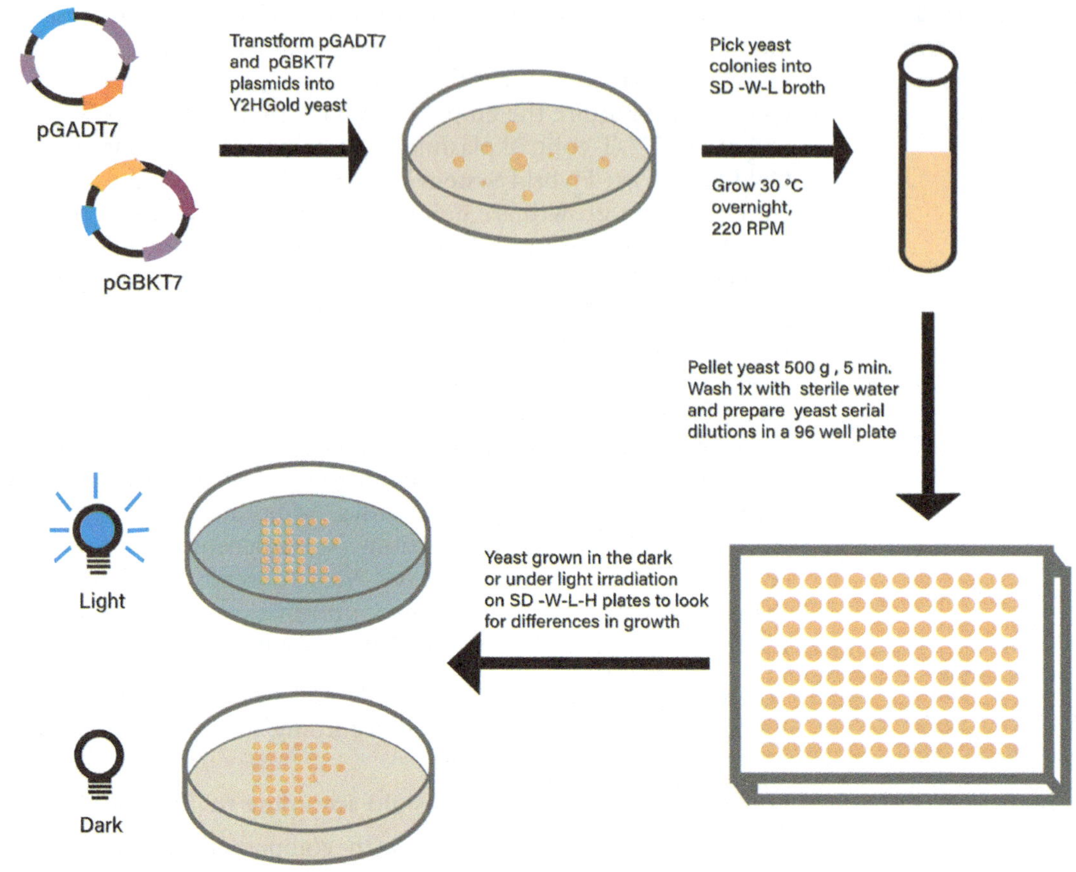

Fig. 3 Schematic for drop test workflow

5. Pick one colony of yeast from each plate into 10 mL SD -W-L broth and grow at 30°C overnight.
6. Centrifuge cells at 500 × g for 5 min.
7. Remove the supernatant and resuspend yeast in 5 mL sterile water.
8. Centrifuge cells at 500 × g for 5 min.
9. Remove the supernatant and resuspend yeast in 1 mL SD -W-L-H broth.
10. Measure the OD600 (*see* **Note 6**).
11. Prepare 1 mL OD600 = 0.4 solutions of Y2HGold expressing each construct and serially dilute these according to Table 2.
12. Plate 10 μL of the diluted yeast solutions onto two SD -W-L-H agar plates and one SD -W-L agar plate using a multichannel pipette (*see* **Note 6**).

Table 2
Drop test assay dilution scheme

Row #	Dilution factor	Volume of yeast (μL)	Volume of do W-L-H- broth (μL)
1	1	200	0
2	2	100	100
3	5	100	150
4	10	100	100
5	50	50	200
6	100	100	100
7	1000	20	180
8	10,000	20	180

13. Wrap one plate loosely with aluminum foil and place the second plate under ~0.1 mW/cm^2 blue light irradiation (*see* **Notes 7 and 8**).
14. Incubate plates for 3–5 days at 30°C.
15. Each day remove any excess moisture with a KimWipe and acquire an image of the plate using a camera or gel imager to track yeast growth over time (*see* **Note 9**).

3.3 Library Screening

This section details how to use the drop test assay to perform library selections to search for photoswitchable protein–protein interactions with desirable properties. This section is divided into three subsections (1) library selections, (2) library screening, and (3) hit validation. A schematic for the workflow can be found in Fig. 4.

3.3.1 Library Selections

This section details how to transform your library into Y2HGold yeast and perform an initial selection to remove variants that misfold and/or show no interaction between the two proteins. The electroporation protocol has been adapted from the work of Benatuil et al. [16].

1. Prepare a library of pGADT7-PSP (*see* **Note 10**).
2. Transform pGBKT7-BP into Y2HGold yeast using the Fast Yeast Transformation™ kit according to the manufacturer's instructions.
3. Plate yeast on SD -W agar plates and grow at 30°C for 3 days.
4. Pick one colony of pGBKT7-BP into a 20 mL SD -W broth and grow overnight at 30°C, 220 rpm.
5. Add the overnight culture to 200 mL SD -W broth and grow until an OD600 of 0.6–0.8 is reached.

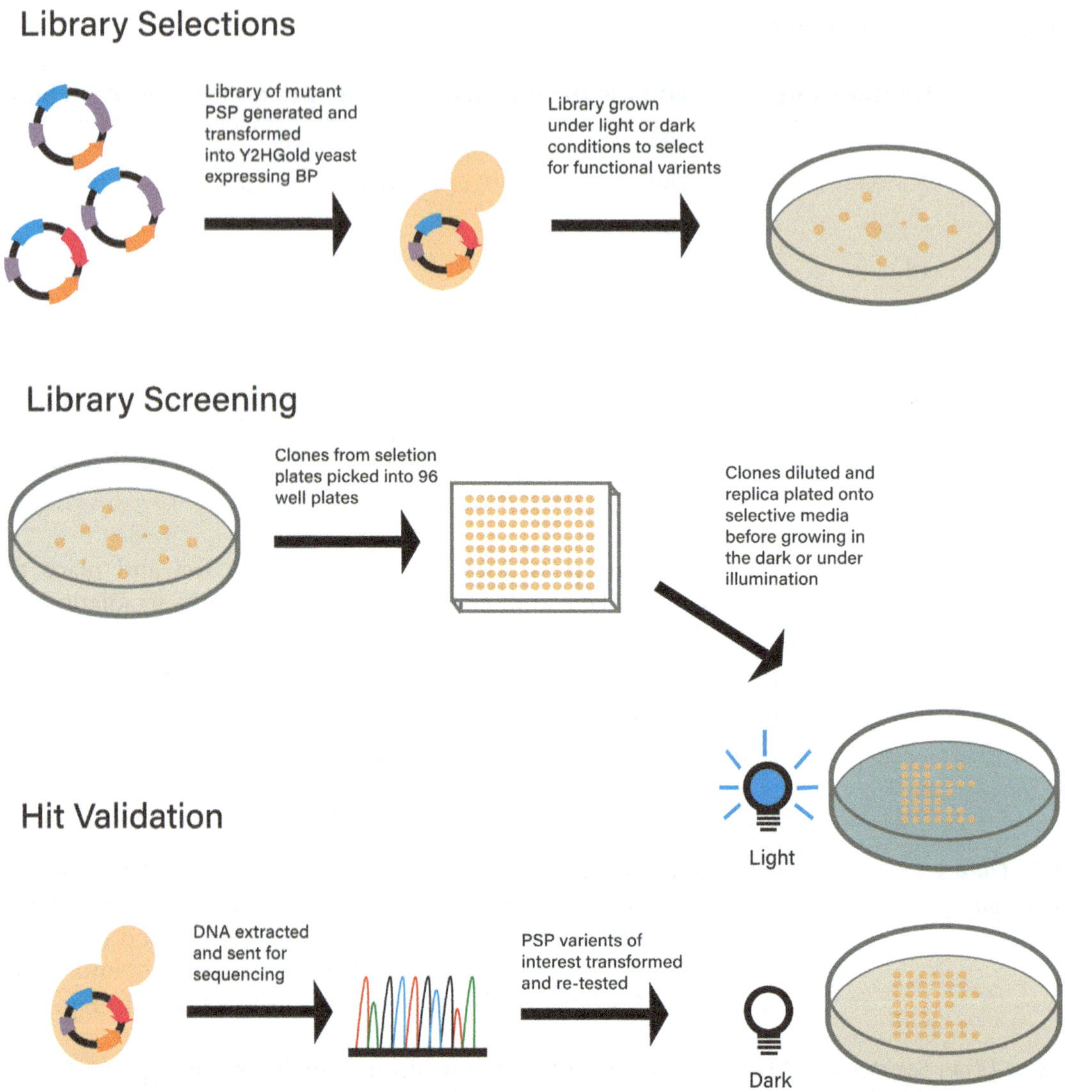

Fig. 4 Schematic of a Y2H-based selection and screening method for evaluating light-dependent protein–protein interactions. The workflow consists of three steps: (1) library selections to remove nonfunctional variants, (2) library screening to identify photoswitchable clones with desirable properties, and (3) hit validation

6. While on ice, split the yeast culture into four 50 mL Falcon tubes and centrifuge for 5 min at 3000 rpm.
7. While on ice, remove the supernatant and resuspend cells in 50 mL ice-cold sterile water. Centrifuge again at 3000 rpm for 5 min.
8. Repeat Step 7.

9. While on ice, resuspend the yeast pellet in 50 mL ice-cold electroporation solution. Centrifuge again at 3000 rpm for 5 min
10. Discard the supernatant and resuspend yeast in 10 mL competence solution.
11. Incubate yeast for 30 min at 30°C and 220 rpm (*see* **Note 11**).
12. Centrifuge yeast for 5 min at 3000 rpm. Discard the supernatant (*see* **Note 12**).
13. While on ice, resuspend yeast pellets with 50 mL ice-cold electroporation solution. Centrifuge yeast solutions at 3000 rpm for 5 min.
14. On ice, resuspend and pool the yeast pellets using 200 μL of electroporation buffer. Top up the final volume to 1.2 mL with electroporation buffer (*see* **Note 13**).
15. In a cold sterile Eppendorf tube, combine 4 μg of the library with 400 μL competent yeast.
16. On ice, mix the yeast DNA solution and load it into a BioRad 0.2 cm gene pulser cuvette (*see* **Note 14**).
17. Incubate on ice for at least 5 min.
18. Electroporate yeast DNA solutions at 2.5 kV, 250 Ω, and 25 μF (*see* **Note 15**). Immediately after electroporation, resuspend yeast with 1 mL YPDA and transfer the yeast to a sterile 50 mL Falcon tube. Wash the cuvette one additional time with YPDA and remove as much of the yeast from the cuvette as possible.
19. Grow the solutions for 1 h at 30°C and 220 rpm.
20. Dilute yeast and plate on multiple SD -W-L-H plates under desired light conditions (*see* **Notes 15 and 16**). Additionally, perform a serial dilution of the library and plate on SD -W-L agar plates to determine library size.
21. Grow plates in the dark or under blue light irradiation until colonies appear.

3.3.2 Library Screening

After selections are performed, the resulting pool of clones will be a mixture of false positives and genuine hits. To isolate clones with genuine photoswitchable interactions, it is necessary to screen the clones for light-dependent growth.

1. From the selection plates, pick any yeast colonies that grew into a 96-well plate filled with 200 μL SD -W-L-H broth per well (*see* **Note 17**).
2. Seal plates with a Breathe-Easy® sealing membrane (Sigma-Aldrich) to prevent mixing between wells.
3. Grow plates overnight at 30°C shaking on a plate shaker.

4. Dilute the overnight cultures tenfold by adding 20 μL overnight culture to 180 μL SD -W-L-H broth in a fresh 96-well plate.
5. Using a multichannel pipette, plate 5 μL from each well onto two, SD -W-L-H agar plates.
6. Wrap one plate loosely with aluminum foil while the second plate is irradiated with 0.1 mW/cm^2 blue light (*see* **Notes 7 and 8**).
7. Grow plates for 1–5 days at 30°C.
8. Image plates each day to search for colonies with light-dependent changes in growth.

3.3.3 Hit Validation

Post screening, it will be necessary to extract the DNA sequences of any clones of interest and perform a final Y2H drop test to confirm that the light-dependent growth observed by any one clone is genuine. Clones that show consistent light-dependent behavior can then be further tested in vitro or in other systems to confirm their validity.

1. Pick one colony of yeast for each clone of interest into 10 mL SD -L broth (*see* **Note 18**).
2. Grow at 30°C 220 rpm overnight.
3. Pellet cells at 500 × g for 5 min and then resuspend in 0.5 mL Qiagen P1 buffer and transfer to a 2 mL screw cap tube.
4. Add 0.4 mL acid-washed glass beads.
5. Lyse cells using a 1 min on/off cycle with glass beads for ten cycles using a HT mini homogenizer (OPS Diagnostics).
6. Transfer the lysate to a fresh tube and add 0.5 mL Qiagen P2 buffer.
7. Leave cells at room temperature for 5 min before adding 0.6 mL Qiagen N3 buffer.
8. Add 40 μL of DNAse free water to elute any bound DNA.
9. Transform 5 μL of the recovered DNA into competent DH10β according to the manufacturer's instructions.
10. Obtain sequences of recovered clones using Sanger sequencing.
11. Re-transform clones into fresh Y2HGold with pGBKT7-BP and verify results obtained with drop tests according to Sect. 3.2.

4 Notes

1. For Gibson assembly, design primers to anneal to your gene of interest with a Tm of ~65°C. Additionally, include 20–30 bp of overhang with homology to the pGADT7 or pGBKT7 MCS.

2. For yeast broths, combine all components except for the carbon source in a suitably sized glass contained and add 90% of the final volume of milliQ water. In a separate container, prepare a 20% dextrose solution. Sterilize the broth by autoclaving and the dextrose solution by filtration and then combine dextrose and broth solutions for a final dextrose concentration of 2%.
3. Autoclave agar for 15 min at 120°C or the shortest time possible to prevent the agar from getting too soft. Alternatively, agar can be autoclaved separately and then combined with the other plate ingredients.
4. When performing drop tests on a potential photoswitchable interaction, there are several key variables that need to be optimized to observe light-dependent changes in growth. It will be necessary to find a light intensity that maximizes activation of the PSP while minimizing light toxicity effects. This can be done by finding the highest light intensity that does not affect the growth of the Y2H positive control between the dark and light irradiated plates. An additional consideration is the light and dark affinities of the protein–protein interaction. For interactions where both the light and dark K_d's are in the nanomolar range, it will be necessary to add competitive inhibitors, such as 3-AT, because the *HIS3* gene expression in both states is sufficient for growth. For example, in several PYP-based tools 5–15 mM 3-AT was required to observe light-dependent growth [9, 12]. However, when using 3-AT, due to its toxicity, the yeast will grow slower and will be more susceptible to light toxicity. As a result, it will be necessary to adjust both the growth times and light intensities used in the experiment. Finally, if the PSP requires a chromophore, this also needs to be considered. In our experience, the chromophore can often be added directly to the agar of the plate for drop test assays and library selections. In the case of PYP, we prepare agar plates with 0.4–40 μM chromophore by adding a 1000× stock solution of the activated chromophore in DMSO. The chromophore is then added directly to the molten agar just prior to plate pouring. We recommend allowing the agar to cool as much as possible prior to chromophore addition as it is not stable at high temperatures.
5. Yeast can be prepared the day of use and stored on ice until they are transformed or can be stored at −80°C and thawed when ready for use.
6. When working with yeast solutions, it is important to vigorously mix the solution prior to any measurement or dilution step. Yeast cells rapidly settle out of the solution and mixing is necessary for accurate measurements and dilutions.

7. For best results, use pre-dried yeast plates which have been preincubated at 30°C for several hours or overnight to remove some of the moisture. This ensures that the yeast solutions are absorbed into the plate and are more likely to form neat droplets rather than running into one another. Furthermore, this reduces the amount of condensation build up as the yeast are grown at 30°C. Excessive condensation build up can disrupt yeast colonies growing on the plate.
8. The necessary light intensity will vary depending on the PSP of interest as well as what wavelength of light is used. The suggested value is based on the interaction between the PSP photoactive yellow protein and its BP BoPD [9]. For LOV domains, we use 0.23–0.25 mW/cm^2 [11].
9. When acquiring images, ensure that the image settings are consistent across days and samples.
10. There are a wide variety of library generation methods that can subsequently be screened using Y2H. Our lab uses Kunkel mutagenesis to make multiple site-specific libraries [17]. Alternatively, error-prone PCR or any other mutagenesis methods can be used. For a variety of library generation methods, we recommend reading reviews by Alejaldre et al. or Currin et al. [18, 19].
11. Growing yeast for too long in the competence solution can result in a loss of viability.
12. It is important to remove as much of the competence solution as possible to prevent sparking during electroporation.
13. This volume will be sufficient for three electroporation that can be done one after the other and then pooled.
14. It is important that the cuvette is cold for optimal electroporation efficiency. We recommend storing cuvettes at −20°C.
15. Time constants are typically between 3 and 5 s. Faster constants may indicate insufficient removal of salts from the competence solution.
16. It will be necessary to titrate the amount of yeast plated. Over-plating of yeast can result in false positives.
17. Use the same light conditions as would be used for the drop test. To ensure more uniform irradiation of the plates consider placing some paper in front for the LED to act as a diffuser.
18. Also, include the positive control from Sect. 3.2 to ensure that any light-dependent phenotypes are not due to light toxicity.

References

1. Fields S, Song O (1989) A novel genetic system to detect protein–protein interactions. Nature 340(6230):245–246
2. Brückner A et al (2009) Yeast two-hybrid, a powerful tool for systems biology. Int J Mol Sci 10(6):2763–2788
3. Fu X et al (2013) Design and selection of a camelid single-chain antibody yeast two-hybrid library produced de novo for the cap protein of porcine circovirus type 2 (PCV2). PLoS One 8(3):e56222
4. Huang Z et al (2020) Creating red light-switchable protein dimerization systems as genetically encoded actuators with high specificity. ACS Synth Biol 9(12):3322–3333
5. Chen W et al (2022) The roles of optogenetics and technology in neurobiology: a review. Front Aging Neurosci 14:867863
6. Method of the Year 2010 (2011). Nat Method 8(1):1
7. Fischer AAM et al (2022) Shedding light on current trends in molecular optogenetics. Curr Opin Chem Biol 70:102196
8. Oh T-J et al (2021) Steering molecular activity with optogenetics: recent advances and perspectives. Adv Biol 5(5):2000180
9. Woloschuk RM et al (2020) Yeast two-hybrid screening of photoswitchable protein–protein interaction libraries. J Mol Biol 432(10):3113–3126
10. Taslimi A et al (2016) Optimized second-generation CRY2-CIB dimerizers and photoactivatable Cre recombinase. Nat Chem Biol 12(6):425–430
11. Lu H et al (2019) A yeast system for discovering optogenetic inhibitors of eukaryotic translation initiation. ACS Synth Biol 8(4):744–757
12. Woloschuk RM et al (2021) Structure-based design of a photoswitchable affibody scaffold. Protein Sci 30(12):2359–2372
13. Tanaka T, Rabbitts TH (2009) Selection of complementary single-variable domains for building monoclonal antibodies to native proteins. Nucleic Acids Res 37(5):e41
14. Harmer ZP, McClean MN (2023) Lustro: high-throughput optogenetic experiments enabled by automation and a yeast optogenetic toolkit. ACS Synth Biol 12(7):1943–1951
15. Chien CT et al (1991) The two-hybrid system: a method to identify and clone genes for proteins that interact with a protein of interest. Proc Natl Acad Sci U S A 88(21):9578–9582
16. Benatuil L et al (2010) An improved yeast transformation method for the generation of very large human antibody libraries. Protein Eng Des Sel 23(4):155–159
17. Kunkel TA (1985) Rapid and efficient site-specific mutagenesis without phenotypic selection. Proc Natl Acad Sci U S A 82(2):488–492
18. Alejaldre L, Pelletier JN, Quaglia D (2021) Methods for enzyme library creation: which one will you choose?. BioEssays 43(8): 2100052
19. Currin A et al (2021) The evolving art of creating genetic diversity: From directed evolution to synthetic biology. Biotechnol Adv 50:107762

Part II

Engineering Viral Delivery Systems

Check for updates

Chapter 6

CRISPR-Engineered Bacteriophage T4 for Foot-and-Mouth Disease Nanoparticle Vaccine Development

Mengling Li, Cen Chen and Pan Tao

Abstract

Peptide-based vaccines offer a safer alternative to inactivated vaccines. However, the immunogenicity of the peptides is usually poor, and therefore, adjuvants or delivery systems are required. Bacteriophage T4, with its intrinsic immunostimulatory properties, provides a promising platform for antigen display. Here, we used CRISPR-Cas genome editing to insert a gene encoding a foot-and-mouth disease virus (FMDV) B-cell epitope ($VP1_{130\text{-}158}$) into the C-terminus of the soc gene in the T4 genome. The T4 phage self-assembly system enables the display of epitopes on the capsid surface in vivo, generating $VP1_{130\text{-}158}$-T4 virus-like particles. To further enhance immune activation, $CD4^+$ T-cell epitopes FMDV $3A_{21\text{-}35}$ or tetanus toxoid $P2_{830\text{-}844}$ were fused downstream of the VP1 epitope. These recombinant T4 phages provide proof-of-concept for the development of safe, epitope-based FMDV vaccines.

Keywords Bacteriophage T4, FMDV vaccines, Virus-like particles, B- and T- cell epitopes

1 Introduction

Foot-and-mouth disease (FMD) is a highly contagious viral infection of cloven-hoofed animals and is designated as a notifiable disease by the World Organization for Animal Health (WOAH) [1, 2]. Vaccination with inactivated virus remains the most effective strategy for the control of FMD [3, 4]. However, vaccine production is limited by the need for biosafety level 3 (BSL-3) facilities, which imposes significant cost and infrastructure burdens [5, 6].

Peptide vaccines containing protective epitopes of the foot-and-mouth disease virus (FMDV) offer a safer alternative to whole-virus formulations [7]. Several antigenic epitopes have been identified on the FMDV capsid proteins, among which the G-H loop of the VP1 protein (residues 140–160) is considered the most critical B-cell epitope due to its ability to elicit neutralizing antibodies [8–10]. This epitope has been incorporated into a commercially available peptide-based FMD vaccine [10].

Huan Peng and S. Brian Kim (eds.), *Biological Macromolecule Engineering Technologies*,
Methods in Molecular Biology, vol. 3032, https://doi.org/10.1007/978-1-0716-5257-2_6,

Peptide vaccines lack pathogen-associated molecular patterns (PAMPs) necessary to activate innate immunity, resulting in limited immunogenicity and a need for adjuvants or delivery platforms to enhance immune responses [11, 12]. Incorporation of $CD4^+$ T cell epitopes, such as residues 21–35 of nonstructural protein 3A or 20–40 of VP1, has been shown to significantly enhance the immunogenicity of FMDV epitope-based vaccines [13–15]. In addition, we previously developed a T4 bacteriophage-based nanoparticle platform through in vitro assembly, which exhibits innate adjuvant properties and enhances antigen-induced immune responses [16, 17]. We recently advanced this system by using CRISPR-Cas genome editing to insert antigen-encoding sequences into the C-terminus of the *Soc* gene in the T4 genome [18, 19]. Antigens are assembled onto the surface of T4 phage capsids, forming antigen-decorated nanoparticles during phage propagation in *Escherichia coli* [18, 19]. This eliminates the need for large-scale purification and post-assembly of Soc-antigen fusion proteins, enabling cost-effective production of antigen-presenting T4 virus-like particles (VLPs) during phage propagation.

Results from our laboratory have demonstrated that B cell epitope of the FMDV VP1 protein ($VP1_{130\text{-}158}$) is displayed on the T4 phage capsid surface through in vivo assembly, forming VP1-T4 recombinant phage nanoparticles that induce a robust immune response without the need for adjuvants. Additionally, incorporating the $3A_{21\text{-}35}$ of the FMDV 3A protein or the $P2_{830\text{-}844}$ of tetanus toxoid into the VP1 epitope enhances the immunogenicity of VP1-T4 phage nanoparticles [20]. In this chapter, we describe the methods to generate VP1- T4 phage nanoparticles as potential next-generation FMDV vaccine candidates.

2 Materials

2.1 Construction of pSoc-VP1 Donor Plasmids

1. *Escherichia coli* cells, plasmid: *E. coli* DH5α cells (Novagen), plasmid pModB-Soc-35mer-56 (prepared in our laboratory).
2. Luria-Bertani (LB) medium: Prepare LB medium by dissolving 10 g tryptone, 5 g yeast extract, and 5 g NaCl in 1 L of distilled water. Sterilize the medium by autoclaving at 121°C for 20 min.
3. 1000× Ampicillin (100 mg/mL): Dissolve 1 g ampicillin in 10 mL of sterile distilled water. Filter-sterilize the solution using a 0.22 μm membrane filter and store at −20°C.
4. Ampicillin LB plates: Prepare LB agar by adding 20 g agar to 1 L of LB medium. Sterilize by autoclaving at 121°C for 20 min. Allow the medium to cool to approximately 50°C before adding 1 mL of 1000× ampicillin stock solution per

liter. Pour the medium into sterile Petri dishes and allow it to solidify at room temperature. Store plates at 4 °C.

5. Enzymes: 2× Phanta Flash Master Mix (Vazyme) was used for high-fidelity PCR for PCR amplification to construct clones. 2× Rapid Taq Master Mix (Vazyme) was used for PCR amplification for identification. ClonExpress Ultra One Step Cloning Kit V2 (Vazyme) was used for homologous recombination-based fragment cloning.
6. 50× TAE: Dissolve 242 g Tris and 37.2 g EDTA in about 800 mL of distilled water. Add 57.1 mL of acetic acid and stir until the solution is homogeneous. Adjust the volume to 1 L with distilled water and adjust the pH to 8.0.
7. Agarose gel electrophoresis buffer: Add 20 mL of 50× TAE stock solution to 1 L of sterile distilled water to prepare 1× TAE buffer.
8. Agarose gel: Add 1 g of agarose to 100 mL of 1× TAE buffer and heat the mixture in a microwave until it boils. Allow the solution to cool to about 50°C, then add 1 μL of GoldView™ dye (Aidlab Biotech). Pour the agarose solution into the electrophoresis gel mold and allow it to solidify.
9. Kit: Gel Extraction Kit (Omega) was used for the isolation of DNA from agarose gel slices. Plasmid Extraction Kit (Omega) was used to extract plasmid from bacterial cultures.

2.2 Generation of Recombinant T4 Phages

1. Phage and *E. coli* cells: *Δsoc* T4 phage (prepared in our laboratory), *E. coli* B834, and *E. coli* P301.
2. Growth media: Luria-Bertani (LB) medium.
3. Antibiotics: 1000× ampicillin (100 mg/mL): Dissolve 1 g ampicillin in 10 mL of sterile distilled water, 1000× spectinomycin (50 mg/mL): Dissolve 0.5 g spectinomycin in 10 mL of sterile distilled water. Filter-sterilize the solution using a 0.22 μ m membrane filter and store at −20°C.
4. Ampicillin-spectinomycin LB plates: Prepare LB agar by adding 20 g agar to 1 L of LB medium. Sterilize by autoclaving at 121°C for 20 min. Allow the medium to cool to approximately 50°C before adding 1 mL 1000× ampicillin and 1 mL 1000× spectinomycin stock solution per liter. Pour the medium into sterile Petri dishes and allow it to solidify at room temperature. Store plates at 4°C.
5. Top agar: Add 3 g of tryptic soy broth (TSB) and 0.75 g of agar to 100 mL distilled water. Mix thoroughly and sterilize by autoclaving at 121°C for 20 min. Keep the sterilized top agar at 42°C until use.
6. Agar plates: Prepare agar plates by adding 20 g agar to 1 L distilled water. Sterilize by autoclaving at 121°C for 20 min. Pour

the medium into sterile Petri dishes and allow it to solidify at room temperature. Store plates at 4°C.

7. Pi-Mg Buffer: Dissolve 26 mM Na_2HPO_4, 22 mM KH_2PO_4, 70 mM NaCl, and 1 mM $MgSO_4$ in distilled water.
8. PCR identification reagents: agarose gel and agarose gel electrophoresis buffer.
9. Enzymes: 2× Phanta Flash Master Mix (Vazyme) was used for high-fidelity PCR for PCR amplification to construct clones. 2× Rapid Taq Master Mix (Vazyme) was used for PCR amplification for identification.

2.3 Purification of Recombinant T4 Phages for Immunizations

1. Phage and *E. coli*: recombinant T4 phages, *E. coli* P301.
2. Growth media: LB medium; M9CA medium: Dissolve 13.3 g of M9CA broth (Sangon Biotech) in 1 L of sterile distilled water. Sterilize the medium by autoclaving at 121°C for 20 min.
3. 1 mg/mL DNase I (Biosharp): Dissolve 10 mg of DNase I in 10 mL sterile distilled water. Aliquot the solution and store at −20°C.
4. Buffers: Pi-Mg buffer: Dissolve 26 mM Na_2HPO_4, 22 mM KH_2PO_4, 70 mM NaCl, and 1 mM $MgSO_4$ in distilled water; Dialysis buffer I: Dissolve 10 mM Tris, 200 mM NaCl, and 5 mM $MgCl_2$ in distilled water, and adjust the pH to 7.5 with HCl; Dialysis buffer II: Dissolve 10 mM Tris, 50 mM NaCl, and 5 mM $MgCl_2$ in distilled water, and adjust the pH to 7.5 with HCl.
5. CsCl stock solution: Dissolve 8 M CsCl, 100 mM Tris, 85 mM NaCl, and 200 mM NH_4Cl in distilled water, adjust the pH to 7.5 with HCl, filter the solution through a 0.22 μm membrane filter, and store in the dark.
6. Dialysis membrane with a molecular weight cutoff (MWCO) of 10 kDa.
7. 1× PBS pH 7.4: Dissolve 2.68 mM KCl, 1.47 mM KH_2PO_4, 136.9 mM NaCl, and 8.06 mM Na_2HPO_4 in distilled water.
8. 10% Ammonium Persulfate (AP) solution: Dissolve 1 g of ammonium persulfate in 10 mL of distilled water. Prepare fresh before use and store at 4°C if necessary.
9. 10% SDS solution: Dissolve 1 g of Sodium Dodecyl Sulfate (SDS) in 10 mL of distilled water. Stir the solution until completely dissolved. Store the solution at room temperature.
10. 15% Polyacrylamide gel: Resolving gel: Mix 2.3 mL distilled water, 5 mL of 30% acrylamide/bis-acrylamide solution (29:1), 2.5 mL of 1.5 M Tris–HCl (pH 8.8), and 100 μL of 10% SDS solution. 100 μL of 10% AP solution and 10 μL TEMED were added immediately before casting. The solution was mixed gently and poured into the casting cassette. Stacking gel: Mix 2.1 mL distilled water, 0.5 mL of 30%

acrylamide/bis-acrylamide solution (29:1), 0.38 mL of 1 M Tris-HCl (pH 6.8), and 30 μL of 10% SDS solution. 30 μL of 10% AP solution and 3 μL TEMED were added immediately before casting. The solution was mixed gently and layered on top of the resolving gel. A comb was inserted, and the gel was allowed to polymerize at room temperature.

11. 1% Bromophenol blue solution: Dissolve 0.1 g of bromophenol blue in 10 mL of distilled water to prepare a 1% (w/v) solution. Store the solution at room temperature.
12. 5× SDS-PAGE sample buffer: Mix 0.6 mL of 1 mol/L Tris-HCl (pH 6.8), 2 mL of 10% SDS, 0.5 mL of β-mercaptoethanol, 1 mL of 1% bromophenol blue, and 0.9 mL of distilled water. Mix thoroughly and store at −20°C.
13. SDS-PAGE running buffer: Dissolve 4.53 g Tris, 28.2 g glycine, and 1.5 g SDS in 1.5 L of distilled water. The solution was mixed thoroughly until fully dissolved.
14. Coomassie brilliant blue staining solution: Dissolve 5 g Coomassie Brilliant Blue R-250 in 250 mL isopropanol and 100 mL glacial acetic acid. The solution was mixed thoroughly and then diluted with distilled water to a final volume of 1 L.
15. SDS-PAGE gels destaining solution: Mix 100 mL glacial acetic acid, 50 mL ethanol, and 850 mL distilled water. The solution was mixed thoroughly before use.
16. 0.45 μm PVDF membrane (Millipore) was used for protein transfer.
17. Western blotting transfer buffer: Dissolve 4.5 g Tris and 21.6 g glycine in 300 mL methanol. The solution was then diluted with distilled water to a final volume of 1.5 L. The buffer was thoroughly mixed, sealed, and kept on ice before use.
18. TBST: Dissolve 2.42 g Tris and 8 g NaCl in 1 L distilled water. 1 mL of Tween-20 was added, and the solution was mixed thoroughly.
19. 5% non-fat milk: Dissolve 1 g non-fat milk powder in 20 mL TBST.
20. HRP-conjugated goat anti-mouse IgG (Abclonal) was used as a secondary antibody.

3 Methods

3.1 Construction of pSoc-VP1 Donor Plasmids

Plasmid pModB-Soc-35mer-56, constructed in our laboratory, contains the *T4* phage *Soc* gene, its flanking homologous arms, and the $VP1_{158\text{-}6\times His}$ antigen gene. It serves as a donor plasmid for targeted gene insertion at the C-terminus of the *Soc* gene, enabling the editing of the phage genome. Four different *VP1* gene sequences were introduced by designing specific primers and performing PCR amplification with plasmid pModB-Soc-35mer-56 as a template. These genes were then precisely inserted into the C-terminus of the *Soc* gene using Gibson assembly, resulting in the generation of four pSoc-VP1$_{158}$ donor plasmids (*see* **Note 1**).

1. The pUC19 vector DNA was amplified from the plasmid pModB-Soc-35mer-56 using 2× Phanta Flash Master Mix and the primers listed below.
   Vector Forward: 5′-TAATAACTCAAGGACTCCTTCGG-GA-3′
   Vector Reverse: 5′-GTAAACAGAACCAGAACCACCAGT-TACTTTCCACAA-3′
2. 10× DNA loading buffer was added to the PCR products. The products were then loaded onto a 1% agarose gel and separated by electrophoresis.
3. The gel containing the expected size band was excised and transferred to a clean 1.5 mL tube. DNA was extracted using the Omega Gel Extraction Kit according to the manufacturer's instructions.
4. The *VP1*$_{158}$ gene was amplified from the plasmid pModB-Soc-35mer-56 using 2× Phanta Flash Master Mix and the primers listed below (*see* **Note 2**).
   S158 FW: 5′-GGTTCTGGTTCTGTTTACAACGGGAATT-GC-3′
   S158 BW: 5′-AAGGAGTCCTTGAGTTATTACCTCGC-CGCCTTCTGAGCCA-3′
5. The *VP1*$_{158C}$ gene was amplified from the plasmid pModB-Soc-35mer-56 and the primers listed below.
   S158 FW: 5′-GGTTCTGGTTCTGTTTACAACGGGAATT-GC-3′
   S158C BW: 5′-AAGGAGTCCTTGAGTTATTAGCACCTC-GCCGCCTTCTGAGCCA-3′
6. The primers S158 FW/S158C-3A BW1 were used to amplify a DNA fragment from plasmid pModB-Soc-35mer-56. This fragment served as the template for sequential rounds of overlap extension PCR to generate the full-length *VP1*$_{158C}$*-3A* gene, using the primers S158 FW/S158C-3A BW2 and S158 FW/S158C-3A BW3 listed below.
   S158 FW: 5′-GGTTCTGGTTCTGTTTACAACGGGAATT-GC-3′
   S158C-3A BW1:
   5′-CATGCCGAAGAATTCAATTGCAGCGCTACCAC-CACCGCACCTCGCCGC-3′

S158C-3ABW2:5′-TTATTTAATAGAGTCATGAACCATG-CCGAAGAATTCAATTGCAGC-3′
S158C-3A BW3: 5′-GAAGGAGTCCTTGAGTTATTAT-TTAATAGAGTCATGAACC-3′

7. The primers S158 FW/ S158C-P2 BW1 were used to amplify a DNA fragment from plasmid pModB-Soc-35mer-56. This fragment served as the template for sequential rounds of overlap extension PCR to generate the full-length $VP1_{158C}$-*P2* gene, using the primers S158 FW/S158C-P2 BW2, S158 FW/S158C-P2 BW3, and S158 FW/S158C-P2 BW4 listed below.
S158 FW: 5′-GGTTCTGGTTCTGTTTACAACGGGAAT-TGC-3′
S158C-P2BW1:5′-GCTACCACCACCGCACCTCGCCGC-CTTCTGAGCCAGCACTTGGAG-3′
S158C-P2 BW2:
5′-TTTAGAATTTGCTTTTATATACTGGCTACCAC-CACCGCACCTCGCCGC-3′
S158C-P2 BW3: 5′-TTATAGTTCAGTTATACCTATAAAT-TTAGAATTTGCTTTTATATACTG-3′
S158C-P2 BW4: 5′-GAAGGAGTCCTTGAGTTATTATA-GTTCAGTTATACC-3′
8. 10× DNA loading buffer was added to the PCR products of $VP1_{158}$, $VP1_{158C}$, $VP1_{158C}$-3A, and $VP1_{158C}$-P2. The samples were then loaded onto a 1% agarose gel and separated by electrophoresis.
9. Repeat Step 3.
10. Each of the $VP1_{158}$, $VP1_{158C}$, $VP1_{158C}$-3A, and $VP1_{158C}$-P2 DNA fragments was ligated with the pUC19 vector using ClonExpress Ultra One Step Cloning Kit V2 according to the manufacturer's instructions.
11. The ligation products were individually transformed into *E. coli* DH5α competent cells and plated onto ampicillin LB plates, and incubated overnight at 37°C.
12. Single colonies were picked, transferred to LB broth containing 100 μg/mL ampicillin, and incubated overnight at 37°C with shaking.
13. PCR amplification of the bacterial culture was performed, and the products were resolved on a 1% agarose gel by electrophoresis. The bacterial culture containing the expected band was sent for sequencing.
14. The bacterial culture containing the expected band was diluted 1:100 into LB broth containing 100 μg/mL ampicillin and incubated overnight at 37°C with shaking.

15. Plasmid DNA was extracted from the bacterial culture using the Omega Plasmid Extraction kit according to the manufacturer's instructions. The plasmids obtained were pSoc-VP1$_{158}$, pSoc-VP1$_{158C}$, pSoc-VP1$_{158C}$-3A, and pSoc-VP1$_{158C}$-P2.

3.2 Generation of Recombinant T4 Phages

Plasmid pLbCas12a-ModB, constructed in our laboratory, encodes the Cas12a effector protein and a spacer sequence targeting the *modB* gene of T4 phage, enabling efficient and precise cleavage of the phage genome. The recombinant T4 phages constructed in this study included VP1$_{158}$-T4, VP1$_{158C}$-T4, VP1$_{158C}$-3A-T4, and VP1$_{158C}$-P2-T4.

1. The plasmids pSoc-VP1$_{158}$, pSoc-VP1$_{158C}$, pSoc-VP1$_{158C}$-3A, and pSoc-VP1$_{158C}$-P2 were each co-transformed into *E. coli* B834 along with pLbCas12a-ModB, and incubated overnight at 37°C on ampicillin-spectinomycin LB plates.
2. A single colony was picked and inoculated into 5 mL of LB medium containing 100 μg/mL ampicillin and 50 μg/mL spectinomycin, and incubated overnight at 37°C with shaking.
3. About 10^4–10^5 pfu of T4Δsoc phages were incubated with 300 μL of *E. coli* (~10^8 cells/mL) in Step 2 at 37°C for 7 min (*see* **Note 3**).
4. 3 mL of top agar was added to the infection mixture, poured onto agar plates, and incubated overnight at 37°C (*see* **Note 4**).
5. A portion of a single phage spot was picked with a sterile pipette tip to amplify the *VP1* genes inserted in the recombinant phage using the following primers, with T4Δsoc phages as controls.
   Fwd: 5′-CTTGGTGAAATGTCACGTGGT-3′
   Rev: 5′-CGCTCTGAATCTAATCCATCA-3′
6. PCR products were loaded onto 1% agarose gels and separated by electrophoresis. Recombinant T4 phages with products of the expected target fragment sizes were labeled. The following steps (Steps 7–10) were conducted to purify the recombinant phages.
7. *E. coli* P301 was streaked onto LB agar plates with an inoculation loop and incubated overnight at 37°C. A single colony was inoculated into 10 mL of LB broth and incubated overnight at 37°C with shaking.
8. 300 μL of *E. coli* P301 was mixed with 8 mL of top agar and poured onto an agar plate. After solidification, the pipette tips were dipped into the correctly identified phages from Step 6 and inoculated onto the top agar containing *E. coli* P301, followed by incubation overnight at 37°C.

9. Step 8 was repeated twice.
10. A single plaque was transferred to 1.5 mL tube containing 1 mL of Pi-Mg buffer and gently vortexed at low speed (*see* **Note 5**).
11. About 300 μL of phage-containing Pi-Mg buffer was incubated with 300 μL of *E. coli* P301 culture at 37°C for 7 min, poured onto agar plates, and incubated overnight at 37°C.
12. Repeat Steps 5–6.
13. Products with expected target fragment sizes were sent for sequencing and further identification.
14. Single phages sequenced correctly in Step 13 were picked into 1 mL of Pi-Mg buffer for subsequent use, respectively.

3.3 Proliferation and Purification of Wild-Type and Recombinant T4 Phages

The recombinant phages propagated and purified in this study included $VP1_{158}$-T4, $VP1_{158C}$-T4, $VP1_{158C}$-3A-T4, and $VP1_{158C}$-P2-T4.

1. *E. coli* P301 was streaked onto LB agar plates with an inoculation loop and incubated overnight at 37°C.
2. A single colony was inoculated into 10 mL of LB broth and incubated overnight at 37°C with shaking.
3. The recombinant phage T4 in Pi-Mg buffer from Step **14** (Sect. 3.2) was serially diluted, incubated with 300 μL of *E. coli* P301 culture at 37°C for 7 min, then poured onto agar plates and incubated overnight at 37°C.
4. A single plaque was transferred to 1.5 mL tube containing 1 mL of Pi-Mg buffer and gently vortexed at low speed.
5. Approximately 300 μL of Pi-Mg buffer containing recombinant T4 phage was mixed with 500 μL of freshly cultured *E. coli* P301 culture and incubated at 37°C for 7 min.
6. The infection mixture was mixed with 5 mL of top agar, poured onto LB plates, and incubated overnight at 37°C.
7. Repeat Step 2.
8. The overnight culture of *E. coli* P301 was diluted 1:100 into 1 L of LB and M9CA medium (a mixture of 500 mL LB medium and 500 mL M9CA medium), and incubated at 37°C with shaking until the cell density reached 2×10^8 cells/mL.
9. The entire top agar layer containing recombinant T4 phages was collected and added to a 2-L flask containing 1 L of *E. coli* P301 culture (2×10^8 cells/mL), and incubated at 37°C with shaking at 200 rpm for 4 h (*see* **Note 6**).
10. The addition of chloroform to the culture mixture, followed by incubation at 37°C with shaking for 30 min, induces

bacterial lysis, resulting in the complete release of the phage progeny.

11. The culture mixture containing phage progeny was filtered through sterile gauze to remove agar, followed by centrifugation at 25,000 g for 30 min at 4°C.
12. The pellet was resuspended with Pi-Mg buffer containing 10 μg/mL DNase I and 500 μL chloroform, and incubated at 37°C with shaking at 200 rpm for 30 min.
13. The mixture was centrifuged at 4300 g for 10 min at 4 °C to remove bacteria, and the resulting phage suspension was collected.
14. Phage particles were collected by centrifugation at 30,000 g for 45 min at 4°C and soaked in 1 mL of Pi-Mg buffer, followed by incubation at 4°C overnight (*see* **Note 7**).
15. A CsCl gradient was prepared for phage purification. First, layer buffers were prepared according to the compositions listed in Table 1. Subsequently, 2 mL each of layer buffer No. 6, No. 5, No. 4, No. 3, No. 2, and No. 1 were sequentially added from bottom to top into a 13.2 mL Beckman centrifuge tube (*see* **Note 8**).
16. The phage pellet obtained in Step 14 was gently resuspended using a pipette, and 1 mL of the resuspended phage sample was carefully layered onto the top of the CsCl gradient. The gradient was then ultracentrifuged at 148,596 g for 1 h at 4°C using a Beckman SW41 Ti rotor.
17. The centrifuge tube was carefully removed from the rotor using forceps and placed in a tube rack. The upper layer containing bacterial debris and impurities was gently aspirated and discarded using a pipette. The visible phage band was then carefully extracted using a pipette and transferred into a clean tube.

Table 1
Layer buffer preparation compositions

Layer no.	Stock CsCl (mL)	H_2O (mL)	Total volume (mL)
1	1	4	5
2	1.5	3.5	5
3	2	3	5
4	2.5	2.5	5
5	3	2	5
6	3.5	1.5	5

18. The phage sample was transferred into a 10-kDa molecular weight cutoff dialysis membrane and dialyzed against Dialysis Buffer I at 4°C for 5 h with gentle stirring using a magnetic stirrer.
19. The dialysis membrane containing the phage sample was transferred into Dialysis Buffer II and dialyzed overnight at 4°C.
20. The phage particles were collected, quantified, and stored at 4°C for subsequent use.
21. The wild-type and recombinant T4 phages were mixed with 5× SDS loading buffer, boiled for 7 min, and loaded onto separate wells of a 15% SDS-PAGE gel for electrophoresis.
22. Proteins separated by SDS-PAGE were transferred onto a PVDF membrane using a wet transfer method. The membrane was then blocked with 5% non-fat milk in TBST buffer at room temperature for 2 h to reduce nonspecific binding.
23. The membrane was washed three times with TBST buffer and then incubated overnight at 4°C with anti-FMDV serum (diluted at 1:6000 with TBST).
24. The membrane was washed three times with TBST buffer, then incubated with HRP-conjugated goat anti-mouse IgG (diluted at 1:8000 with TBST) at room temperature for 1 h.
25. The membrane was washed five times with TBST buffer, and the protein bands were visualized using the ChemiDoc Touch Imaging System (BioRad). The following steps (Steps 26–29) were performed to quantify the number of phage particles and the amount of antigen displayed on the phage capsid surface. Bovine serum albumin (BSA, 1 μg/μL) was used as a standard.
26. Recombinant T4 phages and BSA were mixed with 5× SDS loading buffer and boiled for 7 min.
27. The boiled phage samples and 1, 2, 3, 4, 5, and 6 μg of BSA were loaded into separate wells of a 15% SDS-PAGE gel and subjected to electrophoresis.
28. The SDS-PAGE gel was transferred to a tray, stained with Coomassie Brilliant Blue solution, and destained with destaining solution.
29. The gel was imaged using a gel imager, and the intensities of phage gp23*, VP1 proteins, and varying amounts of BSA band were quantified using Image-Pro Plus software. The phage particle concentration and the amount of VP1 antigen per capsid were calculated by constructing a BSA standard curve (*see* **Note 9**).

3.4 Animal Immunizations

The exact quantities of wild-type and recombinant T4 phages propagation were determined based on how many animals to be used. The dose used in the experiment was 7.5 μg of antigen per animal.

1. Based on the amount of antigen displayed per phage capsid calculated in Steps 26–29 under Sect. 3.3, recombinant phages containing 7.5 μg of antigen were resuspended in 50 μL of PBS.
2. Animals were immunized via intramuscular injection with 7.5 μ g of antigen displayed on the T4 phage capsid. Antigen-specific antibody and cellular immune responses were evaluated as described previously [20].

4 Notes

1. We used the COOH-terminal of T4 Soc, rather than the NH2-terminal, as the insertion site for antigen display. Previous studies have shown that the COOH-terminal of T4 Soc is more suitable for in vivo display than the NH2-terminal [19].
2. We previously constructed the plasmid pModB-Soc-35mer-56, which contains the $VP1_{158}$ gene fragment, the phage T4 *Soc* gene, and its flanking homologous arms [19].
3. Immediate mixing upon introducing the phage into *E. coli* cultures is essential to prevent localized high multiplicity of infection.
4. Top agar was added to the infection mixture and immediately inverted and mixed thoroughly to ensure uniform distribution.
5. Phage T4, a long-tailed bacteriophage, is susceptible to tail structure damage from vigorous vortexing, which can markedly reduce its ability to infect host cells.
6. *E. coli* cells were aliquoted onto a cell counting chamber and counted under a microscope to determine the cell concentration.
7. The phage pellet was soaked overnight in Pi-Mg buffer to ensure even distribution of the phage in the buffer, avoiding potential damage to the phage caused by physical resuspension methods such as pipetting or vortexing.
8. To form a CsCl density gradient, CsCl solutions of varying concentrations were slowly layered along the wall of the ultracentrifuge tube without stirring or shaking to prevent mixing of solutions with different concentrations.
9. The phage particle concentration and the copy number of VP1 proteins were determined using gp23* (49 kDa, 930 copies per capsid) as internal control.

Acknowledgments

This work was supported by grants from the National Key R & D Program of China (2023YFD1802505).

References

1. Choudhury SM, Ma X, Li Y, Nian X, Luo Z, Ma Y, Zhu Z, Yang F, Cao W, Zheng H (2021) FMDV leader protein interacts with the NACHT and LRR domains of NLRP3 to promote IL-1β production. Viruses 14(1):22. https://doi.org/10.3390/v14010022
2. Di Giacomo S, Bucafusco D, Schammas JM, Pega J, Miraglia MC, Barrionuevo F, Capozzo AV, Perez-Filgueira DM (2022) Assessment on different vaccine formulation parameters in the protection against heterologous challenge with FMDV in cattle. Viruses 14(8):1781. https://doi.org/10.3390/v14081781
3. Grant CFJ, Carr BV, Kotecha A, van den Born E, Stuart DI, Hammond JA, Charleston B (2017) The B cell response to foot-and-mouth disease virus in cattle following sequential vaccination with multiple serotypes. J Virol 91(9):e02157. https://doi.org/10.1128/JVI.02157-16
4. Harvey Y, Jackson B, Carr BV, Childs K, Moffat K, Freimanis G, Tennakoon C, Juleff N, Seago J (2022) An improved αvβ6-receptor-expressing suspension cell line for foot-and-mouth disease vaccine production. Viruses 14(3):621. https://doi.org/10.3390/v14030621
5. Hardham JM, Krug P, Pacheco JM, Thompson J, Dominowski P, Moulin V, Gay CG, Rodriguez LL, Rieder E (2020) Novel foot-and-mouth disease vaccine platform: formulations for safe and DIVA-compatible FMD vaccines with improved potency. Front Vet Sci 7:554305
6. Lu Z, Yu S, Wang W, Chen W, Wang X, Wu K, Li X, Fan S, Ding H, Yi L et al (2022) Development of foot-and-mouth disease vaccines in recent years. Vaccines 10(11):1817. https://doi.org/10.3390/vaccines10111817
7. Soria I, Quattrocchi V, Langellotti C, Pérez-Filgueira M, Pega J, Gnazzo V, Romera S, Schammas J, Bucafusco D, Di Giacomo S et al (2018) Immune response and partial protection against heterologous foot-and-mouth disease virus induced by dendrimer peptides in cattle. J Immunol Res 2018:3497401
8. Fernandez-Sainz I, Gavitt TD, Koster M, Ramirez-Medina E, Rodriguez YY, Wu P, Silbart LK, de Los Santos T, Szczepanek SM (2019) The VP1 G-H loop hypervariable epitope contributes to protective immunity against foot and mouth disease virus in swine. Vaccine 37:3435–3442
9. Bittle JL, Houghten RA, Alexander H, Shinnick TM, Sutcliffe JG, Lerner RA, Rowlands DJ, Brown F (1982) Protection against foot-and-mouth disease by immunization with a chemically synthesized peptide predicted from the viral nucleotide sequence. Nature 298:30–33
10. Wang G, Liu Y, Feng H, Chen Y, Yang S, Wei Q, Wang J, Liu D, Zhang G (2018) Immunogenicity evaluation of MS2 phage-mediated chimeric nanoparticle displaying an immunodominant B cell epitope of foot-and-mouth disease virus. PeerJ 6:e4823
11. Jackson DC, Lau YF, Le T, Suhrbier A, Deliyannis G, Cheers C, Smith C, Zeng W, Brown LE (2004) A totally synthetic vaccine of generic structure that targets toll-like receptor 2 on dendritic cells and promotes antibody or cytotoxic T cell responses. Proc Natl Acad Sci U S A 101:15440–15445
12. Malonis RJ, Lai JR, Vergnolle O (2020) Peptide-based vaccines: current progress and future challenges. Chem Rev 120:3210–3229
13. Li Q, Wubshet AK, Wang Y, Heath L, Zhang J (2023) B and T cell epitopes of the incursionary foot-and-mouth disease virus serotype SAT2 for vaccine development. Viruses 15(3):797. https://doi.org/10.3390/v15030797
14. Lei Y, Shao J, Zhao F, Li Y, Lei C, Ma F, Chang H, Zhang Y (2019) Artificially designed hepatitis B virus core particles composed of multiple epitopes of type A and O foot-and-mouth disease virus as a bivalent vaccine candidate. J Med Virol 91:2142–2152
15. Lee MJ, Jo H, Park SH, Ko MK, Kim SM, Kim B, Park JH (2020) Advanced foot-and-mouth disease vaccine platform for stimulation of simultaneous cellular and humoral immune responses. Vaccines 8(2):254. https://doi.org/10.3390/vaccines8020254
16. Li M, Guo P, Chen C, Feng H, Zhang W, Gu C, Wen G, Rao VB, Tao P (2021) Bacteriophage T4 vaccine platform for next-generation influenza vaccine development. Front Immunol 12:745625
17. Tao P, Mahalingam M, Kirtley ML, van Lier CJ, Sha J, Yeager LA, Chopra AK, Rao VB (2013) Mutated and bacteriophage T4 nanoparticle arrayed F1-V immunogens from Yersinia pestis as next generation plague vaccines. PLoS Pathog 9:e1003495

18. Li M, Chen C, Wang X, Guo P, Feng H, Zhang X, Zhang W, Gu C, Zhu J, Wen G et al (2023) T4 bacteriophage nanoparticles engineered through CRISPR provide a versatile platform for rapid development of flu mucosal vaccines. Antiviral Res 217:105688
19. Dong J, Chen C, Liu Y, Zhu J, Li M, Rao VB, Tao P (2021) Engineering T4 Bacteriophage for in vivo display by type V CRISPR-cas genome editing. ACS Synth Biol 10:2639–2648
20. Chen C, Zhang N, Li M, Guo A, Zheng Y, Humak F, Qian P, Tao P (2023) Recombinant bacteriophage T4 displaying key epitopes of the foot-and-mouth disease virus as a novel nanoparticle vaccine. Int J Biol Macromol 258:128837

Chapter 7

Endoskeleton-Enhanced Virus-Like Particles (VLPs): Internal Cross-Linking of Physalis Mottle Virus and Stability Tests

Zhuohong Wu, Sean K. Hsu, Stephen L. Craig and Nicole F. Steinmetz

Abstract

Virus-like particles (VLPs) often exhibit favorable chemical, thermal, and mechanical stability compared to other biologic materials. Nonetheless, enhancing their stability remains crucial for advancing VLP-based materials toward practical and commercial use. To address this, we introduced an "endoskeleton-armoring" approach aimed at reinforcing VLP stability. Using the VLP from physalis mottle virus (PhMV) as a model system, we created an internal polymer backbone by covalently linking the coat proteins within the capsid using a maleimide–PEG_{15}–maleimide crosslinker. This contrasts with native VLPs, which rely solely on non-covalent interactions for structural integrity. The resulting armored VLPs demonstrated markedly improved thermal stability, enhanced resistance to various denaturants (including surfactants, extreme pH, chemical agents, and organic solvents), and superior mechanical robustness. In this chapter, we detail the methods to achieve the "endoskeleton-armoring" approach using PhMV as proof of concept, which can be applied to other VLPs and protein cages.

Keywords Virus-like particles (VLPs), Protein cage, Stabilization, Crosslinking, Single-molecule force spectroscopy

1 Introduction

Virus-like particles (VLPs) are non-infectious nanostructures formed from viral capsid proteins that self-assemble into organized and monodisperse nanoparticles. They serve as promising platforms in virology, vaccinology, bionanotechnology, and materials science. A key challenge in their broader application lies in maintaining their structural and functional integrity during synthesis, storage, and operation under stress conditions. Although exterior polymer conjugation has been explored to enhance VLP stability, it limits further functionalization due to surface occupancy. Alternatively, internal polymer filling has been used to improve loading, but its role in structural stabilization remains underexplored.

To enhance stability of VLPs, we developed an "endoskeleton-armoring" approach by covalently crosslinking coat proteins (CPs)

Huan Peng and S. Brian Kim (eds.), *Biological Macromolecule Engineering Technologies*,
Methods in Molecular Biology, vol. 3032, https://doi.org/10.1007/978-1-0716-5257-2_7,

inside the capsid using a PEG linker, specifically a homo-bifunctional maleimide-PEG_{15}-maleimide linker to crosslink CPs through solvent-exposed cysteine residues on the internal surface of the capsids (Fig. 1). As a testbed, we used the VLPs.

VLP from physalis mottle virus (PhMV), a ~30 nm icosahedral nanoparticle made of 180 copies of identical coat proteins (CPs); the VLPs can be recombinantly produced in *E. coli*. Each CP offers a solvent-accessible cysteine (C75) directed toward the interior, allowing selective internal modification. The capsid also features ~1.0–1.4 nm pores that permit cargo access to the interior [1–3].

Here, we detail the methods to achieve CP crosslinking resulting in endoskeleton-enhanced PhMV, termed EE-VLPs. We describe

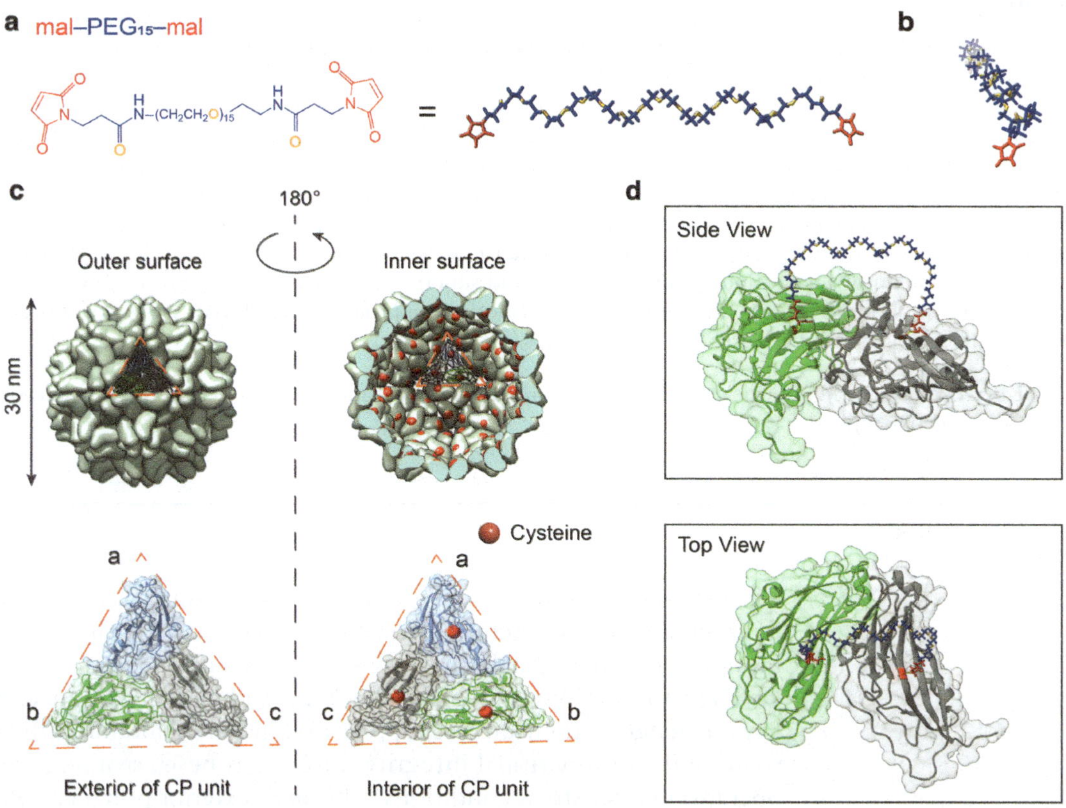

Fig. 1 (**a**) Chemical structure and projected 3D structure of the maleimide–PEG_{15}–maleimide crosslinker (polymer model created by MolView.org). (**b**) Side view to show its helical conformation in water. (**c**) Structure of the Physalis mottle virus (PhMV) virus-like particle (VLP) and its coat protein (CP); the VLP has an icosahedral T = 3 symmetry and is assembled from 180 identical copies of a CP; three CP for a triangular unit with a-c (blue, green, grey) denoting their position; "a" in blue clusters at the fivefold axis while "b" in green and "c" in grey is at the two- and threefold axis. The internal solvent-exposed cysteine is highlighted in red. The outer size is estimated at ~30 nm (PhMV structure created by UCSF Chimera, PDB: 1E57). (**d**) Structural model of PhMV coat proteins with viral endoskeleton via our endoskeleton-armored strategy, highlighted by top and side views. Note: the protein–PEG in this work is a model purely for illustration and does not depict the realistic conformation. (Reprinted with permission from [4]) Copyright 2024 American Chemical Society

the methods used to assay the structural integrity of the EE-VLPs and crosslinking efficiency, namely native and denaturing gel electrophoresis, size-exclusion chromatography (SEC), dynamic light scattering (DLS), transmission electron microscopy (TEM), and assays where the EE-VLPs are subjected to various chemical, thermal, and mechanical stressors. Their structural integrity was assessed via turbidimetry, TEM, circular dichroism (CD) spectroscopy, and single-molecule force spectroscopy (SMFS).

2 Materials

All solutions are prepared using ultrapure water (MilliQ water). Buffers and VLPs are stored at 4°C, and antibiotics and isopropyl β-D-thiogalactopyranoside (IPTG) are stored at −20°C. Unless specifically noted, chemicals are purchased through Sigma.

2.1 Double Plasmid Transformation into E. coli *and Preparation of the Starting Culture*

All reagents are prepared fresh.

1. Plasmids: A pRSETa plasmid (Novagen, USA) containing a 564-bp PhMV coat protein gene (NCBI Gene ID 940246) and a carbenicillin-resistance gene (Fig. 2). A pET plasmid (Novagen, USA) containing the same 564-bp PhMV coat protein gene (NCBI Gene ID 940246) and a kanamycin-resistance gene (Fig. 2) (*see* **Note 1**).
2. *E. coli* BL21(DE3) cells.
3. SOC medium: Super Optimal broth with Catabolite repression medium.

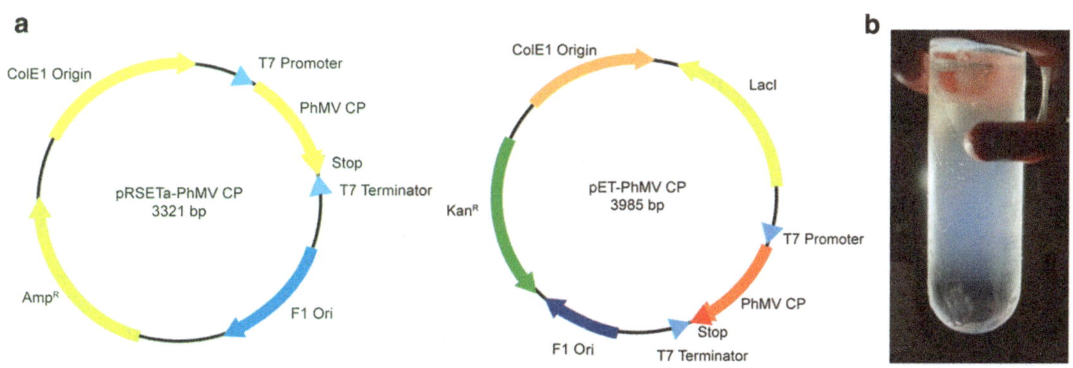

Fig. 2 Plasmids for PhMV VLP cloning and expression. (**a**) pRSETa-PhMV CP is the entry vector possessing the PhMV coat protein (NCBI Gene ID 940246), with ampicillin resistance. pET-PhMV CP is the entry vector possessing the same PhMV coat protein (NCBI Gene ID 940246), with kanamycin resistance. T7 is the promoter to initiate the transcription of the gene of PhMV CP. (**b**) Examples of sucrose gradients of PhMV VLP post-spin. To see the native PhMV, a flashlight is placed at the bottom of the tubes and directed upwards. (Reprinted with permission from [5]) Copyright 2024 American Chemical Society

4. 50 mg/mL carbenicillin.
5. 50 mg/mL kanamycin.
6. Selective LB agar plate: Sterilize 40 g LB agar in 1 L ultrapure water, pour into a 25 mL Petri dish, cool to ~50°C, and supplement with 25 μL of 50 mg/mL carbenicillin and 25 μL of 50 mg/mL kanamycin. The plates can be stored at 4°C for a few months.
7. LB medium: Lysogeny Broth medium.
8. Glycerin.
9. Use polypropylene tubes unless stated otherwise.
10. Heat block (e.g., ThermalMixer F2.0).
11. Icebox containing ice.
12. A UV–Vis spectrometer (e.g., Implen Diluphotometer) and matching disposable plastic cuvettes to monitor bacterial growth via optical density at 600 nm (OD_{600}).

2.2 Expression and Purification of PhMV VLPs

All reagents are prepared fresh.

1. Isopropyl-β-D-1-thiogalactopyranoside (IPTG, 0.5 M): Dissolve in ultrapure water and store at −20°C.
2. Phosphate-buffered saline (PBS): Dilute ten times from PBS 10× with ultrapure water. The pH should be 7.4–7.8.
3. Sodium citrate buffer (SCB, 50 mM, pH 5.5): per 1 L add 10.34 g sodium citrate dihydrate and 2.85 g citric acid; adjust pH with HCl/NaOH.
4. n-butanol.
5. Chloroform.
6. Magnesium sulfate: Dissolved magnesium sulfate in ultrapure water with a final concentration of 1 M stock.
7. Terrific Broth (TB): To prepare 1 L of TB media, add 47 g of dehydrated culture medium to 900 mL of distilled water. Add 10 mL of 1 M magnesium sulfate to yield a final concentration of 10 mM. Add 4 mL of glycerol, mix well, and add distilled water to 1 L. Adjust the pH value to 7.2 with HCl and NaOH and sterilize by autoclaving for at least 15 min at 121°C.
8. Lysogeny Broth (LB) medium: Stir to suspend 25 g powder in 1 L distilled water, autoclave for 15 min at 121°C to sterilize, and supplement with 25 μL of 50 mg/mL carbenicillin and 25 μL of 50 mg/mL kanamycin. The medium can be stored at room temperature unless aggregates are found. Autoclave before the next use.
9. 0.22 μm cellulose acetate syringe filter membranes, nonsterile.
10. Inoculation loop.

11. Use a polypropylene tube unless stated otherwise.
12. 2.5 L Flask
13. Flask screw cap
14. Sucrose gradient (10%–40% w/v): Two methods can be used to prepare the sucrose gradient.
    a) The first method uses a Gradient Master (BioComp Instruments, Inc.) to form the gradient. 15 mL of 40% (w/v) sucrose in SCB buffer is added to the bottom half of the ultraclear centrifuge tube, and 15 mL of 10% (w/v) sucrose in SCB buffer is carefully added on top. The tubes are then run in the gradient maker using the 10%–40% (w/v) program provided by BioComp.
    b) Alternatively, prepare a 40% (w/v) solution of sucrose in SCB buffer, freeze it in ultraclear centrifuge tubes at −20°C, and thaw it at room temperature. A gradient will form during the thawing process (*see* **Note 2**).
15. Orbital shaker with controlled temperature/speed and fixed rack for 2.5 L flask.
16. UV–Vis spectrometer (e.g., Implen Diluphotometer) and matching disposable plastic cuvettes (to monitor bacterial growth via optical density at 600 nm with 1 cm pathlength (OD_{600}).
17. Sonicator (e.g., Soniprep 150, MSS150, MSE with an MSE-SH100 process timer).
18. Suggested centrifuges, rotors, and tubes.
    a) Beckman Coulter Optima MAX-TL Ultracentrifuge.
    b) Beckman Coulter TLA 55 Fixed-angle Rotor.
    c) Beckman Coulter Polypropylene Microfuge Tube (Product No. 357448)
    d) Beckman Coulter Optima L-90 K Ultracentrifuge.
    e) Beckman Coulter JLA16.250 Rotor.
    f) Beckman Coulter Polypropylene Microfuge Tube (Product No. 356011)
    g) Beckman Coulter SW 32 Ti Rotor.
    h) Beckman Coulter Ultraclear Centrifuge Tubes (Product No. 344058)
    i) Beckman Coulter Type 50.2 Ti Rotor.
    j) Beckman Coulter 26.3 mL, Polycarbonate Bottle Assembly (Product No. 355616)
    k) Beckman Coulter Avanti J-E High Performance Centrifuge.

l) Beckman Coulter JLA 10.500 Rotor.
m) Beckman Coulter Polypropylene Centrifuge Bottle Assembly (Product No. 361691).
n) Beckman Coulter Microfuge 16 Centrifuge.

2.3 Synthesis of Endoskeleton-Enhanced PhMV Nanoparticles

1. Maleimide-PEG_{15}-maleimide: MW 1k, dissolved in dimethyl sulfoxide with vigorous vortexing at a final concentration of 400 mg/mL. Avoid exposure to light and prepare upon use (*see* **Note 3**).
2. 20% sucrose cushion (w/v, in PBS buffer).
3. PBS.
4. Use 1.5 mL microcentrifuge tube for ultra-centrifugation.
5. Use a polypropylene tube unless stated otherwise.

2.4 Characterization of EE-PhMV

All solutions or agarose gels are prepared fresh for each characterization. Pre-cast NuPAGE gel can be stored at room temperature.

2.4.1 Denaturing SDS-PAGE Analysis

1. SDS-PAGE gel: 10 lanes, 12% Bis-Tris NuPAGE SDS Gel.
2. Use a polypropylene tube unless stated otherwise.
3. Pre-stained Protein Standard: SeeBlue Plus2 Pre-stained standard.
4. Loading buffer: NuPAGE LDS Sample Buffer (4×).
5. Reducing buffer: NuPAG Sample Reducing Agent (10×).
6. Running buffer: 50 ml 20× MOPS buffer, bring the final volume to 1 L with ultrapure water.
7. GelCode Blue Stain Reagent.
8. Destaining buffer: 20% (v/v) methanol, 10% (v/v) acetic acid in ultrapure water.
9. Gel Electrophoresis system.
10. Heat block (e.g., Thermal Mixer F2.0).
11. 1000-Watt Microwave (e.g., Vissani).

2.4.2 Electrophoretic Mobility Shift Assay

1. Gel loading dye purple (6×)
2. TBE Buffer (Tris-borate-EDTA): Mix 50 mL of UltraPure 10× TBE buffer with 450 ultrapure water to bring to 500 mL. Final concentration: 0.1 M Tris-borate-EDTA, pH 8.3.
3. 0.8% (w/v) agarose gel: Mix 4 g of Agarose with 496 mL TBE, heat in the microwave for 2 min for three times to dissolve agarose, and cool at room temperature in a gel mold for the experiment.
4. Staining solution: 0.25% (w/v) Coomassie Brilliant Blue R-250, 30% (v/v) methanol, 10% (v/v) acetic acid in ultrapure water.

5. Destaining solution: 40% (v/v) methanol, 10% (v/v) acetic acid in ultrapure water.
6. Kimwipe.
7. Use a polypropylene tube unless stated otherwise.
8. Gel Electrophoresis system.

2.4.3 Transmission Electron Microscopy

1. 2% (w/v) uranyl acetate.
2. Carbon type-B TEM grid (Ted Pella).
3. Glow discharge cleaning system (Ted Pella).
4. Transmission electron microscope (JEOL JEM 1400 Plus operating at 80 kV with a Gatan 4k digital camera).

2.4.4 Size Exclusion Chromatography (SEC)

1. AKTA explorer FPLC system (GE Lifesciences).
2. Superose 6 10/300GL size exclusion column.
3. 100 μL coil.
4. 0.22 μm cellulose acetate syringe filter membranes.
5. 100 μL syringe.

2.4.5 Dynamic Light Scattering

1. Dynamic light scattering (DLS) instrument (Malvern Instruments Ltd., UK).
2. Disposable single sealed cuvettes.

2.4.6 UV–Vis Absorbance Spectroscopy

1. Infinite M PLEX microwell plate reader.
2. Non-Treated Surface 96-well plates.

2.4.7 Temperature-Dependent Circular Dichroism (CD) Spectroscopy

1. CD spectroscopy with temperature control (e.g., Aviv model 215 CD spectrometer).
2. 2 mm path length quartz cuvette.

2.5 Stability Test

All reagents are prepared fresh with ultrapure water.

1. SDS: Sodium dodecyl sulfate (1% w/v).
2. NaCl: Sodium Chloride (1, 2, and 4 M).
3. pH Stressor (pH 4.0, hydrochloric acid supplemented with sodium chloride to 100 mM in ionic strength).
4. pH Stressor (pH 1.2, hydrochloric acid supplemented with sodium chloride to 100 mM in ionic strength).
5. pH Stressor (pH 12, sodium hydroxide supplemented with sodium chloride to 100 mM in ionic strength).
6. Urea (1, 3, and 6 M).

7. GdnHCl: Guanidine hydrochloride (1–6 M).
8. MeOH: Methanol (v/v 25%, 50%, and 90%).
9. EDC: 1-Ethyl-3-(3-dimethylaminopropyl)carbodiimide (dissolved in DMSO).
10. Use a polypropylene tube unless stated otherwise.
11. Sharp Microlever Silicon probes (Bruker, Camarillo, CA).
12. Piranha solution (H_2SO_4: H_2O_2 = 3:1) for 30 min (Extreme caution should be taken when working with piranha).
13. Silicon wafer (~1 × 1 cm).
14. Sharp Microlever silicon probes (MSNL) are purchased from Bruker (Camarillo, CA), and are prepared by soaking in piranha solution for ~15 min at room temperature. Silicon surfaces are prepared by soaking ~30 min in hot piranha solution, followed by washing with ultrapure water and drying under a stream of nitrogen. The force curves used for analysis are obtained with rectangular-shaped cantilevers (205 μ m × 15 μm, nominal tip radius ~2 nm, nominal spring constant k ~0.02 N/m, frequency ~15 kHz).
15. Amine-functionalized surface silicon substrates are prepared by incubating the substrates described in Step 14, above, in a 2% v/v solution of *N*-[3-(trimethoxysilyl)propyl]ethylenediamine in methanol for 30 min. The substrates are then rinsed with methanol and annealed at 110°C for 30 min.
16. Amine-functionalized AFM probes are prepared by incubating the probes described in Step 14, above, in a 2% (v/v) solution of *N*-[3-(trimethoxysilyl)propyl]ethylenediamine in methanol for 10 min. The probes are then rinsed with methanol and annealed at 110°C for 30 min.
17. Matlab (MathWorks, Natick, MA).
18. AFM: Atomic force microscopy, e.g., Bruker (previously Digital Instruments) Multimode AFM head mounted on top of a piezoelectric positioner (Physik Instrumente, GmbH).

3 Methods

3.1 Double Plasmid Transformation into E. coli and Preparation of the Starting Culture

1. Thaw *E. coli* BL21 competent cells on wet ice (*see* **Note 4**).
2. Pre-chill 1.5 mL microcentrifuge tubes on wet ice.
3. When ready, gently mix the cells, then prepare 50 μL aliquots of BL21 cells in the chilled 1.5 mL tubes.
4. Dilute the stock plasmids (see Fig. 2) to 100 ng/μL with DI water.
5. Add 1 μL of the diluted plasmids to the 50 μL of BL21 cells.

6. Incubate the cells on ice for 20 min.
7. Prepare a 42°C water bath (Thermal Mixer F2.0) (*see* **Note 5**).
8. Heat-shock the cells for exactly 45 s in the 42°C-warm water bath. Do not mix or shake the 1.5 mL tube.
9. Immediately place the cells on ice for 2 min.
10. Add 200 μL of room-temperature SOC medium.
11. Shake the 1.5 mL tube at 250 RPM for 1 h at 37°C.
12. Spread 100 μL of the transformed cells onto an agar plate supplemented with 50 μg/mL kanamycin and 50 μg/mL carbenicillin (*see* **Note 6**).
13. Invert the plate and incubate for 16–24 h at 37°C.
14. Add 5 mL LB medium supplemented with 50 μg/mL kanamycin and 50 μg/mL carbenicillin to a 50 mL tube.
15. Pick one single colony and add it to the 50 mL tube to produce the starting culture.
16. Shake the starting culture at 225 RPM for 18 h at 37°C.
17. Dilute the cells to OD_{600} = 0.9, then mix 1000 μL of cells with 300 μL glycerin in a cryotube.
18. Place the cryotubes into a cryo-foam pod in −80°C for at least 12 h, then transfer the cryotubes to cryogenic freezer boxes for long-term preservation at −80°C.

3.2 Expression and Purification of PhMV VLPs

1. Add a small amount of transformed BL21 cells from the starting culture (see Sect. 3.1) or from the cryo-preserved sample (*see* **Note 7**) to a 50 mL tube supplemented with 50 μg/mL of carbenicillin and 50 μg/mL kanamycin. The lid is half-capped.
2. Incubate at 225 RPM for 18 h at 37°C.
3. Add 10 mL to 500 mL of TB supplemented with 50 μg/mL of carbenicillin and 50 μg/mL kanamycin in a 2.5 L flask, then incubate at 250 RPM for 18 h at 37°C.
4. Monitor the OD_{600} over a time course of ~8 h.
5. When OD_{600} = ~1.8 (approximately at 8 h), add 0.5 mL of IPTG stock to yield a final 0.5 mM IPTG (*see* **Note 8**).
6. Decrease the temperature to 30°C and incubate for 18 h.
7. To collect the cells, transfer the cells to two 500 mL centrifuge tubes, and centrifuge (7500 × *g*, 10 min, 4°C).
8. Resuspend the cell pellet using 50 mM SCB (~35 mL each pellet).
9. To extract the VLPs, add ~40 mL cell suspension to 50 mL tubes and incubate on wet ice.
10. Sonicate to lyse the cells while keeping the 50 mL tube on wet ice. Set the sonication to 10 min, 30% amplitude, 5 s ON, 2 s OFF.

11. Centrifuge (10,000 × *g*, 10 min, 4°C) the samples and then transfer the supernatant to new 250 mL tubes (*see* **Note 9**)
12. Centrifuge (27,000 × *g*, 30 min, 4°C; Beckman Coulter JLA-16.250 rotor) the samples and then transfer the supernatant to polycarbonate tubes and ultra-centrifugation in a fixed-angle rotor (111,000 × *g*, 3 h, 4°C; 35,000 RPM using 50.2 Ti rotor).
13. Add 3 mL SCB to each pellet to resuspend overnight at 4°C on an orbital shaker.
14. Remove the insoluble materials in a tabletop centrifuge (15,000×*g*, 10 min, 4°C; e.g., Beckman Coulter Microfuge 16); discard the pellet and collect the supernatant.
15. Prepare chloroform and n-butanol with a 1:1 ratio of volume.
16. To the supernatant, add half volume of 1:1 chloroform/n-butanol (e.g., 6.25 mL of 1:1 chloroform/n-butanol to 12.5 mL of supernatant) (*see* **Note 10**), and mix.
17. Centrifuge (5000 × *g*, 10 min, 4°C) and collect the top aqueous layer suspension while discarding the bottom waxy (organic) layer.
18. Remove the insoluble materials in a tabletop centrifuge (15,000 × *g*, 10 min, 4°C); discard the pellet and collect the supernatant.
19. Add 2.5 mL sample atop a sucrose gradient (10%–40% w/v) (*see* Sect. 2.2 and **Note 11**).
20. Centrifuge in a swing bucket rotor (96,000 × *g*, 3 h, 4°C; 28,000 RPM when using Beckman Coulter SW 32 Ti rotor).
21. Collect the light scattering fractions that contain the VLPs (see Fig. 2) and dilute with SCB in a 1:1 volume ratio.
22. To collect the VLPs, an ultra-centrifuge in a fixed-angle rotor for 3 h at 160,000×*g* and 4°C (42,000 rpm when using Beckman Coulter type 50.2 Ti rotor).
23. Discard the supernatant and resuspend the pellet in 2 mL PBS overnight on an orbital shaker.
24. Remove the insoluble materials in a tabletop centrifuge (15,000 × *g*, 10 min, 4°C); discard the pellet and collect the supernatant.
25. Filter the sample using a 0.22 μm filter to obtain purified VLPs.
26. Test concentration using the bicinchoninic acid (BCA) assay using bovine serum albumin (BSA) as a standard.
27. Test purity using SEC with a Superose 6 10/300GL column by monitoring the absorbance at 280 nm (Fig. 3).

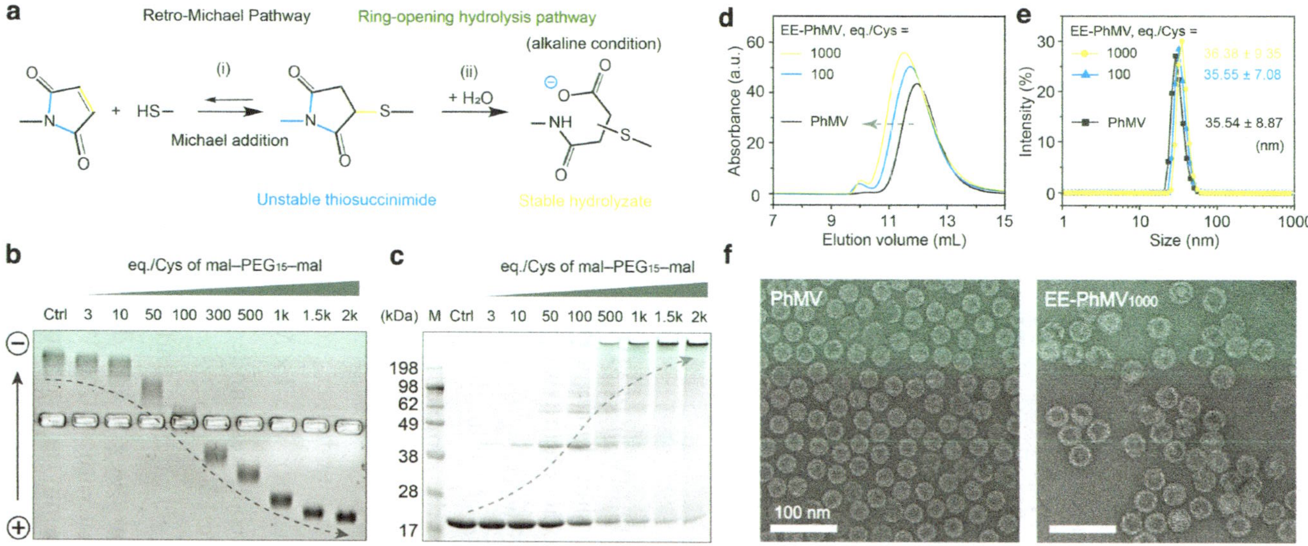

Fig. 3 Characterization of Endoskeleton-Enhanced (EE-) PhMV. (**a**) Michael addition of the thiol to the maleimide results in the maleimide–thiol conjugate, a succinimide thioether. The reaction is fast, but the conjugate is relatively unstable and can undergo further reaction via one of two pathways: (i) it can undergo a retro-Michael conversion back to the original thiol and maleimide; (ii) it can undergo irreversible ring-opening hydrolysis to yield a stable hydrolysate (succinamic acid thioether) under alkaline condition, which prevents elimination of the maleimide–thiol conjugate and generates negatively charged/deprotonated carboxylate. This charge change serves as a proxy of the degree of conjugation to the VLPs. (**b**) Electrophoretic mobility shift assay (EMSA) and (**c**) denaturing sodium dodecyl sulfate–polyacrylamide gel electrophoresis (SDS-PAGE) of PhMV and EE-PhMV under various eq./Cys. The distance of negative shifting toward the anode positively correlates to the value of eq./Cys. (**d**) Fast protein liquid chromatography (FPLC) data indicate that the endoskeleton leads to swelling of the VLPs, which agrees with (**e**) dynamic light scattering (DLS) data. FPLC Detection wavelength is 260 nm. (**f**) TEM images demonstrate a good dispersity and integrity of native PhMV and EE-PhMV$_{1000}$ (eq./Cys = 1000). (Reprinted with permission from [4]) Copyright 2024 American Chemical Society

3.3 Synthesis of Endoskeleton-Enhanced PhMV Nanoparticles

1. At room temperature, in a 2-mL tube, 100 μL of PhMV stock (10 mg/mL, in PBS buffer) is first added to 770 μL of PBS buffer with a gentle vortex.
2. 400 mg/mL of maleimide–PEG_{15}–maleimide is freshly made in DMSO with vigorous vortexing, resulting in the maleimide–PEG_{15}–maleimide stock solution.
3. 130 μL of the maleimide–PEG_{15}–maleimide stock solution is added dropwise to PhMV while gently vortexing.
4. The reaction is then incubated at 4°C for 16 h in the dark or using an amber tube.
5. To a 1.5 mL ultra-centrifuge tube (*see* **Note 12**), 1 mL of EE-PhMV nanoparticle mixture is layered atop 100 μL of 20% sucrose cushion (w/v, in PBS buffer), and is pelleted by ultra-centrifugation (124,000×*g*, 90 min, TLA-55 rotor).
6. Resuspend the pellet overnight in 2 mL PBS at 4°C.
7. Determine the final concentration by bicinchoninic acid (BCA) assay for further characterization.
8. Test sample with DLS, TEM, and FPLC (see Sects. 3.4.3–3.4.5) (Fig. 3).

3.4 Characterization of EE-PhMV

3.4.1 Denaturing SDS-PAGE Analysis

1. Mix 5 μL EE-PhMV or PhMV samples (1 mg/mL) with 8 μL PBS, 5 μL loading buffer (4×), and 2 μL reducing buffer (10×). The total volume of working solutions is 20 μL.
2. Denature the working solution by heating it using a heat block at 95°C for 5 min; then place samples on ice.
3. Load 15 μL of the sample into a ten-lane SDS-PAGE gel (*see* **Note 13**) placed in 1× MOPS buffer in the gel electrophoresis apparatus.
4. Load 10 μL SeeBlue ladder protein in a separate lane.
5. Fill the gel chamber with 1× MOPS buffer.
6. Run gel electrophoresis under 180 V for 60 min, with a limited current of 120 mA.
7. Uninstall the gel carefully, and place it in a container.
8. To pre-destain the gel, add distilled water until gel is submerged, heat it with a 1000-W microwave for 40 s, and then incubate for 5 min on an orbital shaker.
9. Decant distilled water carefully, add GelCode Blue Stain Reagent until gel is submerged, heat it at 1000 W with a microwave for 40 s, and then incubate for 5 min for staining on an orbital shaker.

10. Pour GelCode Blue Stain Reagent into a waste container, add distilled water until gel is submerged, heat it with a 1000-W microwave for 40 s, add a Kimwipe to the solution, and then incubate for 5 min for destaining on an orbital shaker.
11. Decant distilled water and wash the gel with distilled water.
12. Image the gel with the gel imager.
13. Analyzed under UV light (Fig. 3).

3.4.2 Electrophoretic Mobility Shift Assay

1. Heat 0.8% (w/v) agarose solution (see Sect. 2.4.2) in a microwave for a few minutes until it is completely dissolved.
2. Cast the warm dissolved 0.8% (w/v) agarose solution in the native gel electrophoresis mold, place a gel comb to create eight wells in the gel to create approximately 4–6 mm depth of well for sample loading (*see* **Note 14**).
3. After the agarose solution is cooled down and solidified as a gel, put the mold and gel into a gel tank and submerge the gel in TBE (1×) buffer.
4. Mix 8 μL EE-PhMV or PhMV sample (1 mg/mL) with 8.7 μL PBS and 3.3 μL gel loading dye purple (6×).
5. Load 15 μL of the working solution into each well.
6. Run gel electrophoresis under 100 V for 40 min, with a limited current of 400 mA.
7. To pre-destain the gel, add destaining solution (see Sect. 2.4.2) until gel is submerged and then incubate for 5 min (*see* **Note 15**).
8. Pour the destaining solution into the waste container (*see* **Note 16**), add staining solution (see Sect. 2.4.2) until gel is submerged, and then incubate on an orbital shaker for 1 h for staining.
9. Pour staining solution into a waste container, add destaining solution until gel is submerged, place a Kimwipe in the solution (*see* **Note 17**), and incubate on an orbital shaker for 3 h for destaining.
10. Pour the destaining solution into a waste container and wash the gel with distilled water.
11. Image the gel with the gel imager.
12. Analyze under UV light (Fig. 3).

3.4.3 Transmission Electron Microscopy

1. Treat the TEM grid with a glow discharge cleaning system for 30 s.
2. Dilute the EE-PhMV sample to a concentration of 0.5 mg/mL in ultrapure water.
3. Float 10 μL droplets of each of the following solutions on parafilm: EE-PhMV, ultrapure water, ultrapure water, 2% (w/v) uranyl acetate.

4. In an upside-down manner, allow the carbon-coated side of the TEM grid to contact the EE-PhMV droplet. Keep it upside-down for 2 min.
5. Blot off the droplet by tapping and drying with the edge of the filter paper.
6. Contact the grid on the first ultrapure water droplet for 30 s.
7. Blot off the droplet by tapping and drying with the edge of the filter paper.
8. Contact the grid on the second ultrapure water droplet for 30 s.
9. Blot off the droplet by tapping and drying with the edge of the filter paper.
10. Contact TEM grid on 2% (w/v) uranyl acetate droplet for 90 s.
11. Blot off the droplet by tapping and drying with the edge of the filter paper.
12. Leave the grid in the original grid container overnight.
13. Use TEM for imaging at 200 kV.

3.4.4 Size Exclusion Chromatography

1. Set up a Superose 6 10/300GL size exclusion column on an AKTA explorer FPLC system (GE Lifesciences). A 100 μL coil is used.
2. Equilibrate columns with PBS.
3. Prepare 150–200 μL of a 1 mg/mL solution of EE-PhMV or PhMV sample (1 mg/ mL) in PBS. Filter sample with 0.22 μm cellulose acetate syringe filter membranes.
4. Inject solution and run column at a flow rate of 0.5 mL/min. Set the absorbance detectors for 260 nm (host RNA) and 280 nm (PhMV protein and PEG polymer) (*see* **Note 18**).

3.4.5 Dynamic Light Scattering

1. Add 8 μL EE-PhMV or PhMV sample (1 mg/mL) to 72 μL of 10 mM NaCl.
2. Measure with DLS with measurements for ten times and 10 s each. Each sample is measured in triplicate.
3. Report on the data in intensity percentage.

3.4.6 UV–Vis Absorbance Spectroscopy

1. Add 100 μL of sample to a 96-well microwell plate.
2. Measure with a microwell plate reader at 600 nm (Fig. 4).

3.4.7 Temperature-Dependent Circular Dichroism Spectra

1. Circular dichroism (CD) spectra are acquired using an Aviv model 215 CD spectrometer with a 2 mm pathlength quartz cuvette.
2. Add 500 μL of EE-PhMV or PhMV sample (1 mg/mL) to 500 μL PBS. The final sample concentration is 0.5 mg/mL.

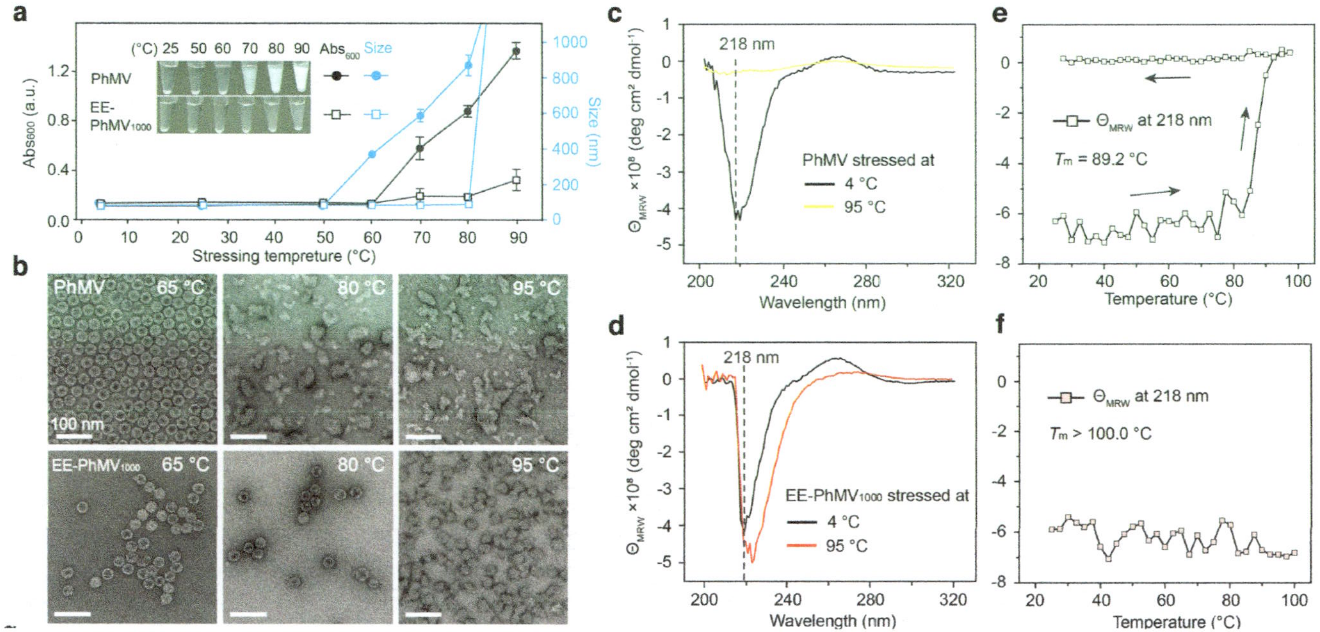

Fig. 4 Comprehensive stability assay of PhMV and EE-PhMV. (**a**) Turbidimetric assay (Abs_{600}, marked in black) and size measurement (marked in blue) for determining the thermal-induced aggregation of PhMV (solid circle) and EE-PhMV (hollow square) in the range of 25–90°C. Both groups appear as microscale aggregates at 90°C. Samples at 4°C are used as a negative control. Inset: photographs of samples after treatment at different temperatures for 15 min. (**b**) TEM images showing the thermal-induced morphological changes of PhMV and EE-PhMV over various temperatures. Circular dichroism (CD) spectra of (**c**) PhMV and (**d**) EE-PhMV after thermal stressing at 4°C (negative control) and 95°C for 15 min. The content of β-structure is quantified at 218 nm, as indicated by dashed lines at 218 nm. Θ_{MRW} represents the mean residue weight. Temperature-dependent CD is conducted by monitoring the change of β-structure of (**e**) PhMV and (**f**) EE-PhMV, ranging from 25–100°C. (Reprinted with permission from [4]) Copyright 2024 American Chemical Society

3. Set the temperature increase rate to 20°C per min, the temperature interval to 2.5°C, and the equilibration time at each point was 2.5 min.
4. Measure and analyze the sample (Fig. 4).

3.5 Stability Tests

3.5.1 Thermal Stability

1. 15 min prior to the thermal stressing experiment, set the temperature of a heat block (e.g., Eppendorf 5333 Master Cycler Thermal Cycler) to 25, 50, 60, 70, 80, or 90°C.
2. In a 2 mL tube, add 10 μL of purified EE-PhMV or PhMV (10 mg/mL) in 90 μL PBS, yielding a final 1 mg/mL sample.
3. Put the tube in the pre-heated heat block for 15 min (*see* **Note 19**).
4. Cool down the sample to room temperature and characterize with UV–Vis, DLS, and CD Spectra (see Sects. 3.4.5–3.4.7).

3.5.2 Comprehensive Stability

1. In a 2 mL tube, add 10 μL of purified EE-PhMV or PhMV (10 mg/mL) to 90 μL stressing solutions, yielding a final 1 mg/mL sample.
2. Incubate the samples with the following time and temperature settings:
   a) 1% (w/v) SDS, 24 h, room temperature.
   b) 1, 2, and 4 M NaCl, 6 h, room temperature.
   c) PBS (pH 7.4), 24 h, room temperature.
   d) pH stressors (pH 1.2, 4.0, and 12), 24 h, room temperature.
   e) Urea (1, 3, and 6 M), 24 h, room temperature.
   f) Guanidine hydrochloride (1, 3, and 6 M), 24 h, room temperature.
   g) NaCl (1, 2, and 4 M), 6 h, room temperature.
   h) Methanol (25%, 50%, and 90%), 2 h, room temperature.

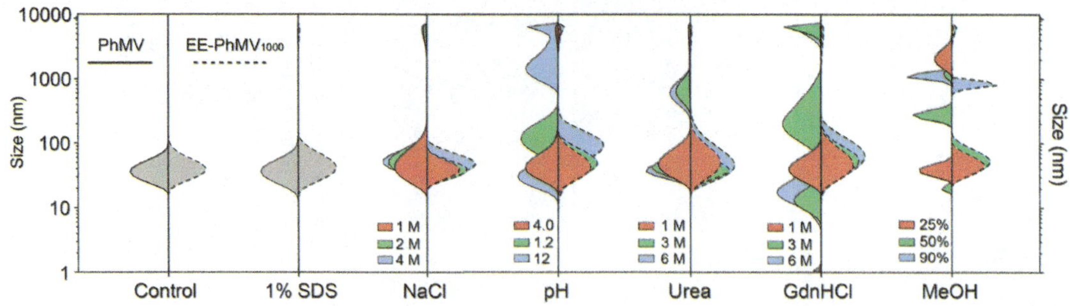

Fig. 5 Size distribution under miscellaneous stressing conditions, including in the presence of surfactant, salt, pH, chemical denaturants, and organic solvents. DLS samples are triplicated and presented by intensity (%). (Reprinted with permission from [4]) Copyright 2024 American Chemical Society

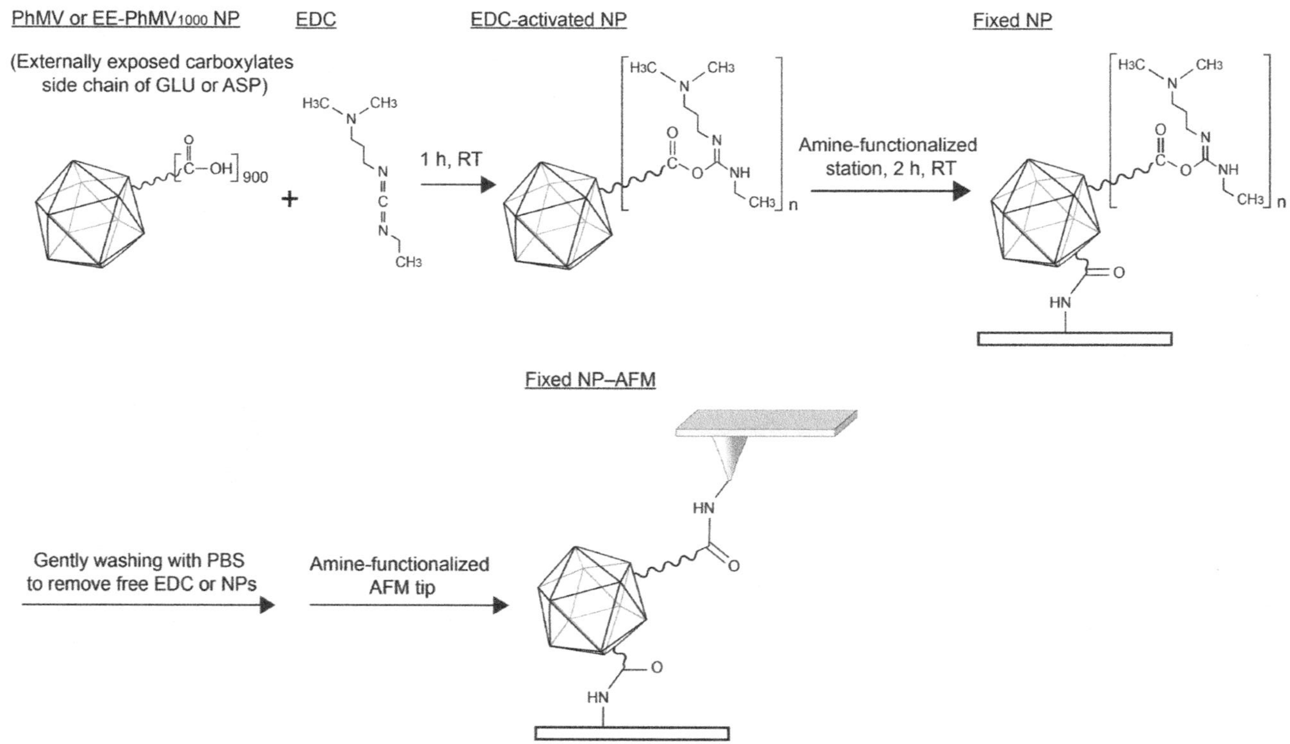

Fig. 6 Conjugation scheme of PhMV or EE-PhMV$_{1000}$ NP for SMFS test. PhMV or EE-PhMV$_{1000}$ exhibits five exterior-exposed carboxylates (namely D58, D83, E39, E47, and C-terminus). These carboxylates are activated by the addition of 1-ethyl-3-(3-dimethylamino-propyl)carbodiimide (EDC) crosslinker to yield o-acylisourea active ester for the following conjugations. First, the activated particles are immobilized on an amine-functionalized station via EDC chemistry, followed by PBS washing to remove free reagents. Then, the immobilized particles are conjugated onto an amine-functionalized AFM cantilever tip by conjugating the o-acylisourea active ester and the amine from the cantilever tip. The force and distance are monitored between VLP and the cantilever. (Reprinted with permission from [4]) Copyright 2024 American Chemical Society

3. Characterize the sample with DLS (see Sect. 2.5) and report data as a population of hydrodynamic radius based on intensity (Fig. 5).

3.5.3 Mechanical Stability

1. To activate the VLP with EDC, at room temperature, add 1 μL EDC (3 μM) to 99 μL PhMV or EE-PhMV sample (1 mg/mL) to yield a final concentration of EDC of 0.03 μM (Fig. 6).
2. Incubate for 1 h at room temperature.
3. To fix EDC-activated VLP, 100 μL of EDC-coupled VLP solution is placed on the amine-functionalized silicon wafer (see Sect. 2.5) and incubated for 2 h to allow immobilization.

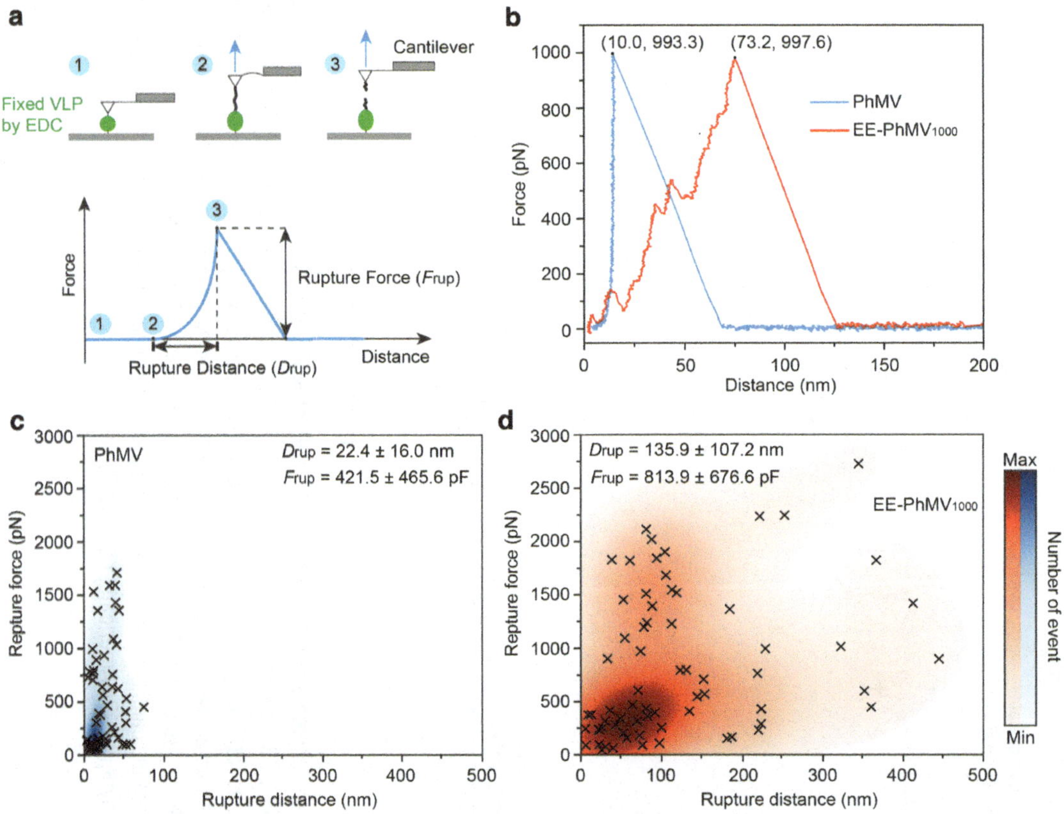

Fig. 7 Mechanical performance of PhMV and EE-PhMV. (**a**) Schematic of single-molecule force spectroscopy (SMSF) study performed by AFM at the "force-extension" operational mode. The exterior-exposed carboxylates are activated by 1-ethyl-3-(3-dimethyl aminopropyl)carbodiimide (EDC) crosslinker, followed by conjugation onto an amine-functionalized station and amine-functionalized AFM cantilever, respectively. Two critical parameters are schematized: rupture distance (D_{rup}) and rupture force (F_{rup}). (**b**) Representative force-extension curve of PhMV and EE-PhMV$_{1000}$. (**c** and **d**) Statistic mapping shows the two-dimensional distribution of rupture forces and distance of PhMV and EE-PhMV$_{1000}$. The D_{rup} and F_{rup} are sixfold and 1.9-fold increased. Number of events = 71. Statistical analysis: a p-value <0.05 is statistically significant. Independent samples t-test analysis reveals $p < 0.001$ for both rupture force and rupture distance between native PhMV and EE-PhMV$_{1000}$. (Reprinted with permission from [4]) Copyright 2024 American Chemical Society

4. Gently rinse with PBS to remove free VLPs and EDC.
5. To conjugate AFM cantilever tip to VLP, at room temperature, the VLP-containing substrate is placed on a piezoelectric stage is brought toward the amine-functionalized cantilever tips (see Sect. 2.5) at a velocity of 1000 nm/s until a contact force of 50 pN is obtained. The piezoelectric stage is subsequently retracted at a velocity of 1000 nm/s. The approach/retract process is repeated at different spots on the substrate in a raster pattern.
6. Collect data using Matlab.
7. Analyze the data with two critical parameters: rupture distance (D_{rup}) and rupture force (F_{rup}) (Fig. 7).

4 Notes

1. Ampicillin-resistance genes commonly encode a β-lactamase enzyme, conferring resistance against antibiotics like carbenicillin, which is more stable during bacterial culture.
2. It should be noted that additional freeze–thaw cycles improve the quality of the gradient.
3. If a different VLP was used, different lengths of maleimide-PEG-maleimide could be tested to maximize the degree of crosslinking.
4. The thawing process can be done at room temperature.
5. Use a heat block and thermometer to monitor the temperature. Reserve 15 min of heating for the temperature to be stable.
6. It is suitable to spread 25 μL of 50 mg/mL antibiotics (kanamycin and carbenicillin) and dry it at 37°C for quick use.
7. The sample is transferred using an inoculation loop. Prepare to move fast; repeated thawing of cryo stock should be prevented. It is recommended that multiple cryo stocks be prepared and stored as backup.
8. OD_{600} = ~1.0 is the typical number indicating that the bacterial growth phase is in the middle or late phase of logarithmic growth. This is an experimental balance of sufficient biomass and metabolic activity. Practically, adding 150 μg culture to a 96-well plate yields approximately 0.5 cm.
9. Add an additional low spin (15,000 × *g*, 10 min) if the pellet breaks.
10. If very cloudy, add a full volume of 1:1 chloroform/butanol.
11. Add 1–2.5 mL sample, depending on the sample concentration.

12. It is important to use the right tube for ultra-centrifugation.
13. Add 10 μL if using a 12-lane SDS-PAGE gel. Ideally, 2.5–3.75 μg coat protein per lane.
14. The position of the gel comb should depend on the estimation of the direction and distance of sample shifting. Gel loading dye purple can be applied to show the frontier of moving samples.
15. The first destaining step is to avoid staining impurity and make a clearer background of the native gel.
16. If a fluorescent dye was used, image the gel under the corresponding fluorescent channel before staining.
17. Add Kimwipe to absorb excess Coomassie Blue dye and therefore facilitate destaining. Kimwipe can be changed in the middle of destaining.
18. Both protein and PEG polymer show a strong absorbance at 280 nm. This can be used to detect the purity of the sample after the purification process. Absorbance at 260 nm comes from nucleic acids contained in PhMV VLP. These nucleic acids are mostly *E. coli* host RNA, not PhMV genetic material, as PhMV VLPs are devoid of genetic material.
19. Evaporation of solvent will further concentrate the sample, causing aggregation, or even dehydrating the samples. It is important to minimize condensation using a heated lid. There are two options:

    a) Thermocycler (e.g., Labnet MultiGene OptiMax).
    b) Heat block with heated lid (MyBlock).

Acknowledgments

This work was supported by the NSF Center for the Chemistry of Molecularly Optimized Networks (MONET), CHE-2116298. This work was performed in part at the San Diego Nanotechnology Infrastructure (SDNI) of University of California, San Diego, a member of the National Nanotechnology Coordinated Infrastructure (NNCI), which is supported by the National Science Foundation (Grant ECCS-1542148). The electron microscopic images are taken in the Cellular and Molecular Medicine Electron microscopy core facility, which is supported in part by the National Institutes of Health Award number S10OD023527.

Conflict of Interest The authors declare the following competing financial interest(s): Dr. Steinmetz is a co-founder of, has equity in, and has a financial interest with Mosaic ImmunoEnginering

Inc. Dr. Steinmetz is a co-founder of, and serves as manager of Pokometz Scientific LLC, under which she is a paid consultant to Mosaic ImmunoEngineering Inc., Flagship Labs 95 Inc., and Arana Biosciences Inc. The other authors declare no potential COI.

References

1. Masarapu H, Patel BK, Chariou PL, Hu H, Gulati NM, Carpenter BL, Ghiladi RA, Shukla S, Steinmetz NF (2017) Physalis mottle virus-like particles as nanocarriers for imaging reagents and drugs. Biomacromolecules 18(12):4141–4153. https://doi.org/10.1021/acs.biomac.7b01196
2. Hooker JM, Kovacs EW, Francis MB (2004) Interior surface modification of bacteriophage MS2. J Am Chem Soc 126(12):3718–3719. https://doi.org/10.1021/ja031790q
3. Sastri M, Kekuda R, Gopinath K, Kumar CT, Jagath JR, Savithri HS (1997) Assembly of physalis mottle virus capsid protein in Escherichia coli and the role of amino and carboxy termini in the formation of the icosahedral particles. J Mol Biol 272(4):541–552. https://doi.org/10.1006/jmbi.1997.1258
4. Wu Z, Bayon JL, Kouznetsova TB, Ouchi T, Barkovich KJ, Hsu SK, Craig SL, Steinmetz NF (2024) Virus-like particles armored by an endoskeleton. Nano Lett 24(10):2989–2997. https://doi.org/10.1021/acs.nanolett.3c03806
5. Barkovich KJ, Wu Z, Zhao Z, Simms A, Chang EY, Steinmetz NF (2023) Physalis mottle virus-like nanocarriers with expanded internal loading capacity. Bioconjug Chem 34(9):1585–1595. https://doi.org/10.1021/acs.bioconjchem.3c00269

Chapter 8

An Integrated Workflow for Phage Selection, Amplification, and Surface Modification

Yingqi Xu, Weixuan Luo, Wenli Tong, Jingxiang Fu, Xinya Wang and Huan Peng

Abstract

Bacteriophages, characterized by strong host specificity and high environmental compatibility, are widely recognized for their immense potential in microbial ecological regulation. They can be utilized for foodborne pathogenic bacteria control and drug-resistant bacterial infections treatment. To address the limitations of traditional phage screening methods when isolating target phages from complex environmental samples, a standardized and efficient technical protocol can be established by integrating optimized techniques, including the double-layer plate method, liquid enrichment combined with centrifugation, and supermicro buffer-mediated phage stability maintenance. In this chapter, we describe protocols for the preparation of key reagents, the implementation of a complete phage screening workflow, phage amplification via liquid/plate/OD_{260} optical density methods, and chemical coupling-based phage surface modification to achieve functionalization, along with critical notes to ensure reproducibility and reliability of the experimental process.

Keywords Bacteriophages, Phage screening, Chemical coupling, Surface modification

1 Introduction

Bacteriophages have important applications in microbial ecological regulation, control of foodborne pathogenic bacteria, and treatment of drug-resistant bacterial infections due to their strong host specificity and high environmental compatibility [1, 2]. However, when screening target bacteriophages from complex environmental samples such as soil and sewage, traditional methods often face problems such as low enrichment efficiency, severe interference by miscellaneous bacteria, and insufficient standardization of operations. Moreover, the physicochemical differences of environmental samples (such as acidity and high organic matter) further increase

Yingqi Xu and Weixuan Luo contributed equally to this work.

Huan Peng and S. Brian Kim (eds.), *Biological Macromolecule Engineering Technologies*,
Methods in Molecular Biology, vol. 3032, https://doi.org/10.1007/978-1-0716-5257-2_8,

the difficulty of screening. An efficient and standardized process is urgently needed [3].

Phage surface properties critically govern their functions and applications. Phage display, a powerful biomolecular tool first reported by Smith in 1985 [4], fuses exogenous genes encoding peptides, proteins, or antibodies with phage structural protein genes, enabling the display of foreign molecules on the virion surface while preserving phage infectivity and replicative capacity. This technology has revolutionized functional biomolecule screening by facilitating rapid isolation of high-affinity ligands from large-scale libraries. Nevertheless, its utility is limited. For instance, the M13 phage's pVIII capsid protein can only accommodate short peptides (≈610 amino acids), severely restricting viral particle modification. In contrast, phage chemical engineering has emerged as an increasingly robust alternative, leveraging the abundant reactive groups on capsid proteins for versatile functionalization. Thus, developing a standardized phage chemical engineering protocol is crucial for advancing functional phage-based therapeutics.

This protocol outlines a streamlined workflow encompassing "sample pretreatment—enrichment—phage screening—phage purification—preliminary identification—phage surface modification." It is specifically tailored for isolating functional bacteriophages from complex environmental matrices and engineering their surface properties, laying a standardized foundation for developing next-generation phage-based antimicrobials.

2 Materials

Prepare all solutions using ultrapure water (prepared by purifying deionized water, to attain a sensitivity of 18 MΩ-cm at 25°C) and analytical grade reagents. Prepare and store all reagents at room temperature (unless indicated otherwise). Diligently follow all waste disposal regulations when disposing of waste materials.

2.1 PB Solution

1. Weigh the NaCl: ~3.974 g. $MgSO_4$(anhydrous): ~1.204 g (or ~2.465 g for $MgSO_4 \cdot 7H_2O$). Tris base: ~1.211 g. Add ultrapure water and stir on a magnetic stirrer until all reagents are fully dissolved.
2. Transfer the dissolved solution to a 1000 mL volumetric flask (rinse the beaker with small amounts of water to recover residues). Insert the pH meter electrode, then add concentrated HCl dropwise while gently swirling until the pH stabilizes at 7.5. Fill the flask with ultrapure water to the 1000 mL mark, seal with a stopper, and invert 8–10 times to mix thoroughly.

2.2 LB Powder

1. Use the ratio of tryptone:yeast extract:NaCl = 2:1:2.

2. Weigh the components. Using an analytical balance, accurately weigh the calculated amounts of tryptone, yeast extract, and NaCl into a clean, dry container.
3. Mix thoroughly. Stir or swirl the container to combine the components. For uniform mixing, transfer the mixture to a sealable bag or container and shake vigorously for 5–10 min, or use a mortar and pestle to grind and blend if clumping occurs.
4. Store the powder. Transfer the well-mixed LB powder to an airtight container, label it with the composition, preparation date, and storage conditions. Store in a cool, dry place away from moisture and direct light.

2.3 2M $CaCl_2$ Solution

1. Weigh and dissolve $CaCl_2$. Use an analytical balance to weigh the calculated amount of $CaCl_2$. Add it to a clean beaker, then add approximately 80% of the final volume of distilled water (e.g., 800 mL for 1 L solution). Stir with a magnetic stirrer until completely dissolved.
2. Adjust to the final volume. Transfer the solution to a volumetric flask of the desired volume. Rinse the beaker with small portions of distilled water and add the rinsate to the flask to avoid loss. Fill the flask to the marked line with distilled water, then cap and invert several times to mix thoroughly.
3. Sterilize and store. Sterilize the solution by autoclaving at 121°C for 15–20 min, or filter through a 0.22 μm sterile filter. Store the sterile solution at 4°C. Label the container with concentration, preparation date, and sterility status.

2.4 Agar Medium

1. Reagents: LB powder, agar, distilled water.
2. Weigh and dissolve components: For 1000 mL soft agar (7%) medium: Weigh 25 g LB powder and 7 g agar. Add to a beaker, pour in 1000 mL distilled water, and stir with a magnetic stirrer until fully dissolved. For 1000 mL 1.5% agar medium: Weigh 25 g LB powder and 15 g agar. Add to a beaker, pour in 1000 mL distilled water, and stir with a magnetic stirrer until fully dissolved.
3. Dissolve and sterilize: Pour 1000 mL distilled water into the beaker, add LB powder and agar, stir until fully dissolved. Transfer to an autoclave-safe container, autoclave at 121°C, 103 kPa for 15–20 min.

2.5 Materials for Chemical Coupling of Bacteriophages

1. EDC hydrochloride (N-(3-Dimethylaminopropyl)-N'-ethylcarbodiimide hydrochloride): A water-soluble carbodiimide crosslinker used to activate carboxyl groups. Sourced from Thermo Scientific (Catalog No. 22980). Store desiccated at –20°C.

2. Sulfo-NHS (N-Hydroxysulfosuccinimide): A water-soluble compound used in conjunction with EDC to form stable amine-reactive Sulfo-NHS esters. Sourced from Thermo Scientific (Catalog No. 24510). Store desiccated at 4°C.
3. Branched polyethylenimine (PEI): Use branched polyethylenimine. The common molecular weight is 25 kDa. It can be prepared as a 1.0 mg/mL aqueous solution or PBS solution (pH 7.4) as a stock solution and stored according to the supplier's recommendations (typically at 4°C).
4. Slide-A-Lyzer MINI Dialysis Devices (20K MWCO, Thermo Scientific): Used for all buffer exchange and purification steps to remove small-molecule reactants and by-products. The 20,000 molecular weight cut-off (MWCO) membrane retains the phage particles while allowing unbound chemicals to dialyze out.
5. Amicon Ultra Centrifugal Filters (100K MWCO, Millipore Sigma, optional): Provided as an alternative or supplementary method for rapid concentration and buffer exchange of the phage samples before or after conjugation.
6. Phage activation solution containing: EDC: 4 mg/mL; Sulfo-NHS: 11 mg/mL in distilled water.

3 Methods

3.1 Phage Screening

3.1.1 Preparation

1. Phage Buffer (PB): Used for leaching soil phages, containing 68 mM NaCl, 10 mM $MgSO_4$, and 10 mM Tris-HCl. The pH should be adjusted to 7.5.
2. LB Powder: Pre-mixed for rapid preparation of phage-containing culture medium. The ratio of tryptone: yeast extract: NaCl is 2:1:2. When in use, add 2.5 g per 100 mL of solution.
3. 2 M $CaCl_2$ Solution: Used to enhance the infection ability of phages. The final concentration is fixed at 2 mM, which requires a 1000-fold dilution.
4. Lower Layer Hard Agar Plates: The mass fraction of agar is 1.5%, and the plates do not need to be too thick.
5. Soft Agar LB Medium: The mass fraction of agar is 0.7%. When in use, melt the medium first and cool it to 45°C (it can be kept in a water bath).
6. Saturated Bacterial Culture: Bacterial culture incubated overnight at 37°C with a rotation speed of 220 rpm.

7. Host Bacteria: Dilute the saturated bacterial culture at a ratio of 1:200 to obtain a bacterial solution with an OD_{600} of approximately 0.6 (see **Note 2**). This bacterial solution is in the logarithmic growth phase, which ensures good infection efficiency and makes it suitable as a host bacterium. The preparation process takes about 3 h, and this step must be performed first every subsequent day to prepare the host bacteria for use on that day.

3.1.2 Sampling and Enrichment

1. Sampling and Processing: Collect soil and water samples using 50 mL centrifuge tubes. For soil samples, add an equal volume of phage buffer (PB) (two tubes each), mix thoroughly, and allow to stand for a certain period (see **Note 1**). Subsequently, centrifuge both soil and water samples at 5000 × *g* for 10 min. The supernatants are then filtered through a 0.22 μm filter (via vacuum filtration) to obtain the phage stock solution.
2. Phage Enrichment: Transfer 20 mL of the phage stock solution into a conical flask (typically a 50 mL conical flask). Add LB powder (1 g for 40 mL system) and 2 M $CaCl_2$ solution at a 1:1000 dilution ratio as per the volume proportion. After filtration through a 0.22 μm filter, inoculate 100 μL of host bacterial culture at a 1:200 ratio (see **Note 2**). Incubate overnight at 37°C with shaking at 220 rpm, followed by centrifugation at 5000×*g* for 10 min to obtain the phage-enriched solution.

3.1.3 Screening

1. Double-Layer Agar Plate Screening and Plating: To obtain clearer plaques, three gradients (100-fold, tenfold, and onefold dilutions) were set up. Specifically, 5, 50, and 500 μL of the phage-enriched solution were mixed with 500 μL of host bacteria, respectively, followed by blending with 4.5 mL of soft agar. The mixture was immediately poured onto the surface of the hard agar plates, and the plates were incubated overnight at 37°C in an inverted position (see Fig. 1).
2. Phage Purification: Clear plaques were picked and streaked on hard agar plates. Subsequently, 500 μL of host bacteria was mixed with 4.5 mL of soft agar, and the mixture was poured onto the streaked plates. The plates were incubated overnight at 37°C in an inverted position. On the next day, clear plaques were picked again, and the above procedure was repeated to isolate a single phage strain.

3.1.4 Phage Titer Assay

1. Mix 500 μL of host bacteria in the logarithmic growth phase with 4.5 mL of 0.7% agar medium, and then pour the mixture onto a 1.5% agar plate.
2. Centrifuge the phage concentrate at 5000 × *g* for 10 min.
3. Collect the supernatant and perform a series of 10-fold serial dilutions.

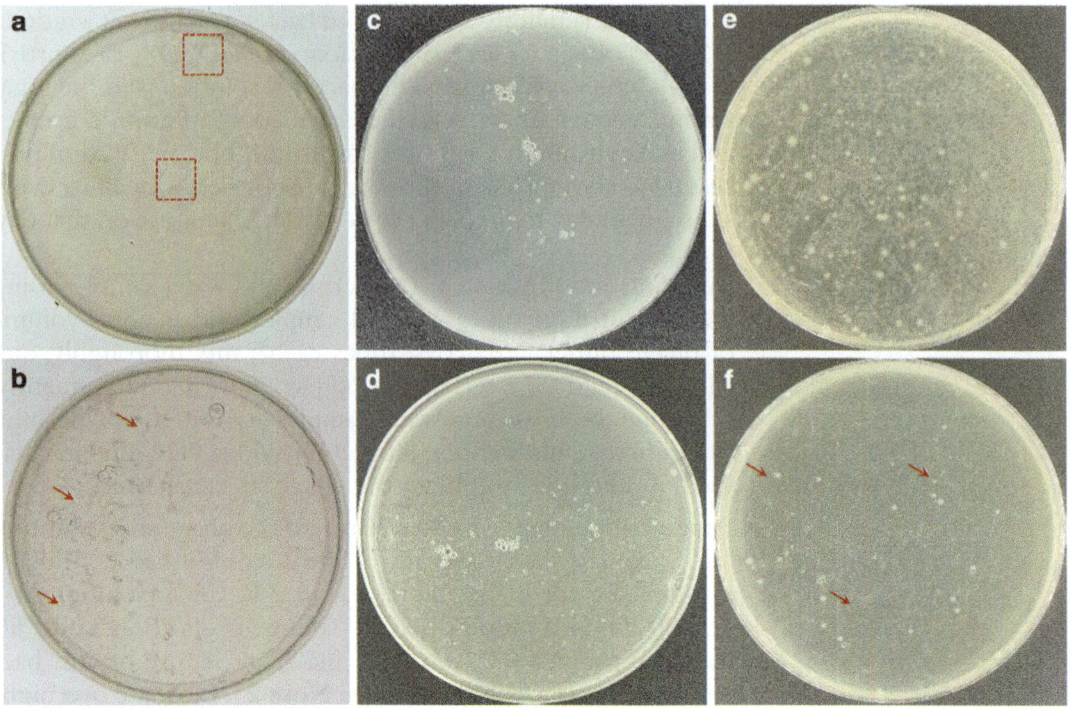

Fig. 1 Plaque morphology of phages against engineered strains on double-layer agar plates. (**a** and **b**). Phages against *Escherichia coli* BL21 strain; (**c**–**f**). Phages against *Bacillus subtilis* BS168 strain. Red arrows and boxes indicate the positions of clearly visible bacterial colonies

Dilution factor	10^0	10^{-1}	10^{-2}	10^{-3}	10^{-4}	10^{-5}	10^{-6}	10^{-7}	Polymer only
BL21									
BS168									

Fig. 2 Spot titer assay of phages. Phages and polymer controls using the modified double-layer agar method

4. Take 5 μL of the diluted solution, spot it onto the double-layer culture plate, and make corresponding labels.
5. After the droplets are completely dried, incubate at 37°C overnight (incubation can be conducted at room temperature), and then count the plaques (see Fig. 2).

This method can be used to determine the phage concentration (plaque-forming units per milliliter, PFU/mL). The phage stock solution is serially diluted in PBS and inoculated onto bacterial medium to count the number of infectious viral particles present in the stock solution.

3.1.5 Transmission Electron Microscopy (TEM) Imaging

1. Dilute the overnight culture of host bacteria at a ratio of 1:40, add it to LB medium, and shake for 2 h to adjust the OD_{600} value of the bacterial solution to 0.6.
2. Take 4 mL of the bacterial solution, centrifuge at 3000 × *g* for 2 min, discard the supernatant, and then resuspend the bacteria in 1 mL of deionized water.
3. Repeat Step 2 three times.
4. Centrifuge the phage concentrate at 5000 × *g* for 10 min.
5. Add phages (10^{11} pfu/mL) to the bacterial solution, incubate for 30 min, then aspirate 20 μL of the mixture and drop it onto a copper grid (placed with the grid side down) for transmission electron microscopy observation (see **Note 4**).

3.1.6 Planktonic Bactericidal Assay of Phages Against Host Bacteria

1. Centrifuge the phage concentrate at 5000 × *g* for 10 min.
2. Based on the concentration of viral particles in the phage-enriched solution determined by the plaque spot titer assay, mix an appropriate volume of the supernatant with host bacteria to achieve a viral particle concentration ten times that of the bacterial concentration in the mixture.
3. Measure the bacterial cell concentration in the mixture every 2 h using spectrophotometry at a detection wavelength of 600 nm (OD_{600}).

3.2 Phage Amplification

3.2.1 Liquid Amplification Method

1. Host Bacteria Preparation: Inoculate host bacteria for amplification into liquid medium at a 1% ratio and culture until the logarithmic growth phase.
2. Phage Preparation: Use a phage solution that has been purified and has a titer ≥107 pfu/mL, or pick a purified phage single plaque and dissolve it in 1 mL of SM buffer.
3. Co-Incubation: Centrifuge the bacteria cultured to the logarithmic phase at 8000×*g*, 4°C for 10 min. Discard the supernatant, then add a certain volume of culture solution or SM buffer to resuspend. Take 500 μL of the resuspended bacterial solution and 5 μL of phage solution for co - incubation for 10 min.
4. Cultivation: Add 5 mL of liquid medium, and culture on a shaker with shaking until all bacteria are lysed and the liquid becomes transparent.

5. Lysis and Separation: Add chloroform to a final concentration of 10%, shake and mix well, then place in an ice-water bath for 20 min until all bacteria are lysed. Centrifuge at 8000 × *g* for 5 min, slowly aspirate the supernatant, taking care not to aspirate chloroform and precipitated impurities (see **Note 3**).
6. Filtration and Preservation: Filter the supernatant through a 0.45 μm filter membrane, collect the filtrate, measure the phage titer, and then store at 4°C.

3.2.2 Plate Amplification Method

1. Host Bacteria Preparation: Culture host bacteria to the logarithmic phase, take 800 μL of logarithmic phase bacterial solution, add 200 μL of phage solution, mix well, and co-incubate for 10 min.
2. Preparation of Double-Layer Plate: Take 12 mL of medium that has been autoclaved and contains 0.5% agar, cool to 45–55°C, add the above-mentioned mixed solution, quickly pour it onto the lower plate pre-paved with 1.5% agar to make a double-layer plate. Let it stand until the upper layer solidifies, then invert and culture in an incubator until the bacteria on the double-layer plate are completely lysed and the plate becomes transparent.
3. Elution: Add 20 mL of physiological saline (or SM buffer), elute on a shaker at 100 r/min for 4 h to release phages into the buffer, and aspirate all the eluent.
4. Lysis and Separation: Add chloroform to a final concentration of 10%, shake and mix well, then place in an ice-water bath for 20 min until all bacteria are lysed. Centrifuge at 8000 × *g* for 5 min, slowly aspirate the supernatant, taking care not to aspirate chloroform and precipitated impurities.
5. Filtration and Preservation: Filter the supernatant through a 0.45 μm filter membrane, collect the filtrate, measure the phage titer, and then store at 4°C.
6. Plate Preparation: Take 50 μL of the infection solution and evenly spread it on an agar plate containing a suitable medium. After drying, place it in a 37°C incubator overnight. At the same time, spread the uninfected host bacteria on an agar plate as a negative control.
7. Counting and Calculation: Count the number of plaques on the agar plate, and calculate the titer of the amplified phages according to the dilution factor. The calculation formula is: Phage titer (pfu/mL) = (Number of plaques × 10 × 50/10 μL) × Dilution factor.

3.2.3 Optical Density Method (OD_{260})

1. Principle: Phage particles have ultraviolet absorption at a wavelength of 260 nm. By measuring the OD260 value, the total

number of particles (including active and inactive ones) can be estimated according to an empirical formula.

2. Formula: Number of phage particles/mL ≈ (OD260 × 6 × 10^16)/(Number of base pairs in the phage genome).

Note: This method cannot distinguish between phages with infectious activity and inactive phages, and is greatly interfered by nucleic acid impurities in the sample. It is only suitable for the rapid estimation of highly purified samples.

3.3 Chemical Coupling of Phages

3.3.1 Activation of Phage Surface Carboxyl Groups

Prepare fresh activation solution, and add activation solution to the blocked phage sample at a 1:10 (v/v) ratio.
Incubate for 20 min at room temperature with gentle mixing.

3.3.2 Removal of Excess Crosslinkers

Transfer the reaction mixture to a Slide-A-Lyzer dialysis device.
Dialyze at 4°C against 1× PBS (pH 7.2): three changes, 40 min each (see **Note 5**).

3.3.3 PEI Conjugation

Transfer dialyzed sample to a 15 mL tube.
Adjust volume to 10 mL with 1× PBS (pH 7.2).
Add PEI to the final concentration of 1.0 mg per 1 mL of phage solution.
Incubate for 2 h at room temperature with gentle mixing.

3.3.4 Purification of PEI-Phage Conjugates

Dialyze the reaction mixture at 4°C against: 1× TBS (pH 7.2): 2 changes, 2 h each;
1× PBS (pH 7.4): three changes, 2 h each.
Alternatively, concentrate using 100 K MWCO centrifugal filters with buffer exchange (see **Note 6**).

3.3.5 Quality Control

Determine phage concentration by UV spectrophotometry (A269/A320).
Verify conjugation efficiency by FTIR, UV–Vis.
Store final Phage conjugates at 4°C in PBS buffer [5–7].

4 Notes

1. The sample collection in Sect. 3.1.2 is important for phage screening, try to get samples from different sources such as sewage, compost, and even animal fecal from diverse environments. Collect environmental samples in sterile containers and process them within 4 h of collection. If immediate processing is not possible, store samples at 4°C for no more than 24 h.

2. Add a proper antibiotic for bacterial growth if the susceptibility of the host bacteria to antibiotics is known.
3. In the liquid amplification method, chloroform is added to a final concentration of 10% for bacterial lysis. Avoid excessive chloroform (≥15%), as it will denature phage capsid proteins and reduce phage activity. After adding chloroform, shake the mixture gently to prevent the formation of chloroform droplets that are difficult to separate by centrifugation.
4. For TEM imaging of phage particles, prepare samples of different concentrations, and stain the samples with uranyl acetate or phosphotungstic acid if necessary.
5. For phage surface chemical coupling, prepare all crosslinking reagents (EDC, Sulfo-NHS, Sulfo-NHS-Acetate) fresh before use. EDC hydrolyzes rapidly in aqueous solutions (half-life <15 min at pH 5.5), and pre-prepared solutions reduce conjugation efficiency by >40%.
6. For long-term storage of PEI-phage conjugates, add 0.02% (w/v) sodium azide to PBS buffer to inhibit microbial contamination.

Acknowledgments

Dr. Peng is grateful for support from the National Natural Science Foundation of China (Grant No. 32201100).

References

1. Abedon ST, Danis-Włodarczyk KM, Alves DR (2021) Phage therapy in the 21st century: is there modern, clinical evidence of phage-mediated efficacy?. Pharmaceuticals 14(11):1157. https://doi.org/10.3390/ph14111157
2. Kutter, E., & Sulakvelidze, A. (Eds.). (2004). Bacteriophages: Biology and Applications (1st ed.). CRC Press. Boca Raton, https://doi.org/10.1201/9780203491751
3. Selim HMRM, Gomaa FAM, Alshahrani MY, Morgan RN, Aboshanab KM. (2025) Phage therapeutic delivery methods and clinical trials for combating clinically relevant pathogens. Ther Deliv. 16(3):247–269. https://doi.org/10.1080/20415990.2024.2426824
4. Smith GP (1985) Filamentous fusion phage: novel expression vectors that display cloned antigens on the virion surface. Science 228(4705):1315–1317. https://doi.org/10.1126/science.2983285
5. Kutter E, Mosig G, Blattner FR, Hendrix RW, Mathews CK (eds) (2004) Bacteriophages: biology and applications. CRC Press
6. Hermanson GT (2013) Bioconjugate techniques, 3rd edn. Academic Press, Amsterdam
7. Sundberg RJ, Martin RB (1974) Interactions of histidine and other imidazole derivatives with transition metal ions. Chem Rev 74(4):471–517

Part III

CRISPR-Driven Genome Engineering

Check for updates

Chapter 9

CRISPR-Cpf1-Mediated T4 Phage Genome Editing for One-Step In Vivo Display of Heterologous Protein

Yiyao Wang, Xinfeng Li, Yusen Huang, Mingyue Zhong and Hang Yang

Abstract

The T4 phage is a robust vector for high-density heterologous protein display. It leverages two non-essential outer capsid proteins, i.e., Soc (~870 copies) and Hoc (~155 copies). These two proteins enable the efficient display of target proteins on the capsid of T4. Here, we detail the workflow for one-step in vivo display of a heterologous protein. Specifically, this method utilizes CRISPR-Cpf1-mediated gene editing technology to insert the sequence of interest (using mCherry as an example) downstream of the Soc encoding gene within the T4 phage genome, resulting in a Soc-fused recombinant protein. This engineering method allows for endogenous expression of the Soc-mCherry recombinant protein within *Escherichia coli* cells during phage replication, after which the recombinant protein spontaneously assembles onto the capsid of the engineered phage. By providing a universal framework suitable for in vivo display, this approach empowers researchers to readily construct tailored T4 nanoparticles for diverse biotechnological applications.

Keywords T4 phage, Phage editing, Phage vector, CRISPR-Cpf1, Surface display

1 Introduction

Phage T4 is a powerful yet underutilized platform for the high-density surface display of heterologous proteins. Its capsid contains two non-essential decoration proteins, the small outer capsid protein (Soc) and the highly antigenic outer capsid protein (Hoc), which bind with high affinity in an ordered manner after capsid assembly [1–3]. Soc, present at ~870 copies per capsid, enables dense peptide or small protein display, whereas Hoc, with ~155 copies, accommodates larger proteins [4, 5]. Phage-based display holds significant promise for diverse applications, including vaccine design [3, 6, 7], targeted drug delivery [8–10], and novel biomaterial development [11, 12].

T4-based antigen in vitro display regularly relies on multiple steps. This includes purification of Soc-deficient phage particles and recombinant Soc-fusion proteins, followed by in vitro assembly and a finalized purification step [13–15]. Although effective, this

Huan Peng and S. Brian Kim (eds.), *Biological Macromolecule Engineering Technologies*,
Methods in Molecular Biology, vol. 3032, https://doi.org/10.1007/978-1-0716-5257-2_9,

process is laborious, time-consuming, and difficult to scale, limiting its utility for high-throughput applications. The large genome size of T4 (~172 kb) [16], which offers substantial capacity for genetic insertion, suggests a compelling alternative: engineering the phage genome to enable direct, in vivo expression of Soc-fusion proteins during viral replication within the host bacteria. This approach integrates the display process into the natural life cycle of phage, thereby circumventing the cumbersome in vitro steps. Such an in vivo display strategy has shown promise in vaccine design [17–19], demonstrating high antigen display efficiency and robust immunoprotective effects.

In this chapter, we outline a one-step in vivo display workflow for heterologous proteins, using mCherry as an example. This approach contains two major aspects: insertion the sequence of interest downstream of the Soc encoding sequence within the T4 genome by CRISPR-Cpf1-mediated genome editing, and the propagation of engineered phage. By establishing a robust and universal framework, this approach will enable researchers to efficiently create customized T4 nanoparticles for diverse biotechnological applications.

2 Materials

All solutions should be prepared with analytical grade reagents and ultrapure water (defined as deionized water that has been purified to a specificity of 18 MΩ-cm at 25°C). Unless specified otherwise, reagents are prepared and stored at room temperature.

2.1 CRISPR-Cpf1-Mediated Genome Editing

1. CRISPR-Cpf1 editing plasmids pEcCpf1 and pcrEG [20].
2. mCherry gene donor plasmid pmCherry-C1.
3. *Escherichia coli* (*E. coli*) DH5α (New England Biolabs, MA).
4. Wild-type (WT) T4 phage.
5. Lysogeny broth (LB) medium.
6. LB agar plates: Supplement LB medium with 15 g/L agar prior to autoclaving.
7. Water agar: Add 0.7 g agar to 100 mL water and autoclave.
8. 1000× kanamycin sulfate (50 mg/mL): Dissolve 0.5 g in 10 mL water, filter-sterilize (0.22 μm), and store at −20°C.
9. 1000× streptomycin sulfate (50 mg/mL): Dissolve 0.5 g in 10 mL water, filter-sterilize (0.22 μm), and store at −20°C.
10. 2× Phanta Master Mix (Vazyme, Nanjing).
11. FastDigest restriction enzymes: *Kpn*I, *Xho*I, *Eco*31I (Thermo Fisher Scientific, MA).
12. T4 DNA Ligase (Thermo Fisher Scientific, MA).
13. Phosphate-buffered saline (PBS): 137 mM NaCl, 2.7 mM KCl, 10 mM Na_2HPO_4, 1.8 mM KH_2PO_4, pH 7.4.

14. Plasmid Mini Kit I (Omega Bio-tek, GA).
15. E.Z.N.A.® Cycle Pure Kit (Omega Bio-tek, GA).
16. ClonExpress Ultra One Step Cloning Kit V2 (Vazyme, Nanjing).
17. 20% L-Arabinose.
18. Primers: mCherry-F: 5′-GGCAGTAGCGGCAGCATGGTG AGCAAGGGCGAGGAG-3′;
mCherry-R: 5′-GAGTCCTTGAGTTACTGCAGTTACTTG-TACAGCTCGTC-3′;
Soc-Up-F: 5′-ATAAATGTAGTGATAGCGGTACCTTCTG-TAGCTTTCCCTCTTT-3′;
Soc-Up-R: 5′-GCTGCCGCTACTGCCACCAGTTACTTTC-CACAAAT-3′;
Soc-Down-F: 5′-CTGCAGTAACTCAAGGACTCCTTCG-GGAGTCCTTTTTTCAT-3′;
Soc-Down-R: 5′-GAGCTGCACATGAACTCGAGTGAATG-CACAACAACTTCTGT-3′;
pcrEG-F1: 5′-TTGACAAAAATGGGCTCGTG-3′;
pcrEG-R1: 5′-ATAGCACGATCAACGGCACT-3′;
crRNA-F: 5′-AGATTGGAAAGTAACTGGTTAATAACT-3′;
crRNA-R: 5′-AATTAGTTATTAACCAGTTACTTTCCA-3′;
pcrEG-F2: 5′-CCGAACGACCGAGCGCAGCGAGTCAGT-3′;
pcrEG-R2: 5′-CAACGTTAATAAGGCCAAGC-3′.

2.2 Propagation and Purification of Edited Phage

1. *E. coli* strain P301.
2. WT T4 phage and engineered T4-mCherry phage.
3. Growth media as described in Sect. 2.1.
4. 7 mg/mL of deoxyribonuclease I (DNase I).
5. HPLC-grade chloroform.
6. Cesium chloride (CsCl) stock solution: Dissolve 200 g CsCl, 1.80 g Tris, 0.74 g NaCl, and 0.16 g NH_4Cl in water, adjust to a final volume of 200 mL, filter-sterilize (0.22 μm), and store protected from light at 4°C.
7. Primers: VFY-F: 5′-CTTAGATTAAACAATCCTGAAACAA AAACTCGGATGAAGA-3′; VFY-R: 5′-GACTTGCTTCTCA CCATAATCATATCGAATCTCTGCATTA-3′.

2.3 SDS-PAGE Analysis of Phage Proteins

1. SDS-PAGE gel: 10% (w/v) polyacrylamide gel (Vazyme, Nanjing).
2. 5×SDS loading buffer: 250 mM Tris-HCl (pH 6.8), 10% (w/v) SDS, 0.5% (w/v) bromophenol blue, 50% (v/v) glycerol, 5% (v/v) β-mercaptoethanol. Add dithiothreitol (DTT) to a final concentration of 100 mM before use.

3. 10×Running buffer (Tris-Glycine-SDS): 250 mM Tris, 1.92 M glycine, 1% (w/v) SDS. Dilute to 1×working solution with water before use.
4. Staining solution: 0.1% (w/v) Coomassie blue R-250, 25% (v/v) methanol, 10% (v/v) acetic acid.
5. Destaining solution: 10% (v/v) acetic acid, 5% (v/v) ethanol.
6. Pi-Mg buffer: 26 mM Na_2HPO_4, 22 mM KH_2PO_4, 70 mM NaCl, and 1 mM $MgSO_4$.

3 Methods

3.1 Construction of the CRISPR-Cpf1 Editing Plasmid pcrEG-crRNA-mCherry

1. Amplify the mCherry encoding sequence from the pmCherry-C1 plasmid using primers mCherry-F/R and the 2× Phanta Max Master Mix (*see* **Note 1**).
2. Using WT T4 phage genomic DNA as the template, amplify the upstream and downstream homology arms with primers Soc-Up-F/R and Soc-Down-F/R, respectively (*see* **Note 2**).
3. Purify all PCR products using the E.Z.N.A.® Cycle Pure Kit according to the manufacturer's instructions.
4. Digest 1 μg of the pcrEG plasmid with restriction enzymes *Kpn*I and *Xho*I for 1 h at 37°C.
5. Purify the linearized plasmid using the E.Z.N.A.® Cycle Pure Kit.
6. Assemble the purified mCherry fragment, homology arms, and linearized pcrEG plasmid using the ClonExpress Ultra One Step Cloning Kit V2 according to the manufacturer's instructions.
7. Transform the assembly reaction into chemically competent *E. coli* DH5α cells. Plate the transformation mix onto LB agar supplemented with streptomycin and incubate overnight at 37°C.
8. Screen individual colonies (*see* **Note 3**) by colony PCR using verification primers pcrEG-F1/R1.
9. Confirm positive constructs, designated pcrEG-mCherry, by Sanger sequencing.
10. Extract the validated pcrEG-mCherry plasmid using the Plasmid Mini Kit I.
11. Anneal the crRNA-F and crRNA-R oligonucleotides by incubating at 95°C for 10 min followed by gradual cooling to room temperature (*see* **Note 4**).
12. Digest the pcrEG-mCherry plasmid with *Eco*31I.

13. Purify the linearized pcrEG-mCherry plasmid using the E.Z.N.A.® Cycle Pure Kit.
14. Ligate the annealed crRNA duplex into the *Eco*31I-digested pcrEG-mCherry backbone using T4 DNA ligase for 4 h at 22°C.
15. Transform the ligation product into DH5α cells, plate on LB-streptomycin agar, and incubate overnight at 37°C.
16. Screen colonies by PCR using verification primers pcrEG-F2/R2 and confirm by Sanger sequencing.
17. Extract the final validated plasmid, designated pcrEG-crRNA-mCherry, using the Plasmid Mini Kit I for subsequent phage genome editing.

3.2 Insertion of mCherry Encoding Sequence into T4 Genome

1. Co-transform the pcrEG-crRNA-mCherry and pEcCpfl plasmids into chemically competent *E. coli* DH5α and then plate the transformation on LB agar containing kanamycin and streptomycin and incubate overnight at 37°C to recover the recombinant strain DH5α/double.
2. Inoculate the DH5α/double strain into 5 mL of LB medium and culture at 37°C with shaking until OD_{600} reaches 0.3. Induce the λ-Red recombinase system by adding 50 μL of 20% L-arabinose (*see* **Note 5**). Continue incubation for 3 h.
3. For the plaque assay, prepare a base layer by pouring 10 mL of molten LB agar into a 90 mm Petri dish and allowing the soft agar to harden.
4. Mix 2 mL of the induced bacterial culture with 6 mL of molten water agar (pre-hold at 45°C) and pour immediately onto the base layer to create a double-layer plate. Allow the water agar to harden.
5. Prepare serial tenfold dilutions of the WT T4 phage stock in PBS, ranging from 10^3 to 10^8 PFU/mL (*see* **Note 6**).
6. Spot 10 μL of each dilution onto the double-layer plate and allow the spots to dry.
7. Incubate the plates at 37°C overnight.
8. Pick a single, well-isolated plaque with a sterile pipette tip and transfer it into 10 μL of PBS.
9. Vortex the suspension vigorously for 1 min.
10. Using 1 μL of the suspension as the template, perform PCR detection with verification primers VFY-F/R, which bind outside the homologous arm regions (*see* **Note 7**). A successful edit is indicated by a larger amplicon (1606 bp for T4-mCherry *vs.* 886 bp for WT T4) (Fig. 1). Confirm PCR-positive samples by Sanger sequencing.

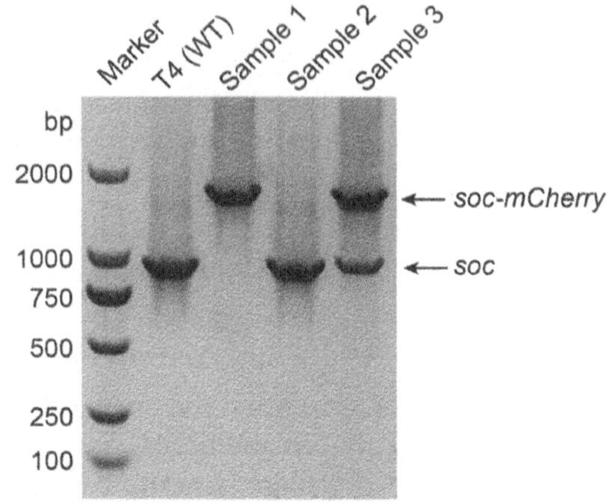

Fig. 1 PCR verification of edited T4-mCherry phage. The size difference between the amplified products (1606 bp for the engineered phage *vs.* 886 bp for WT T4) confirms successful genetic modification. Sample 1 represents the band of successfully edited phage, Sample 2 represents the unedited phage, and Sample 3 represents a heterogeneous population consisting of a mixture of both edited and wild-type phages

11. To ensure genetic homogeneity and eliminate any wild-type or mixed phage populations, perform two additional rounds of plaque purification and PCR verification (*see* **Note 8**).
12. Following the final purification, harvest the phage from the plate by scraping the top agar into 20 mL of PBS. Vortex for 1 min and centrifuge at 8000 × *g* for 5 min to remove agar debris.
13. Filter the supernatant through a 0.22 μm membrane and store at 4°C.

3.3 Propagation and Purification of Edited T4-mCherry Phage

1. Inoculate 5 mL of the overnight culture of *E. coli* P301 into 500 mL of fresh LB medium. Incubate at 37°C with shaking at 160 rpm until the OD_{600} reaches 0.6–0.7 (~2×10^8 CFU/mL).
2. Infect the culture with phage T4-mCherry at a multiplicity of infection (MOI) of 0.1. Continue incubation with shaking for 5–8 h until lysis occurs.
3. Clarify the lysate by centrifugation at 8000×*g* for 10 min at 4°C to pellet cell debris. Transfer the supernatant to a clean flask.
4. Precipitate phage particles by adding 29.2 g of NaCl and 50.0 g of PEG 8000 to the supernatant. Stir until completely dissolved and incubate overnight at 4°C.
5. Collect the precipitate by centrifugation at 9500×*g* for 15 min at 4°C. Discard the supernatant.

6. Resuspend the pellet in 6 mL of Pi-Mg buffer. Add an equal volume of chloroform (6 mL), vortex vigorously for 30 s, and centrifuge at 1500×*g* for 10 min.
7. Carefully recover the upper aqueous phase that contains the phage particles.
8. Treat the aqueous phase with 43 µL of DNase I (7 mg/mL) and incubate at 37°C for 30 min with shaking to digest residual nucleic acid.
9. Prepare a discontinuous CsCl density gradient in an ultracentrifuge tube by carefully layering solutions of decreasing density as outlined in Table 1 (*see* **Note 9**).
10. Gently transfer the DNase-treated phage sample onto the top of the gradient solution. If necessary, top up the centrifuge tube with Pi-Mg buffer. Centrifuge using a Beckman SW41 Ti rotor (or equivalent) at 180,000 × *g* for 1 h at 4°C (*see* **Note 10**).
11. After centrifugation, a distinct pink band corresponding to the T4-mCherry phage will be visible, separate from the milky white WT T4 band (Fig. 2) (*see* **Note 11**). Carefully aspirate and discard the upper layers. Collect the colored phage band using a pipette.
12. Transfer the collected phage into a 1000 kDa MWCO dialysis bag and dialyze extensively against Pi-Mg buffer at 4°C with at least two buffer changes over 8 h to remove residual CsCl.
13. Collect the dialyzed, purified phage suspension, determine the titer by plaque assay, and store at 4°C.
14. Analyze the purified phage particles by SDS-PAGE (Fig. 3) to confirm the presence and approximate stoichiometry of the Soc-mCherry fusion protein.

Table 1
Preparation of CsCl gradient layers for phage purification

Layer no.	H_2O (mL)	Stock CsCl (mL)	Total volume (mL)
1	8	2	10
2	7	3	10
3	6	4	10
4	5	5	10
5	4	6	10
6	3	7	10

Fig. 2 Phage isolation by CsCl density gradient ultracentrifugation. The milky white band near the bottom in the left tube, as indicated by the arrow, represents the WT T4 phage, and the pink band near the bottom in the right tube, as indicated by the arrow, represents the T4-mCherry phage

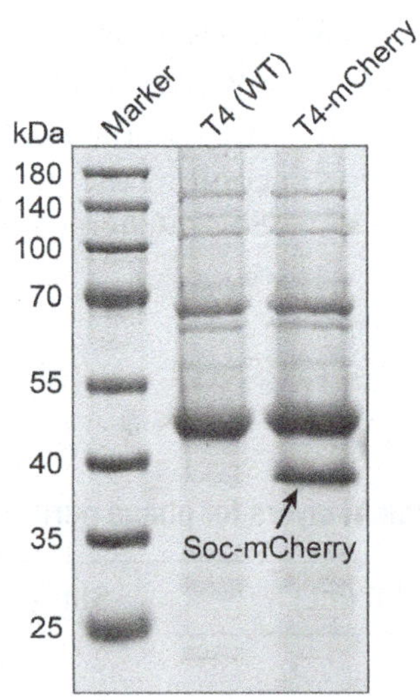

Fig. 3 SDS-PAGE analysis of the components of T4 and T4-mCherry phages. The Soc-mCherry recombinant protein is indicated by the arrow

4 Notes

1. The DNA template for mCherry amplification is the commercial plasmid pmCherry-C1. While it contains a humanized codon sequence that is functional in *E. coli*, further codon optimization for *E. coli* may enhance the expression level of the fusion protein.
2. The upstream and downstream homology arms must be designed to be specific to the intended genomic locus and should be computationally verified to avoid potential off-target sites within the T4 genome.
3. The pcrEG vector contains a constitutively expressed superfolder GFP (sfGFP) gene flanked by the *Kpn*I and *Xho*I sites. Successful cloning of the insert (upstream arm - mCherry - downstream arm) results in the displacement of the sfGFP gene. Consequently, positive recombinant colonies will lack green fluorescence, providing a simple visual screening method during colony selection.
4. The crRNA was designed as a 23-base spacer sequence with perfect complementarity to the target site in the T4 genome. The synthesized oligonucleotides (crRNA-F and crRNA-R) incorporate this spacer and include 4-nucleotide overhangs designed to be compatible with the sticky ends generated by *Eco*31I digestion, ensuring efficient and directional ligation into the linearized vector backbone.
5. The λ-Red recombinase genes are encoded on the pEcCpf1 plasmid under the control of an arabinose-inducible promoter. Induction with L-arabinose is critical for the expression of these proteins, which greatly enhances the efficiency of homologous recombination repair and is therefore essential for achieving high genome editing efficiency.
6. When setting up the plaque assay for isolating edited phages, a precise dilution series is crucial for obtaining well-isolated plaques. Due to the inherently low efficiency of homologous recombination and the concurrent CRISPR-Cpf1-mediated cleavage of unedited genomes, using an excessively diluted phage stock may result in no plaque formation. It is advisable to test a wide range of dilutions.
7. Primers for PCR verification (e.g., VFY-F/R) must be designed to bind to genomic sequences outside the regions used for the homology arms. This strategy ensures that the amplicon is generated from the engineered phage genome and not from any residual editing plasmid DNA carried over from the bacterial host, thereby preventing false-positive results.

8. Phage plaques picked from the first round of selection are often heterogeneous and may contain a mixture of wild-type and engineered genomes. To isolate a genetically pure, clonal population of T4-mCherry phage, at least two additional rounds of plaque purification and PCR screening should be performed.
9. To achieve a well-defined discontinuous CsCl density gradient, each solution must be layered very slowly and carefully along the inner wall of the ultracentrifuge tube. This technique minimizes mixing between layers and is essential for forming sharp interfaces, which are critical for effective separation of phage particles during ultracentrifugation.
10. After loading the phage sample onto the CsCl gradient, ensure the ultracentrifuge tube is filled with buffer (e.g., PBS or Pi-Mg). This prevents the tube from collapsing under the high centrifugal forces during ultracentrifugation.
11. Concentrated preparations of T4 phage Bacteriophage T4 (e.g., $\geq 5 \times 10^{12}$ PFU/mL) typically exhibit a milky white, opalescent appearance. At slightly lower concentrations (e.g., ~1×10^{12} PFU/mL), the suspension may appear translucent with a faint bluish tint.

Acknowledgments

This work was supported by the National Natural Science Foundation of China (grant 32161133003).

References

1. Sathaliyawala T, Islam MZ, Li Q et al (2010) Functional analysis of the highly antigenic outer capsid protein, Hoc, a virus decoration protein from T4-like bacteriophages. Mol Microbiol 77:444–455
2. Rao VB, Black LW (2010) Structure and assembly of bacteriophage T4 head. Virol J 7:356
3. Jaroszewicz W, Morcinek-Orłowska J, Pierzynowska K et al (2022) Phage display and other peptide display technologies. FEMS Microbiol Rev 46:fuab052
4. Gamkrelidze M, Dbrowska K (2014) T4 bacteriophage as a phage display platform. Arch Microbiol 196:473–479
5. Rao VB, Fokine A, Fang Q et al (2023) Bacteriophage T4 head: structure, assembly, and genome packaging. Viruses 15:527
6. Hess KL, Jewell CM (2020) Phage display as a tool for vaccine and immunotherapy development. Bioeng Transl Med 5:e10142
7. Palma M (2023) Aspects of phage-based vaccines for protein and epitope immunization. Vaccines 11:436
8. Sioud M (2019) Phage display libraries: from binders to targeted drug delivery and human therapeutics. Mol Biotechnol 61:286–303
9. Rao VB, Zhu J (2022) Bacteriophage T4 as a nanovehicle for delivery of genes and therapeutics into human cells. Curr Opin Virol 55:101255
10. Wang H, Yang Y, Xu Y et al (2024) Phage-based delivery systems: engineering, applications, and challenges in nanomedicines. J Nanobiotechnol 22:365
11. Farr R, Choi DS, Lee SW (2014) Phage-based nanomaterials for biomedical applications. Acta Biomater 10(4):1741–1750
12. Martins IM, Reis RL, Azevedo HS (2016) Phage display technology in biomaterials engineering: progress and opportunities for

applications in regenerative medicine. ACS Chem Biol 11:2962–2980
13. Tao P, Mahalingam M, Zhu J et al (2018) A bacteriophage T4 nanoparticle-based dual vaccine against anthrax and plague. mBio 9:e01926-18
14. Li M, Guo P, Chen C et al (2021) Bacteriophage T4 vaccine platform for next-generation influenza vaccine development. Front Immunol 12:745625
15. Chen C, Zhang N, Li M et al (2024) Recombinant bacteriophage T4 displaying key epitopes of the foot-and-mouth disease virus as a novel nanoparticle vaccine. Int J Biol Macromol 19:128837
16. Miller ES, Kutter E, Mosig G et al (2003) Bacteriophage T4 genome. Microbiol Mol Biol Rev 67:86–156
17. Li M, Chen C, Wang X et al (2023) T4 bacteriophage nanoparticles engineered through CRISPR provide a versatile platform for rapid development of flu mucosal vaccines. Antiviral Res 217:105688
18. Zhu J, Sha J, Batra H et al (2025) A modular bacteriophage T4 nanoparticle platform enables rapid design of dual COVID-19-Flu mucosal vaccines. Small Sci 5:2400580
19. Dong J, Chen C, Liu Y et al (2021) Engineering T4 bacteriophage for in vivo display by type V CRISPR-cas genome editing. ACS Synth Biol 10:2639–2648
20. Zhu X, Wu Y, Lv X et al (2022) Combining CRISPR–Cpf1 and recombineering facilitates fast and efficient genome editing in *Escherichia coli*. ACS Synth Biol 11:1897–1907

Chapter 10

Methods for Applying Prime Editing and Inverse Prime Editing in Mammalian Cell Culture

Yasaman Mahdavi-Amiri and S. Brian Kim

Abstract

Prime editing (PE) is a powerful method for introducing point mutations into the genomes of living organisms. PE utilizes a Cas9 nickase fused to an engineered Moloney Murine Leukemia Virus reverse transcriptase (MLV-RT), paired with an extended guide RNA known as pegRNA, which contains a primer binding site (PBS) and a reverse transcriptase template (RTT) complementary to the non-target strand DNA. Recently described inverse prime editing (iPE) also employs reverse transcriptase and pegRNAs; however, it utilizes an RNA template complementary to the target strand, resulting in the polymerization of target-strand DNA in the opposite direction. In this work, we provide a practical protocol for using either PE or iPE to introduce point mutation(s) or small-to-medium sized insertions and deletions in cell culture, and demonstrate how to assess editing efficiency via flow cytometry and next-generation sequencing. We include recommendations for prime editor selection and straightforward guidelines for pegRNA design, intended for researchers unfamiliar with genome editing technologies.

Keywords Genome editing, CRISPR, Cas9, Prime editing, Cell culture, Next-generation sequencing, Inverse prime editing, Flow cytometry

1 Introduction

The advent of genome editing technologies has transformed our ability to interrogate and manipulate genetic information with unprecedented precision [1]. The earliest widely adopted tool, CRISPR–Cas9, enables targeted double-stranded DNA breaks (DSBs) that harness cellular repair pathways to generate insertions, deletions, or sequence replacements [2]. While powerful, the reliance on error-prone non-homologous end joining (NHEJ) or inefficient homology-directed repair (HDR) limits the precision and predictability of CRISPR–Cas9–based editing, especially in non-dividing or recalcitrant cell types [3, 4].

Formerly Y. Bill Kim.

Huan Peng and S. Brian Kim (eds.), *Biological Macromolecule Engineering Technologies*,
Methods in Molecular Biology, vol. 3032, https://doi.org/10.1007/978-1-0716-5257-2_10,

To overcome these constraints, next-generation genome editing systems have been developed that bypass the need for DSBs. Base editors, for example, couple catalytically impaired Cas proteins with deaminase enzymes to enable targeted single-base conversions without inducing DSBs [5, 6]. While highly efficient and precise, these systems are restricted to a narrow subset of possible substitutions and cannot facilitate complex edits requiring combinations of mutations or small insertions and deletions [2].

Prime editing (PE) expands the editing toolkit by offering a versatile strategy for introducing a wide range of genetic alterations, including substitutions, insertions, and deletions, without the need for DSBs or donor DNA templates [2, 7]. This technology employs a Cas9 nickase fused to a reverse transcriptase, guided by a prime editing guide RNA (pegRNA) that both directs Cas9 to the target site and encodes the desired edit within its RNA template (Fig. 1). Consequently, PE provides greater flexibility than base editing. Several PE-based strategies have been developed to enhance precision, specificity, efficiency, and cargo capacity [8–12].

Similarly, inverse prime editing (iPE) offers an alternate arrangement of molecular components compared to PE, resulting in distinct mechanistic differences at the target loci [13–15]. During iPE, the pegRNA encodes a sequence complementary to the target

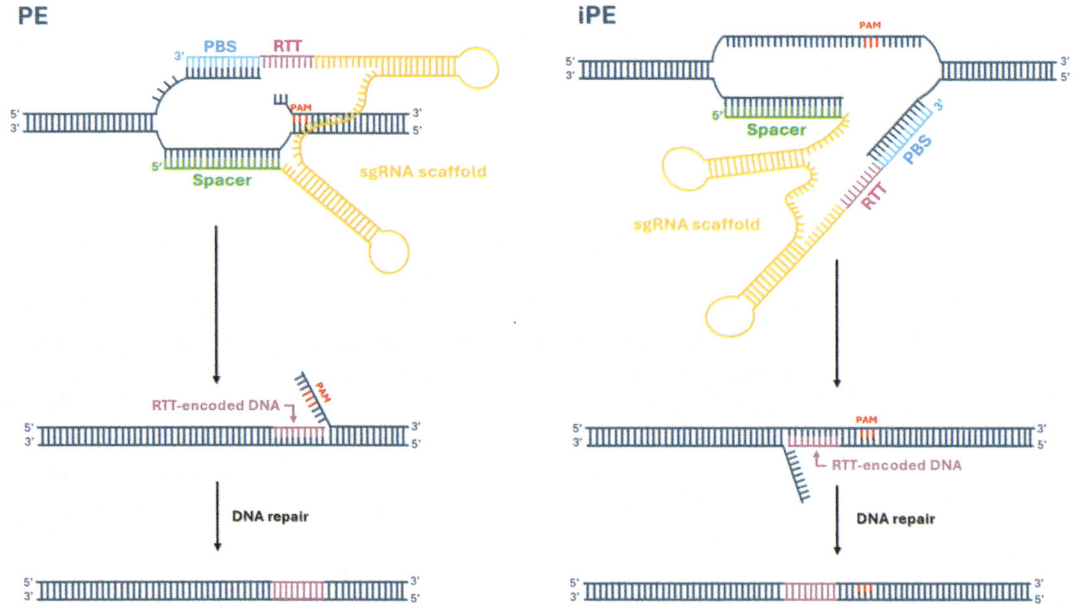

Fig. 1 Comparison of prime editing (PE) and inverse prime editing (iPE) mechanisms. PE uses a pegRNA containing a PBS and RTT complementary to the non-target strand to extend and rewrite that strand. In contrast, iPE employs a pegRNA with extension sequences complementary to the target strand, enabling reverse transcription of the target strand DNA. This inverted architecture allows iPE to introduce mutations within or upstream of the spacer sequence, complementing the editing range of PE

strand DNA. When coupled with a Cas9 nickase that nicks the target strand, this results in the extension of the target strand DNA. This mechanism provides an opportunity to install mutations within or upstream of the spacer sequence (Fig. 1). Because PE is limited in its ability to install mutations upstream of the target site, iPE offers a complementary, non-overlapping editing window. This distinction is particularly advantageous when there is limited availability of desirable PAMs or high-efficiency spacer sequences.

Here, we describe the design and testing of both inverse and standard prime editing in cultured cells via transient transfection of the required components. We place particular emphasis on practical strategies for pegRNA construction, delivery, and the evaluation of editing outcomes. By placing these prime editing strategies within the broader landscape of genome editing technologies, our aim is to provide a methodological framework that enables researchers to effectively implement and benchmark these rapidly evolving tools.

2 Materials

2.1 Consumables/Plasticware

- 96-well and 48-well tissue culture plates (e.g., Corning Costar™).
- 30-100 μm mesh filter.
- 96-well U-bottom plates compatible with a flow cytometer.

2.2 Cell Culture Reagents

- Dulbecco's Modified Eagle Medium (DMEM) (Gibco, 10566-016).
- Opti-MEM Reduced-Serum Medium (Gibco, 31985-070).
- Fetal bovine serum (FBS) (Gibco, 12483-020).
- Phosphate-buffered saline (PBS) (Gibco, 10010031).
- TrypLE Express Enzyme (Gibco, 12605-028).
- Lipofectamine 2000 (Invitrogen, 52887).

2.3 Molecular Biology Reagents

- T4 DNA ligase and reaction buffer (New England Biolabs, M0202).
- Restriction enzymes (e.g., BsaI, New England Biolabs, R0134).
- High-fidelity DNA polymerase (e.g., Q5 High-Fidelity DNA Polymerase, New England Biolabs, M0491).
- Agarose and DNA gel electrophoresis supplies.
- Gel-PCR 2in1 Kit- HiPure DNA Kit (GeneBio Systems, D211101).

- E.Z.N.A. Tissue DNA Kit (Omega Bio-Tek, D3396).
- Qubit dsDNA quantification kit (Invitrogen, Q32851).

2.4 Equipment

- CO_2 incubator set to 37°C.
- Biosafety cabinet.
- Vacuum aspiration system.
- Thermocycler.
- Centrifuge.
- Gel electrophoresis apparatus.
- UV or blue-light gel imaging system.
- Flow cytometer (e.g., Cytek Aurora).
- NanoDrop spectrophotometer (Thermo Fisher Scientific).
- Qubit 4 fluorometer (Thermo Fisher Scientific).

3 Methods

3.1 Plasmid Selection and Design

3.1.1 Protein Constructs

1. Obtain prime editor plasmid(s) from Addgene [16] or other repositories (Table 1). We recommend starting out with PE4max (Addgene ID 174828) as it is the most validated prime editor architecture with sufficiently high efficiency.

3.1.2 pegRNA Constructs

The prime editing guide RNA (pegRNA) comprises four key components: a 20-nucleotide spacer for target specification, a scaffold for Cas9 nickase binding, a primer binding site (PBS) complementary

Table 1
Selection of widely used Prime Editor plasmids and corresponding Addgene repository identifiers

Prime editor	Addgene ID
PEmax	174820
PE4max	174828
PE6a	207851
PE6b	207852
PE6c	207853
PE6d	207854
PE7	214812

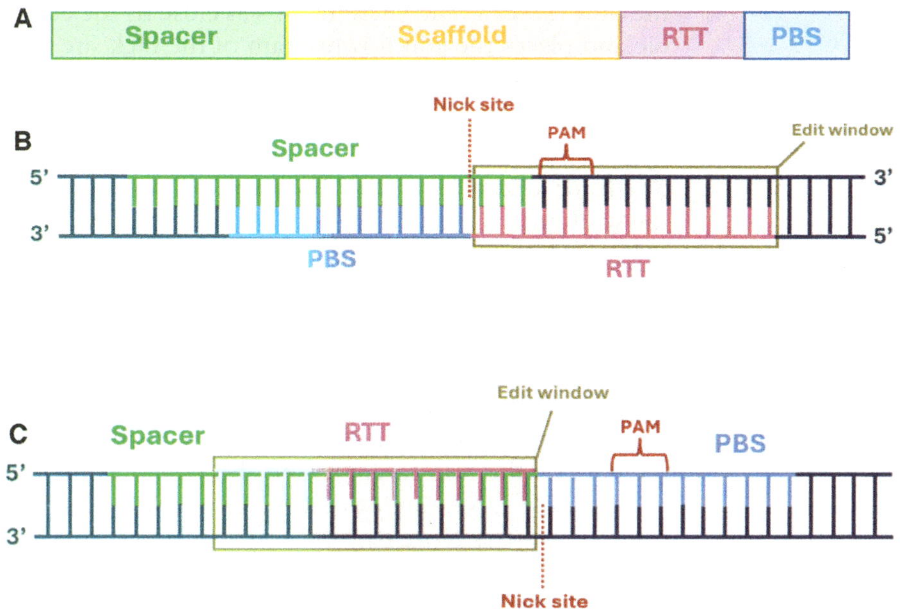

Fig. 2 Structure and sequence source of pegRNA components for prime editing and inverse prime editing. (**A**) Schematic of a prime editing guide RNA (pegRNA) showing its four elements: the 20-nucleotide spacer (green) that defines the genomic target site, the Cas9 scaffold sequence (orange), the reverse transcriptase template (RTT, pink) that encodes the desired edit, and the primer binding site (PBS, blue) that anneals to the nicked DNA strand to initiate reverse transcription. (**B**) Standard prime editing design: The RTT (pink) and PBS (blue) are complementary to the non-target strand. The RTT corresponds to the region immediately downstream of the nick site and includes the desired edit, while the PBS corresponds to the region immediately upstream of the nick site. (**C**) Inverse prime editing design: The RTT (pink) and PBS (blue) are complementary to the target strand. The RTT corresponds to the region downstream of the nick site overlapping the spacer region, and the PBS corresponds to the region upstream of the nick site, including the PAM and away from the spacer

to the genomic DNA, and a reverse transcriptase template (RTT) sequence that encodes the desired mutation (Fig. 2A). To enhance editing efficiency in both PE and iPE, an additional sgRNA may be provided to create a nick on the DNA strand opposite the edited site. This strategy is thought to prevent cellular mismatch repair from reverting the edited strand to the original sequence. For simplicity, this protocol does not describe the use of a nicking sgRNA, although it can be implemented to maximize efficiency.

pegRNA Design for Prime Editing

The pegRNA cassette is typically transcribed from a U6 promoter. The relative positioning of its components with respect to the target mutation can be identified from Fig. 2B.

Clone the sequence into a vector suitable for prime editing. The design guidelines are given below:

- **Protospacer adjacent motif (PAM) and spacer:** Identify a PAM compatible with the chosen Cas9 variant (*see* **Note 1**) on the non-target strand such that it's as close to the edit as possible and places the edit downstream of the nick site. If feasible, choose one that overlaps the edit so the edit can disrupt the PAM (*see* **Note 2**). 20 nt upstream of PAM will be the spacer sequence (*see* **Notes 3 and 4**).

- **Scaffold:** Use the standard Cas9 sgRNA scaffold sequence as the scaffold.
- **RTT:** The RTT sequence (typically 10–20 nt in length) is represented by the region downstream of the nicked site complementary to the non-target strand. RTT includes the desired edit (*see* **Note 5**).
- **PBS:** The PBS sequence (typically 8–15 nt in length) is represented by the region upstream of the nicked site complementary to the non-target strand. PBS enables the formation of an RNA/DNA duplex that serves as a substrate for reverse transcriptase (*see* **Note 6**).
- **Terminal structural RNA motif (optional):** Appending a structured pseudoknot at the 3' end of pegRNA generates an enhanced pegRNA (ePegRNA) [12], whichprotects it from nuclease degradation, enhancing PE efficiency. Such motifs include "tevoPreQ1" and "mpknot".

Alternatively, instead of manually designing each component, several online tools can generate pegRNA sequences automatically based on your target site and desired edit (e.g., PRIDICT [17], PrimeDesign [18], pegFinder [19], and pegIT [20]).

pegRNA Design for Inverse Prime Editing

The pegRNA design for iPE also utilizes RTT and PBS sequences that are appended to a standard sgRNA. However, the target strand and nucleotide compositions are distinct from those used in PE. In iPE, the PBS hybridizes with the target DNA strand, and reverse transcription incorporates the RTT into the target strand upstream of the PAM sequence as shown in Fig. 2C. As an alternative, it is possible to encode RTT and PBS in a separate transcriptional unit, typically done to achieve a circular RNA construct to enhance their intracellular stability. The sequences are similarly cloned into a vector suitable for mammalian expression (e.g., U6-promoter-driven pegRNA expression vector). The design guidelines are given below:

- **Protospacer adjacent motif (PAM) and spacer:** Identify a PAM compatible with the chosen Cas9 variant (*see* **Note 1**) on the non-target strand such that it's as close to the edit as possible and places the edit upstream of the nick site. 20 nt upstream of PAM will be the spacer sequence (*see* **Notes 3 and 4**).
- **Scaffold:** Use the standard Cas9 sgRNA scaffold sequence as the scaffold.

- **RTT:** The RTT sequence (typically 10–20 nt in length) is represented by the region upstream of the nicked site complementary to the target strand DNA. RTT includes the desired edit (*see* **Note 5**).
- **PBS:** The PBS sequence (typically 8–15 nt in length) is represented by the region downstream of the nicked site complementary to the target strand DNA. PBS enables the formation of an RNA/DNA duplex that serves as a substrate for reverse transcriptase (*see* **Note 6**).

Cloning of pegRNA Inserts into Expression Vectors

pegRNA constructs can be cloned into a U6-driven expression vector using a previously reported Golden Gate protocol [7]. The overall cloning workflow is illustrated in Fig. 3, which summarizes the key steps involved in pegRNA assembly. Briefly, the vector backbone is digested with a type II-S restriction enzyme, creating compatible overhangs for insertion. Three sets of annealed oligonucleotides are then prepared:

- **Spacer oligos** encoding the target-specific guide sequence.

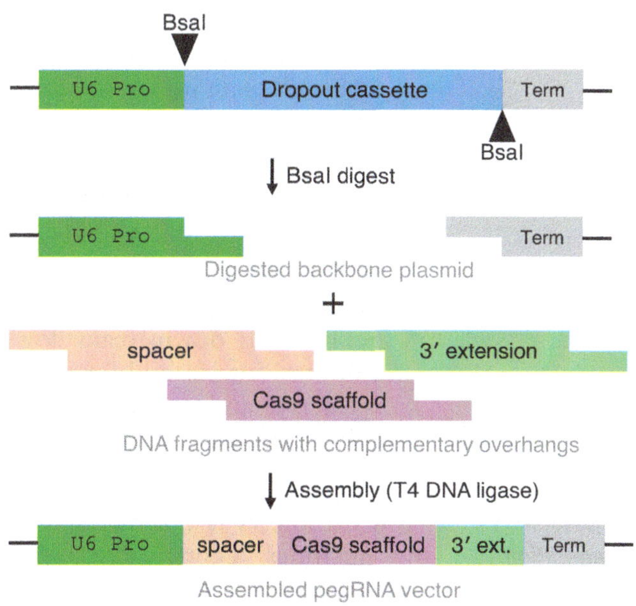

Fig. 3 Schematic overview of pegRNA cloning using the Golden Gate assembly method [7]. pegRNA constructs can be cloned into a U6-promoter-driven expression vector using a Golden Gate protocol. The vector backbone is first digested with the type IIS restriction enzyme BsaI to generate compatible overhangs. Three sets of annealed oligonucleotides are then prepared: (1) spacer oligos encoding the target-specific guide sequence, (2) 3′ extension oligos containing the primer-binding site (PBS) and reverse transcriptase template (RTT), and (3) 5′ phosphorylated scaffold oligos to enable ligation. These oligos are assembled into the digested backbone using T4 DNA ligase, yielding the final pegRNA expression construct

- **3′ extension oligos** containing the primer-binding site and reverse transcriptase template.
- **Scaffold oligos**, which are 5′ phosphorylated to allow ligation.

These annealed oligos are assembled into the digested backbone using T4 DNA ligase, resulting in the final pegRNA expression construct. Alternatively, complete pegRNA constructs can be custom-synthesized and cloned directly into expression vectors by commercial providers such as Twist Bioscience [21].

3.2 Testing PE and iPE via Transient Transfection of Cell Lines Using Lipid Nanoparticles

To perform editing, plasmids encoding both the pegRNA and the selected prime editor must be delivered into the target cells. Because both systems utilize identical protein components, the only distinction between PE and iPE is the pegRNA design. While this protocol uses HEK293T cells as a model system, it is compatible with other adherent cell lines. We recommend optimizing transfection conditions for each specific cell line prior to conducting genome editing experiments.

1. Culture HEK293T cells at 37°C in a humidified incubator with 5% CO_2 in DMEM supplemented with 10% FBS, 1% penicillin-streptomycin, and 1% GlutaMAX.
2. Seed ~25,000 cells into a 96-well plate or ~45,000 cells into a 48-well plate using antibiotic-free DMEM FBS media.
3. Transfect cells when they reach approximately 75% confluency (*see* **Note 7**), typically 24 h after seeding. A common lipid nanoparticle used for transient transfection is Lipofectamine 2000. Modify the manufacturer's protocols for lab-specific conditions.
4. Transfect cells with plasmids encoding the prime editor and pegRNA. If downstream flow cytometry analysis is planned, a fluorescent plasmid for tracking transfection efficiency (e.g., mCherry) and a reporter plasmid designed to measure prime editing activity (see Sect. 3.3) can be included.

3.3 Flow Cytometry Sample Preparation and Analysis

Flow cytometry is a powerful technique for assessing prime editing outcomes. A commonly used assay is the Prime Editor Activity Reporter (PEAR) system [22], which measures editing efficiency via fluorescent readout. In PEAR, a mutation at a splice site prevents GFP expression. Successful prime editing restores the GFP sequence, resulting in green fluorescence. The proportion of successfully edited cells can be determined by quantifying GFP-positive populations.

1. After incubation for 2–3 days, prepare samples for flow cytometry.
2. Remove all media, then wash the cells with PBS.

3. Add 30 μL (96-well) or 50 μL (48-well) TrypLE per well. Incubate for ~15 min at 37°C.
4. Add 170 μL (96-well) or 150 μL (48-well) PBS + 2% FBS (or Opti-MEM; *see* **Note 8**). Pipette up and down thoroughly to disperse clumps.
5. Transfer through a 30–100 μm mesh filter into a U-bottom flow cytometry plate.
6. Keep plates on ice until analysis.
7. Analyze samples on a flow cytometer.
8. Use unstained and single-color controls to set compensation.
9. Use Side Scatter Area (SSC-A) vs. Forward Scatter Area (FSC-A) to identify the live cell population and exclude debris.
10. Gate for the singlet population by plotting Forward Scatter Height (FSC-H) vs. Forward Scatter Area (FSC-A).
11. Define negative and positive populations based on the relevant fluorescent reporter(s).
12. Calculate editing efficiency as the percentage of reporter-positive cells within the relevant population of interest.

Example: Fig. 4 shows the gating strategy for cells co-expressing mCherry as a transfection marker and a reporter that expresses GFP only after successful prime editing. In this case, editing efficiency is calculated as the percentage of GFP-positive cells within the mCherry-positive population.

3.4 Prime Editing Assessment at Endogenous Loci Using Sequencing Methods

Prime editing efficiency can also be evaluated at endogenous genomic loci. In this approach, cells are transfected with the desired prime editing components targeting a specific genomic site. After the experiment, genomic DNA is extracted for downstream analysis. The target locus is amplified by PCR, and the editing efficiency is determined through sequencing-based methods such as Sanger sequencing or next-generation sequencing (NGS), followed by bioinformatic quantification of precise edits.

3.4.1 Genomic DNA Isolation

Using a commercial DNA extraction kit, extract genomic DNA from HEK293T cells that were transfected with editor components and incubated for approximately 3-5 days. This protocol is adapted from the E.Z.N.A. Tissue DNA Kit (Omega Bio-Tek):

1. Prepare BL Buffer Mix (per each well in a 48-well plate: 100 μL BL Buffer, 100 μL PBS, 12 μL Proteinase K solution). Prepare a master mix according to the number of wells, with 10% excess.
2. Remove the culture medium from each well. Add 200 μL BL Buffer Mix and incubate at 37°C for 15 min.

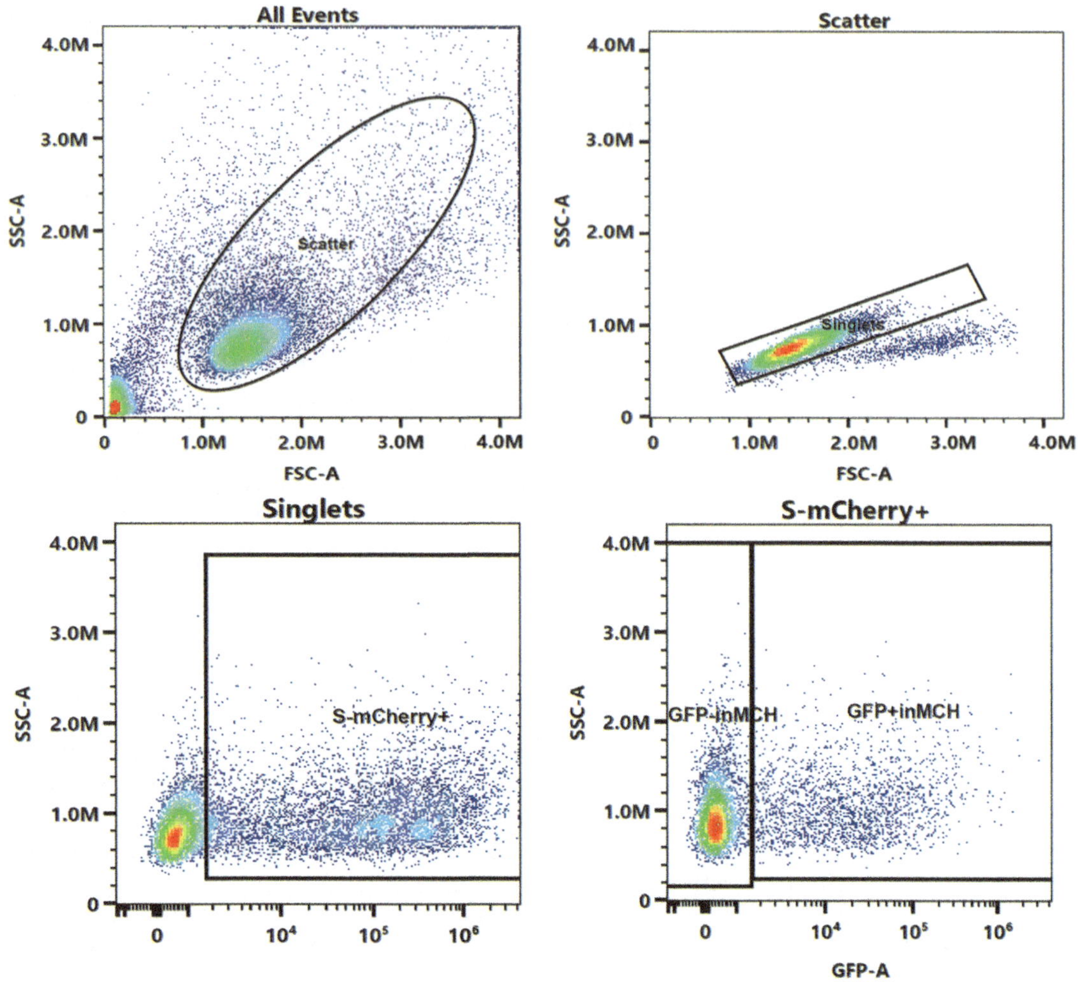

Fig. 4 Flow cytometry for quantifying prime editing efficiency. Cells co-express mCherry as a transfection marker and a GFP reporter that is activated only upon successful prime editing. The main cell population is first gated based on forward and side scatter, followed by singlet selection. mCherry-positive cells are then identified, and GFP expression is measured within this population. Editing efficiency is calculated as the percentage of GFP-positive cells within the mCherry-positive gate

3. Transfer the contents of each well to tubes or PCR strips and incubate at 70°C for 10 min.
4. Add 110 μL of 100% ethanol to each sample and vortex to mix.
5. Transfer the mixture to a DNA spin column. Centrifuge at 13,000 × *g* for 1 min and discard the filtrate.
6. Add 400 μL HBC Buffer to the column, centrifuge for 1 min, and discard the filtrate.
7. Place the column into a new collection tube. Add 700 μL DNA Wash Buffer to the column and centrifuge for 30 s. Discard the filtrate and centrifuge for 1 min to dry the column.

8. Transfer the spin column to a clean microcentrifuge tube. Add 50–100 μL nuclease-free water directly to the membrane. Let it sit for 2 min. Centrifuge for 30 s to elute genomic DNA.

3.4.2 Sanger Sequencing Analysis to Quantify Editing Efficiency

1. Design primers flanking the target editing site to amplify a 200–600 bp fragment with the edited base(s) centrally located.
2. Perform PCR amplification using high-fidelity DNA polymerase and genomic DNA as template.
3. Verify PCR product size by agarose gel electrophoresis and purify using column-based cleanup methods.
4. Obtain Sanger sequencing traces using the forward and/or reverse primer.
5. Analyze the Sanger sequencing traces with a deconvolution software (e.g., ICE [23], EditR [24]) (Fig. 5) to calculate the editing efficiency.

3.4.3 Quantifying Genome Editing Using Amplicon Next-Generation Sequencing (NGS) Analysis

This protocol is compatible with Illumina-based sequencing machines, though other NGS technologies can similarly be used to generate amplicon reads. For labs without easy access to NGS facilities, library samples can be sent to GENEWIZ-Azenta Life Sciences [25], which offers small-scale sequencing solutions (e.g., Amplicon-EZ).

1. Prepare NGS Libraries
   - Library preparation involves two consecutive PCR reactions (Fig. 6):
   - First PCR—Amplifies the target locus from genomic DNA using gene-specific primers.
   - Second PCR—Adds unique sample-specific barcodes (for demultiplexing) and Illumina adapters.

   - After PCR, pool barcoded products and purify them using gel extraction.
   - Quantify purified libraries using Qubit.

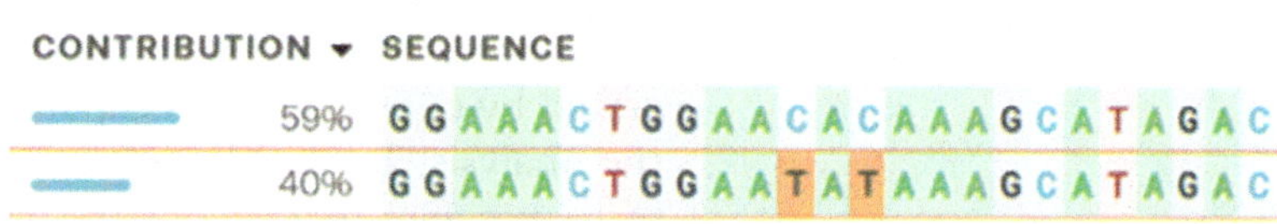

Fig. 5 ICE analysis of Sanger sequencing data to quantify prime editing efficiency. Sanger results (.ab1 files) from both the wild-type (WT) sample and the edited sample were uploaded into the ICE analysis platform. The intended edit was provided as a short "donor sequence" containing the desired base change along with several flanking bases on each side. Using these inputs, ICE calculates the percentage of sequences carrying the desired edit

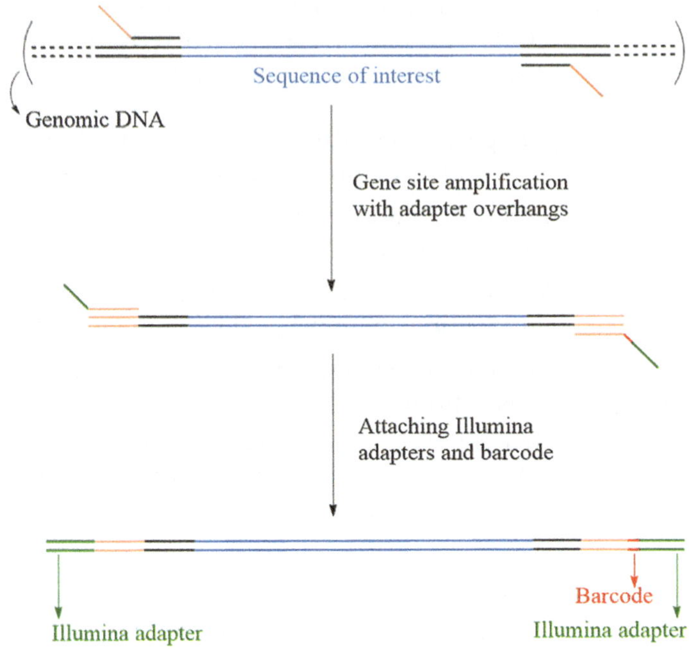

Fig. 6 Workflow for library preparation for next-generation sequencing (NGS). Genomic DNA from transfected cells is first used as a template for PCR amplification of the target locus using gene-specific primers with adapter overhangs. A second PCR then incorporates Illumina adapters and sample-specific barcodes. The resulting amplicons contain the sequence of interest flanked by full Illumina adapter sequences and unique barcodes, enabling pooling and sequencing on an Illumina platform

2. Run the appropriate Illumina sequencing kit. Either single reads or paired-end reads are suitable.
3. Data Analysis

   - Analyze the NGS sequencing reads using CRISPResso2 [26]. CRISPResso2 is the most commonly used method for the quantification of genome editing events from NGS. Ensure CRISPResso2 is properly installed before proceeding with the analysis. Often, many samples are combined into one library. Here, we apply the batch analysis mode ("CRISPRessoBatch") to facilitate the analysis of many samples at the same time. Different outputs can be generated by constructing compatible batch files and running the analysis. Batch files are tab-separated values (.tsv) or text (.txt) files that specify the analysis parameters for each sample in the library.
   - To quantify all of the alleles present in the sample and quickly calculate the indel amount around the spacer, an example "allelebatch.txt" file can be constructed that specifies the following parameters:

name: Your sample name.

fastq_r1: Name of the fastq.gz file.
amplicon_seq: Amplicon sequence without adapters.
guide_seq: Cas9 spacer sequence.
min_average_read_quality: 30.
quantification_window_size: Typically 10.

quantification_window_center: −3 for Cas9 (the cut location relative to PAM).

- This can be run on the computer using the following example command, which will utilize all the processors in your computer and requires the presence of at least 5000 reads for output generation:
CRISPRessoBatch --n_processes max --min_reads_for_inclusion 5000 --batch_settings allelebatch.txt

- Indel frequencies can be calculated from the CRISPResso2 output file: "**CRISPRessoBatch_quantification_of_editing_frequency.txt**" using the formula:
% Indel = 100% × [Insertions + Deletions] / [Reads_aligned_all_amplicons]

- CRISPResso2 can be run with the "prime editing" mode to facilitate quantification of precise edits. Here, a separate batch file, an example "pebatch.txt," can be constructed that specifies the following parameters:

name: Your sample name.
fastq_r1: Name of the fastq.gz file.
amplicon_seq: Amplicon sequence without adapters.
prime_editing_pegRNA_spacer_seq: Cas9 spacer sequence.
prime_editing_target_seq: Edited amplicon sequence.
prime_editing_pegRNA_extension_seq: Sequence of RTT and PBS (use the reverse complement for iPE).

min_average_read_quality: 30.

- This can be run on the computer using the following example command, which will utilize all the processors in your computer and requires the presence of at least 5000 reads for output generation:
CRISPRessoBatch --n_processes max --min_reads_for_inclusion 5000 --batch_settings pebatch.txt

- Precise editing frequencies (PE or iPE) can be calculated from the CRISPResso2 output file: "**CRISPRessoBatch_quantification_of_editing_frequency.txt**" using the formula:
% Edit = 100% × [Prime-edited Reads_aligned] / [Reference Reads_aligned_all_amplicons]

4 Notes

1. PAM is a short DNA sequence adjacent to the target site, required for Cas9 binding. SpCas9 recognizes NGG PAMs, while engineered variants (SpG, SpRY) accept NGN or NRN/NYN motifs. Correct PAM selection is essential for target accessibility.
2. Introducing mutations within the PAM after editing prevents Cas9 from re-cutting the corrected sequence.
3. Spacer sequences should have 40%–60% GC content, avoid long homopolymer runs, and minimize secondary structures. Use design tools (e.g., Benchling [27]) to select guides with high specificity scores.
4. Evaluate the on-target activity and off-target risk of candidate spacers using design tools (e.g., Benchling [27]). Reject spacers with high off-target scores at exonic loci.
5. RTT should be 10–20 nt to include the edit and provide homology for reverse transcription.
6. PBS length is typically 13 nt but may be optimized between 8 and 15 nt based on melting temperature (Tm) and efficiency.
7. Higher or lower confluency can reduce transfection efficiency and/or cell viability.
8. Opti-MEM can help prevent rapid re-adhesion of cells before analysis.

Acknowledgments

This work was supported by the Natural Sciences and Engineering Research Council of Canada (NSERC) Discovery Grant (DG).

References

1. Hsu PD, Lander ES, Zhang F (2014) Development and applications of CRISPR-Cas9 for genome engineering. Cell 157:1262–1278
2. Villiger L, Joung J, Koblan L et al. (2024) CRISPR technologies for genome, epigenome and transcriptome editing. Nat Rev Mol Cell Biol 25:464–487
3. Liao H, Wu J, VanDusen NJ et al. (2024) CRISPR-Cas9-mediated homology-directed repair for precise gene editing. Mol Ther Nucleic Acids 35(4):102344
4. Mao Z, Bozzella M, Seluanov A, Gorbunova V (2008) Comparison of nonhomologous end joining and homologous recombination in human cells. DNA Repair (Amst) 7:1765–1771
5. Komor AC, Kim YB, Packer MS et al. (2016) Programmable editing of a target base in genomic DNA without double-stranded DNA cleavage. Nature 533:420–424
6. Rees HA, Liu DR (2018) Base editing: precision chemistry on the genome and transcriptome of living cells. Nat Rev Genet 19:770–788
7. Anzalone AV, Randolph PB, Davis JR et al. (2019) Search-and-replace genome editing without double-strand breaks or donor DNA. Nature 576:149–157

8. Anzalone AV, Gao XD, Podracky CJ et al. (2021) Programmable deletion, replacement, integration and inversion of large DNA sequences with twin prime editing. Nat Biotechnol 40:731–740
9. Yarnall MTN, Ioannidi EI, Schmitt-Ulms C et al. (2022) Drag-and-drop genome insertion of large sequences without double-strand DNA cleavage using CRISPR-directed integrases. Nat Biotechnol 41:500–512
10. Chen PJ, Hussmann JA, Yan J et al. (2021) Enhanced prime editing systems by manipulating cellular determinants of editing outcomes. Cell 184:5635–5652.e29
11. Pandey S, Gao XD, Krasnow NA et al. (2024) Efficient site-specific integration of large genes in mammalian cells via continuously evolved recombinases and prime editing. Nat Biomed Eng 9:22–39
12. Nelson JW, Randolph PB, Shen SP et al. (2022) Engineered pegRNAs improve prime editing efficiency. Nat Biotechnol 40:402–410
13. Liang R, Wang S, Cai Y et al. (2025) Circular RNA-mediated inverse prime editing in human cells. Nat Commun 16:5057
14. Yang C, Fang Q, Li M et al. (2025) Prime editor with rational design and AI-driven optimization for reverse editing window and enhanced fidelity. Nat Commun 16:5144
15. Kim YB, Pierce EB, Brown M et al (2022) A novel mechanistic framework for precise sequence replacement using reverse transcriptase and diverse CRISPR-Cas systems. bioRxiv 2022.12.13.520319
16. Addgene: A better way to share science. https://www.addgene.org/. Accessed 20 Nov 2025
17. PRIDICT.it - pegRNA Prediction. https://pridict.it/. Accessed 19 Nov 2025
18. GitHub - pinellolab/PrimeDesign: Software tool for the flexible design of pegRNAs and ngRNAs for prime editing!. https://github.com/pinellolab/PrimeDesign. Accessed 19 Nov 2025
19. pegFinder: pegRNA designer for Prime Editing. http://pegfinder.sidichenlab.org/. Accessed 19 Nov 2025
20. pegIT - pegRNAs for your desired edit. https://pegit.giehmlab.dk/. Accessed 19 Nov 2025
21. Vectors | Twist Bioscience. https://www.twistbioscience.com/products/genes/vectors. Accessed 19 Nov 2025
22. Simon DA, Tálas A, Kulcsár PI et al. (2022) PEAR, a flexible fluorescent reporter for the identification and enrichment of successfully prime edited cells. eLife 11:e69504
23. EditCo - CRISPR Performance Analysis. https://ice.editco.bio/#/. Accessed 19 Nov 2025
24. Kluesner MG, Nedveck DA, Lahr WS et al. (2018) EditR: a method to quantify base editing from sanger sequencing. CRISPR J 1:239–250
25. Amplicon-EZ | GENEWIZ from Azenta. https://www.genewiz.com/public/services/next-generation-sequencing/amplicon-sequencing-services/amplicon-ez. Accessed 19 Nov 2025
26. Clement K, Rees H, Canver MC et al. (2019) CRISPResso2 provides accurate and rapid genome editing sequence analysis. Nat Biotechnol 37:224–226
27. CRISPR Guide RNA Design Tool | Benchling. https://www.benchling.com/crispr. Accessed 19 Nov 2025

Check for updates

Chapter 11

Constructing Drive-and-Process (DAP) CRISPR Guide RNA Arrays for Multiplexed Base- and Prime-Editing

Devin A. Golla, Tyler C. Daniel, Logan Haugh and Xue Gao

Abstract

Advancements in base- and prime editing technologies in recent years have offered researchers a plethora of options for introducing precise insertions, deletions, or substitutions into targeted genomic loci. These precision editors' applications have rapidly expanded to the modeling and treatment of polygenic diseases, as well as into the growing field of functional genomics. This has created a need for compact, modular expression systems capable of producing multiple guide RNAs (gRNAs) from a single transcript while preserving high editing efficiency. To address this, we have developed the drive-and-process (DAP) array, a modular architecture composed of alternating gRNA and tRNA units. The DAP array design exploits the cell's endogenous tRNA processing machinery to cleave each tRNA from the array and release individual gRNAs, thereby enabling simultaneous editing at multiple loci following hybridization with Cas9. Here, we outline key design considerations and experimental steps required for the construction and deployment of DAP arrays as multiplex base- or prime editing tools in human cells (e.g., HEK293T). We also highlight how the DAP array can be leveraged to enable efficient processing of gRNAs along with other RNAs of similar size, such as shRNA, potentially broadening its usage in addressing complex biological questions and therapeutic applications that require coordinated genetic perturbation.

Keywords CRISPR, Base editing, Prime editing, Multiplex editing, DAP array, tRNA processing

1 Introduction

Precise and programmable genome-editing technologies built upon the CRISPR-Cas9 architecture have emerged as powerful tools for investigating complex relationships among genes, and in the modeling and treatment of disease. Base editing (BE) enables highly efficient single-base conversions by fusing a deaminase to a catalytically impaired Cas9 nickase [1], while prime editing (PE) combines a Cas9 nickase with a reverse transcriptase (RT) and a prime editing guide RNA (pegRNA) to install small, user-defined sequence modifications, including substitutions, insertions, or deletions, at chosen genomic sites [2]. Both techniques avoid the introduction of double-stranded breaks (DSBs) at the target site, making them safer

Huan Peng and S. Brian Kim (eds.), *Biological Macromolecule Engineering Technologies*,
Methods in Molecular Biology, vol. 3032, https://doi.org/10.1007/978-1-0716-5257-2_11,

and more precise than conventional Cas9 nuclease-based technologies that rely on DSB formation and homology-directed repair for introducing defined edits [3]. BE and PE have been used to perform specific genetic perturbations across a wide range of applications, including in disease modeling [4], in vivo disease correction [5], and agricultural engineering [6]. However, until recently, the utility of these tools was limited by the challenge of performing multiple edits simultaneously. The *Streptococcus pyogenes* Cas9 (SpCas9) used in many of the most common BE and PE systems, including engineered variants like TadCBEd [7], ABE8e [8], PE6 [9], and PE7 [10], is inherently unable to process individual mature single guide RNAs (sgRNAs) from a single transcript [11]. Moreover, although multiple sites can be targeted if each individual sgRNA is expressed from its own corresponding promoter, such an approach may be difficult to employ in viral delivery systems such as AAV, which are restricted by stringent payload size limitations [12].

The drive-and-process (DAP) array was developed to overcome these limitations, taking advantage of the cell's endogenous tRNA processing machinery to release mature gRNAs from a single transcript [13]. In a DAP array, gRNAs are interspersed between repeated tRNA sequences that are recognized and cleaved by the enzymes RNAse P and RNAse Z, which process the 5′ and 3′ ends of the tRNA motifs, respectively (Fig. 1). This processing mechanism releases each gRNA from the array at comparable expression levels, enabling simultaneous BE or PE activities at multiple loci. Additionally, the tRNA at the most 5′ position can also function as an internal RNA Pol III promoter for the transcription of the array [14], reducing the construct size by eliminating the need for a separate promoter. Previous studies have applied such principles to construct tRNA-based multiplex arrays for Cas9-mediated gene targeting in plants [15, 16], achieving indel formation at multiple genomic sites. Importantly, the RNAse P- and Z-mediated tRNA processing mechanism is conserved widely across species [17, 18], offering the potential to employ such a strategy in different cell types. Extending these principles, the DAP array adapts the tRNA-gRNA architecture for base-editing and prime-editing applications in mammalian cells,

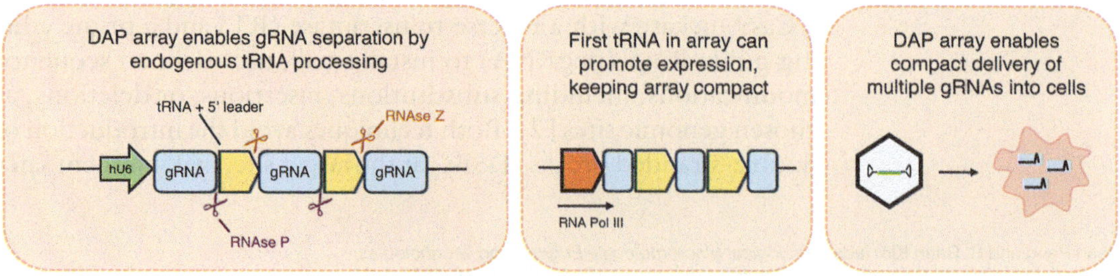

Fig. 1 Schematic of DAP multiplex array components, and summary of advantages of the technology

marking a crucial step toward scalable and multiplexed genome engineering for the study of complex human disease.

The DAP array is well-suited for use in customizable multiplex genome editing applications, as up to 16 individual base editing loci or three individual prime editing loci have been validated to be targetable by a single DAP array at comparable editing efficiency and with comparable off-target profiles to single-gRNA systems [13]. Moreover, the array's construction, requiring only two template plasmids which contain the array's modular "building blocks", can be easily modified to incorporate different guide RNA architectures or other motifs as necessary for a particular application. By using template plasmids corresponding to the characteristic sgRNA scaffold of a chosen CRISPR effector, DAP arrays can be adapted for use with various platforms beyond Cas9, including other programmable RNA-guided endonucleases as they continue to be discovered.

Due to the repetitive nature of the DAP array's components, efficient array assembly can be challenging. Here, we seek to streamline this experimental workflow by describing the process of constructing the DAP array for two representative applications: multiplexed three-locus base editing using SpCas9-based BE and multiplexed two-locus prime editing using SpCas9-based PE3 [2]. Following the design and validation of each individual gRNA, the DAP vector can be assembled using a standard Golden Gate cloning method, in which each gRNA is encoded by primers at the junction of fragments, minimizing nonspecific primer binding to homologous fragments and providing unique 4-nucleotide overhangs for ordered assembly. We also outline procedures for transfecting the assembled array into HEK293T cells and quantifying editing efficiency at each targeted site. Although only two use cases are detailed here, the protocol is effector-agnostic and can be generalized to enable guide RNA multiplexing for a variety of other genome editing systems, by adjusting the sgRNA scaffold as required [19].

2 Materials

2.1 Designed DNA Components for Cloning

1. Plasmid DNA containing the intended backbone and insert sequences. (This protocol describes the use of hCtRNA_FT and hCtRNA_VT plasmids, available from Addgene, #186715 and #186716, respectively.)
2. Commercially synthesized oligo DNA (IDT)—see Sect. 3 for primers used in this example protocol, including those for plasmid assembly and those for genomic PCR.

2.2 Commercial Reagents

1. RNAse/DNAse-free water (Invitrogen).
2. 2× Phanta Max Master Mix (Dye Plus) (Vazyme - catalog number P525-01).

3. Agarose powder.
4. TAE buffer.
5. DNA ladder, such as DL5000 and DL1000 DNA Marker (Abclonal - catalog numbers RK30191 and RK30194, respectively).
6. Ethidium Bromide (Fisher Scientific - catalog number BP1302-10).
7. DNA cleanup buffers QG and PE (Qiagen - catalog numbers 19063 and 19065, respectively).
8. Silica spin columns for DNA extraction, such as EconoSpin® All-In-One DNA Only Mini Spin Column (Epoch Life Sciences, catalog number 1920-050).
9. T4 DNA Ligase & Ligase Buffer (NEB - catalog number M0202S).
10. BsaI-HF® v2 restriction enzyme (NEB - catalog number R3733S) (*see* **Note 1**).
11. Stbl3 chemically competent *E. coli*, such as Mix & Go! *E. coli* Transformation Kit (Zymo Research - catalog number T3001).
12. Ampicillin.
13. LB media (25 g LB powder/L water).
14. Agar.
15. DNA miniprep buffers P1, P2, and N3 (Qiagen - catalog numbers 19051, 19052, and 19064, respectively). P1 buffer requires preparation with RNAse A (Macherey-Nagel - catalog number 740505).
16. HEK293T cells (ATCC - catalog number CRL-3216), or other cell line (*see* **Note 2**).
17. Dulbecco's Modified Eagle Medium (DMEM), high glucose, GlutaMAX™ Supplement, pyruvate (Life Technologies - catalog number 10569010). Supplemented with 10% fetal bovine serum (FBS) (Life Technologies - catalog number A5256701) and 1% Penicillin-Streptomycin ("Pen/Strep") (5,000 U/mL) (Life Technologies - catalog number 15070063).
18. CO_2-humidified incubator.
19. Trypsin solution, such as TrypLE™ Express Enzyme (1X), phenol red (Life Technologies - catalog number 12605010).
20. Tissue culture-treated dishes and 96-well plates.
21. Cell counter, such as Countess 3 Automated Cell Counter (Invitrogen), and required reagents such as Trypan Blue (Invitrogen - catalog number T10282), according to manufacturer's instructions.

22. Lipofectamine 2000 transfection reagent (Thermo Fisher - catalog number 11668027).
23. Opti-MEM™ I Reduced Serum Medium (Life Technologies - catalog number 31985062).
24. Cell lysis buffer (prepared from: 10 mM Tris–HCl (pH 7.5), 0.05% SDS (w/v)).
25. Proteinase K (25 μg/mL) (NEB – catalog number P8107S), stored at −20°C.

3 Methods

3.1 Design and Construction of gRNA-tRNA Fragments

1. Identify a preferred CRISPR genome editor to use, specific to the intended application. This choice of editor will dictate the sgRNA scaffold used in the DAP array template. As most base- and prime-editing systems currently utilize SpCas9, the majority of applications will use Cas9 sgRNA [20]. In such applications, the published Addgene plasmids hCtRNA_FT ("FT" referring to "fragment template") and hCtRNA_VT ("VT" referring to "vector template") may be directly used; otherwise, further cloning is required prior to assembly (*see* **Note 3**).
2. Choose a tRNA sequence to be processed in the DAP array via RNAse P and RNAse Z-mediated cleavage. For most human applications, hCys-GCA-tRNA ("hCtRNA") is known to enable efficient processing of the DAP array [13]. For applications using this hCtRNA, the Addgene plasmid hCtRNA_FT may be directly used; otherwise, further cloning is required prior to assembly (*see* **Note 3**).
3. Design base or prime editing guide RNAs to target sites of interest.
   a. An assortment of gRNA design software optimized for base and prime editing applications, with the option to alter various parameters for increased efficiency, can be found for free online. While gRNAs can be designed by hand, we recommend the following for Cas9-based guide design:
      1. For base editing, we have found success using the "Design and analyze guides" function in the "CRISPR" menu of Benchling [21].
      2. For prime editing guides, we recommend PE-Designer [22] or pegIT [24]. The presented application assumes the use of PE3 (which includes a pegRNA and a "nicking" gRNA at a nearby site). Both gRNAs necessary

Table 1
sgRNA protospacer sequences for Base Editing example

Protospacer #1	5′ GAGTCCGAGCAGAAGAAGAA 3′
Protospacer #2	5′ GGAATCCCTTCTGCAGCACC 3′
Protospacer #3	5′ GTCATCTTAGTCATTACCTG 3′

Table 2
pegRNA and nicking sgRNA guide sequences for Prime Editing example

Nicking gRNA #1	5′ GTCAACCAGTATCCCGGTGC 3′
pegRNA #1 protospacer	5′ GGCCCAGACTGAGCACGTGA 3′
pegRNA #1 extension	5′ TCTGCGATCACGTGCTCAGTCTG 3′
Nicking gRNA #2	5′ TCAACCATTAAGCAAAACAT 3′
pegRNA #2 protospacer	5′ GTCATCTTAGTCATTACCTG 3′
pegRNA #2 extension	5′ AACGAACACATCAGGTAATGACTAAGATG 3′

for PE3 can often be designed by the same software. The result of designing the pegRNA will be two components: a "protospacer" (guide sequence) and an "extension" comprising the primer binding site (PBS) and reverse transcriptase template (RTT).

3. *See* **Note 4** for additional recommendations regarding optimized guide RNA design.

b. Three previously validated BE sgRNA protospacer sequences, and two sets of previously validated PE guide RNA components [13] have been provided for illustrative purposes (Tables 1 and 2).

3.2 Assembling gRNA-tRNA Fragments into DAP Array

Individual DAP fragments can be inserted in a desired order into the hCtRNA_VT vector via Golden Gate assembly. Due to the ability of Type IIS restriction enzymes to cleave outside their respective recognition sites [23], utilizing unique sgRNA protospacers at each site in the DAP array generates unique overhangs at each restriction site, enabling the overall DAP array plasmid to be assembled in an ordered fashion, with only one of each sgRNA protospacer present in the final array. Details on the DAP array assembly procedure are provided below, with certain modifications and troubleshooting tips presented as an addendum (*see* **Note 5**).

3.2.1 Base Editing

1. Using a Golden Gate assembly tool (such as Benchling's assembly wizard [21] or New England Biolabs' NEBridge tool [25]), select the following components for assembly (Fig. 2):

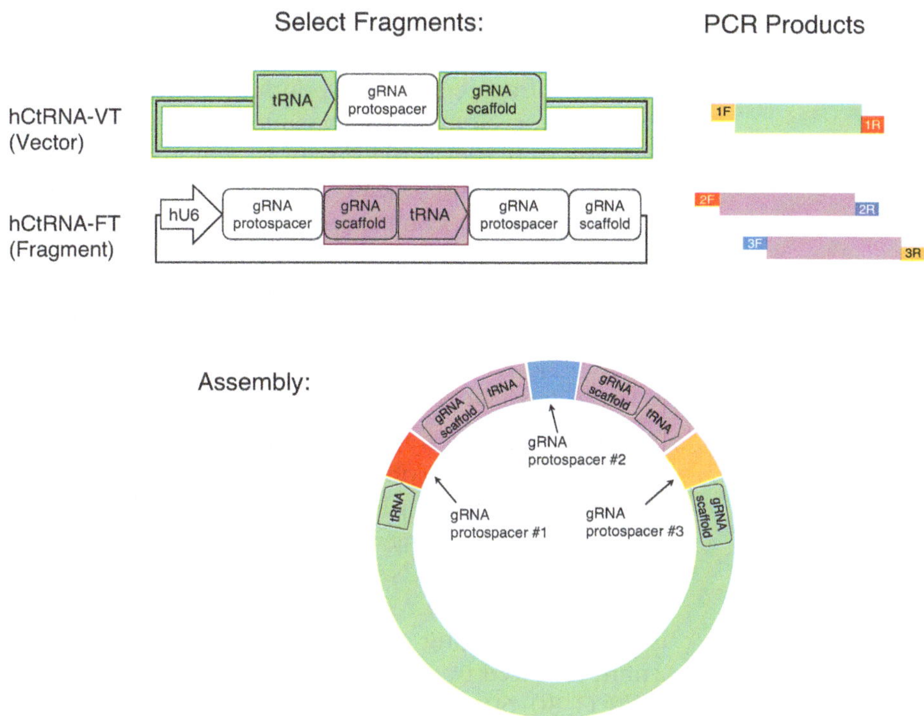

Fig. 2 Illustration of assembly of fragments from starting vectors into DAP array for SpCas9 base editing application. Selected regions of hCtRNA_VT (vector) and hCtRNA_FT (insert) are amplified via PCR, with primers encoding each designed gRNA protospacer; then, fragments are attached via Golden Gate assembly to yield a complete DAP array, of variable size depending on the number of hCtRNA_FT fragments included

a. Backbone: hCtRNA_VT nucleotides 671-650, using the Addgene plasmid's numbering scheme (*see* **Note 6**). This backbone fragment should contain the whole plasmid except for the region labeled "gRNA scaffold" and the 20 nucleotide guide RNA immediately preceding it.

b. Junction 1 between backbone and fragment: Protospacer #1 sequence (Table 1), or the first ~20nt sgRNA guide sequence designed in Sect. 3.1, Step 3 to be inserted.

c. Fragment 1: hCtRNA_FT nucleotides 854-1004 inclusive (this encompasses the Cas9 sgRNA scaffold, the hCtRNA, and a 3 bp 5′ leader sequence).

d. Junction 2 between fragments: Protospacer #2 sequence (Table 1), or the second ~20nt sgRNA guide sequence designed in Sect. 3.1, Step 3 to be inserted.

e. Fragment 2: Identical to Fragment 1.

f. Junction 3 between fragment and backbone: Protospacer #3 sequence (Table 1), or the last ~20nt sgRNA guide sequence designed in Sect. 3.1, Step 3 to be inserted.

For as many base editing sgRNAs as one desires to include in the DAP array, repeat Fragment 1, with each junction between fragments corresponding to the desired sgRNA protospacer. Notably, the maximum size of a single DAP base editing array previously shown to edit efficiently at all target loci is 16 sgRNAs [13]; any longer is theoretically possible, but could potentially complicate assembly or efficient expression.

2. Commercially synthesize complementary oligo primer pairs specified by the assembly tool, which will encode sgRNA protospacers and overhangs compatible with BsaI-digested component fragments (*see* **Note 7**). Primers for the assembly of this example construct are provided in Table 3.
3. For each intended fragment to amplify, assemble 20 μL PCR reactions consisting of: 10 μL Phanta Max master mix, 7 μL nuclease-free water, 1 μL each of the forward and reverse oligo primer diluted to 10 ng/μL, and 1 μL of the template plasmid (hCtRNA_VT, hCtRNA_FT, or equivalent custom plasmid), diluted to 10 ng/μL. Run PCR reaction on a thermocycler, using the following steps: initial denaturation at 95°C for 2 min, followed by 25–35 cycles of (95°C, 20 s; at 60°C, 30 s; at 70°C, 30–60 s/kb), followed by a final extension at 70°C for 2 min (*see* **Note 8**).
4. Verify the lengths of each fragment by running on a ~1% agarose gel (containing ethidium bromide at a final concentration of ~0.5 μg/mL), and purify fragments using Qiagen DNA cleanup buffers, following the manufacturer's protocol.
5. Assemble Golden Gate reaction as follows: 8 μL DNA mixture from previous step; 1 μL T4 Ligase Buffer, 0.5 μL BsaI or equivalent Type IIS restriction enzyme, 0.5 μL T4 DNA Ligase enzyme. Run reaction on a thermocycler, using the following steps: 15 cycles of [37°C for 5 min, 16°C for 5 min], followed by 60°C for 5 min, then 12°C for 5 min (*see* **Note 9**).

Table 3
Primers for Base Editing example

BE_VT_fw	GGTTGGGTCTCGAGTCATTACCTGgttttagagctagaaatagcaagttaaaataaggctagtcc
BE_VT_rev	GGTTGGGTCTCGCTGCTCGGACTCagggggcacccggatttgaa
BE_FT_fw1	GGTTGGGTCTCGGCAGAAGAAGAAgttttagagctagaaatagcaagttaaaataaggctagtcc
BE_FT_rev1	GGTTGGGTCTCGAGAAGGGATTCCagggggcacccggatttgaa
BE_FT_fwd2	GGTTGGGTCTCGTTCTGCAGCACCgttttagagctagaaatagcaagttaaaataaggctagtcc
BE_FT_rev2	GGTTGGGTCTCGGACTAAGATGACagggggcacccggatttgaa

Benchling's Assembly wizard, using the "Golden Gate Assembly" preset, was used to generate these primers. The following settings were used: Pre-recognition site length: 5 bp; Pre-recognition site bases: GGTTG; Min T_m for binding region: 60°C; Min length for binding region: 18 bp; Max T_m difference for primer pairs: 5°C; Type IIS enzyme: BsaI.

6. Transform 5 µL mixture into 10 µL chemically competent *E. coli* by heat shock. (This can be done immediately after the Golden Gate reaction, or after storage of the Golden Gate reaction product overnight at 4°C). Plate transformed E. coli on ampicillin-containing LB-agar plates, and incubate for 16 h at 37°C. Store at 4°C after 16 h to prevent overgrowth.
7. Pick individual colonies from plates, and transfer to round-bottom 15 mL test tubes containing 5 mL LB media supplemented with 5 µL of 100 µg/mL ampicillin. Shake at 37°C for 16 h.
8. Extract plasmid from overnight liquid cultures, using Qiagen DNA miniprep buffers, following manufacturer's protocol.
9. Verify sequence of resulting plasmid by Sanger sequencing. If the combined length of insertion (all fragments) is >700 bp, utilizing a whole-plasmid next-generation sequencing (NGS) service may be a more economical approach. *See* **Note 10** for recommended experiment setup, after DAP array sequences have been verified.

3.2.2 Prime Editing

1. Using a Golden Gate assembly tool, select the following components for assembly (Fig. 3):
   a. Backbone: hCtRNA_VT nucleotides 747-650, using the Addgene plasmid's numbering system. Note that here, in contrast to the base editing protocol, the gRNA scaffold region will be excluded from this fragment.
   b. Junction 1 between backbone and fragment: Nicking gRNA #1 sequence (Table 2), or the first ~20nt sgRNA nicking guide sequence designed in Sect. 3.1, Step 3 to be inserted.
   c. Fragment 1: hCtRNA_FT nucleotides 854-1004 (this encompasses the Cas9 sgRNA scaffold, the hCtRNA, and a 3 bp 5′ leader sequence).
   d. Junction 2 between fragments: pegRNA #1 protospacer (Table 2), or the 20 nt protospacer sequence of the first pegRNA designed in Sect. 3.1 Step 3.
   e. Fragment 2: hCtRNA_FT nucleotides 854-929 (this encompasses **just** the Cas9 sgRNA scaffold).
   f. Junction 3: pegRNA #1 extension, or the extension sequence of the pegRNA designed in Sect. 3.1, Step 3.
   g. Fragment 3: hCtRNA_FT nucleotides 930-1004 (this encompasses **just** the hCtRNA and 3 bp 5′ leader sequence).
   h. Junction 4: Nicking gRNA #2 sequence (Table 2) or equivalent.
   i. Fragment 4: Same as Fragment 1.

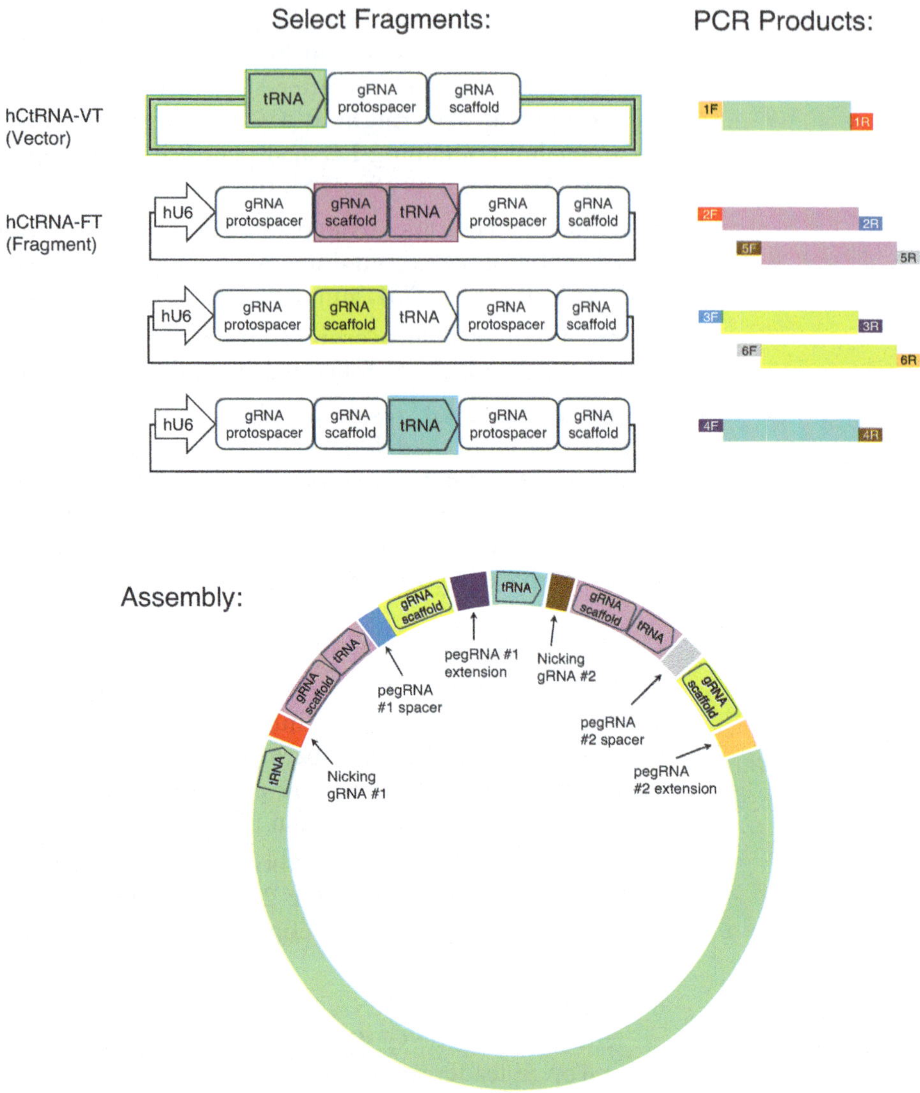

Fig. 3 Illustration of assembly of fragments from starting vectors into DAP array for SpCas9 prime editing application. Selected regions of hCtRNA_VT (vector) and hCtRNA_FT (insert) are amplified via PCR, with primers encoding each designed gRNA protospacer. For the prime editing application, note that multiple fragments derived from hCtRNA_FT, each containing different combinations of components, will be amplified. After PCR, fragments are attached via Golden Gate assembly to yield a complete DAP array, of variable size depending on the number of hCtRNA_FT fragments included

j. Junction 5: pegRNA #2 protospacer (Table 2) or equivalent.

k. Fragment 5: Same as Fragment 2.

l. Junction 6: pegRNA #2 extension (Table 2) or equivalent.

For as many pegRNA-nicking gRNA pairs as one desires to include in the DAP array, repeat Junction, Fragment 1,

Table 4
Primers for Prime Editing example

PE_VT_fw	GGTTGGGTCTCGAGGTAATGACTAAGATGtttttttaagcttgggccgctcgag
PE_VT_rev	GGTTGGGTCTCGATACTGGTTGACagggggcacccggatttgaa
PE_FT_fw1	GGTTGGGTCTCCGTATCCCGGTGCgttttagagctagaaatagcaagttaaaataaggctagtcc
PE_FT_rev1	GGTTGGGTCTCCTCAGTCTGGGCCagggggcacccggatttgaa
PE_FT_fw2	GGTTGGGTCTCCCTGAGCACGTGAgttttagagctagaaatagcaagttaaaataaggctagtcc
PE_FT_rev2	GGTTGGGTCTCCCACGTGATCGCAGAgcaccgactcggtgccac
PE_FT_fw3	GGTTGGGTCTCGCGTGCTCAGTCTGagaggggtatagctcagtggtagagc
PE_FT_rev3	GGTTGGGTCTCGGCTTAATGGTTGAagggggcacccggatttgaa
PE_FT_fw4	GGTTGGGTCTCGAAGCAAAACATgttttagagctagaaatagcaagttaaaataaggctagtcc
PE_FT_rev4	GGTTGGGTCTCGGACTAAGATGACagggggcacccggatttgaa
PE_FT_fw5	GGTTGGGTCTCGAGTCATTACCTGgttttagagctagaaatagcaagttaaaataaggctagtcc
PE_FT_rev5	GGTTGGGTCTCGACCTGATGTGTTCGTTgcaccgactcggtgccac

Benchling's Assembly wizard, using the "Golden Gate Assembly" preset, was used to generate these primers. The following settings were used: Pre-recognition site length: 5 bp; Pre-recognition site bases: GGTTG; Min T_m for binding region: 60°C; Min length for binding region: 18 bp; Max T_m difference for primer pairs: 5°C; Type IIS enzyme: BsaI

Junction, Fragment 2, Junction, Fragment 3. A final Fragment 3 may be omitted at the end of the array. Notably, the maximum size of a DAP prime editing array previously shown to edit efficiently at all loci is three modules (3 pegRNA + 3 nicking gRNA); any longer is theoretically possible, but could potentially complicate assembly or efficient expression. Other modifications to improve editing efficiency may also be included (*see* **Note 5**).

2. Commercially synthesize complementary oligo primer pairs specified by the assembly tool, which will encode sgRNA protospacers and overhangs compatible with BsaI-digested component fragments (*see* **Note 7**). Primers for the assembly of this example construct are provided in Table 4.

Steps 3–9 from Sect. 3.2.1 (Base Editing example) remain the same. As previously discussed, variations of this protocol may enable the substitution of different CRISPR effectors, guide RNA architectures, or tRNAs within the conserved DAP array architecture. One example of such a variation, the recently developed "mvGPT" system [19], is described in **Note 11**.

3.3 Transient Transfecting into HEK Cells

1. From a HEK293T cell culture in DMEM supplemented with FBS and Pen/Strep at ~80% confluence, remove DMEM, then remove cells from treated culture plate using trypsin solution (adding approximately 1 mL trypsin per 10 μL original volume

of DMEM), by incubating for 3 min. at 37°C. After incubating, add 3 mL fresh DMEM per 1 mL trypsin to neutralize trypsin reaction.

2. Seed cells onto a 96-well TC-treated plate at a density of 0.75×10^4 cells per well.
3. 16 h after seeding, perform transfection (using 225 ng/well BE or PE effector plasmid, 75 ng/well DAP array plasmid, diluted in Opti-MEM) with Lipofectamine 2000 reagent or a comparable lipofection kit, following the manufacturer's protocol.
4. Approximately 96 h after transfection, perform DNA extraction:
    1. Prepare a mixture of previously prepared cell lysis buffer (described in Materials) and proteinase K (in a ratio of 1000:1 buffer:enzyme). Remove remaining media from wells.
    2. Add 100 µL of lysis buffer mixture to each well. Incubate for 15 min at 37°C.
    3. Collect cell lysates from wells, and heat using a thermocycler to 95°C for 10 min to inactivate proteinase K.
    4. Store at 4°C until ready to use.

3.4 Analysis of Editing Results

1. Design forward and reverse genomic PCR primers for endogenous sites of interest. Primers for sequencing the genomic sites specific to the gRNAs used here are provided in Table 5. Genomic DNA amplicons should be ≤500bp in length for best PCR efficiency.
2. Perform genomic PCR; this can be done with the same protocol as specified above in Sect. 3.2, Step 3, scaled down to a 10 µL system. Note that separate PCR reactions using the same cell lysate will need to be performed for each individual genomic site of

Table 5
Sequencing primers for genomic sites of interest

Site		Sequencing primer
BE_sgRNA1	Fw	CCAGTGGCTGCTCTGGGGG
	Rev	CGATGTCCTCCCCATTGGCCTG
BE_sgRNA2	Fw	AGTTGCCCAGAGTCAAGGAACACGG
	Rev	GACGTAGGTAGTGCTTGAGACCGCC
BE_sgRNA3 or PE_pegRNA2	Fw	TGTGCAGACAAACGGAACTCAACC
	Rev	CCAACATACAGAAGTCAGGAATGCTTGA
PE_pegRNA1	Fw	ATGTGGGCTGCCTAGAAAGG
	Rev	CCCAGCCAAACTTGTCAACC

interest for which there is a guide in the DAP array. PCR should be performed within a week of DNA extraction for best results.

3. Run PCR products through ~1% agarose gel to verify length, then extract DNA using Qiagen DNA cleanup buffers, following manufacturer's protocol.
4. Perform sequencing (Sanger or NGS) of genomic PCR products, using primers designed previously.
5. Verify editing efficiency from sequencing trace data, using an online tool such as EditR [26] if necessary.

4 Notes

1. hCtRNA_FT and hCtRNA_VT plasmid-derived components are compatible with BsaI type IIS restriction enzyme, as these plasmids do not contain an endogenous BsaI recognition site. Depending on how these plasmids may be modified, if a BsaI recognition site is introduced and cannot be removed (including in the sequence of the desired gRNA protospacers), an alternative type IIS restriction enzyme (such as BsmBI) may be substituted.
2. Currently, the DAP array has been validated in HEK293T, K562, Jurkat, and cord-blood-derived hematopoietic stem and progenitor cells (HSPC) [13], as well as HeLa and HepG2 cells [19]. A number of other papers also indicate that DAP array multiplexing technology can work, yielding varying editing efficiencies, in plant applications [27], fungal applications [28], and porcine fetal fibroblasts [29].
3. For some applications, sgRNA scaffold (otherwise known as "tracrRNA") sequences specific to other RNA-guided endonucleases, including SaCas9-based PE [30, 31] or OgeuIscB-based BE [32–34], or other tRNA sequences, may be substituted into the existing hCtRNA_FT and hCtRNA_VT plasmid system. Such substitutions can be performed by standard Golden Gate assembly on these plasmids prior to DAP array construction; these sgRNA scaffold or tRNA fragments of interest can be obtained from plasmids readily available on Addgene, or from commercially synthesized dsDNA fragments.
4. For base editing guides, it is recommended to try protospacers with different protospacer-adjacent motifs (PAMs), if there are multiple available nearby, to target one's edit of interest. Ensure that your edit of interest is encompassed by the active window of your base editor [35]. For prime editing guides, it is recommended to try different protospacers, PBS lengths, and RTT lengths to achieve the greatest possible

editing efficiency [36]. In our experience, guide RNA optimization may require significant development time if the gene target of interest has not been previously edited in other literature, particularly for novel prime editing guide RNAs.

5. (5.1) Previous research indicates that in multiplex PE applications, if more than one (pegRNA+nicking gRNA) module is desired to be included in the DAP array, an interval sequence containing a 7-bp poly-T termination signal between one module's pegRNA extension and the next module's hCtRNA 5′ leader can be included, which may increase editing efficiencies observed at both intended sites. Such a sequence (reported in [13]) can be inserted in this position with additional cloning, after the intended DAP array is constructed. (5.2) Splitting the backbone fragment into 2 separate fragments, with the split site located within the ampicillin resistance (AmpR) gene, may help increase the rate of successful cloning by enabling stricter selection of successfully assembled plasmids.
6. This notation refers to the sequence of the whole circular plasmid, exclusive of nucleotides 651-670. If Addgene's numbering scheme is not conserved upon downloading the plasmid sequence, Nucleotide 1 can be assigned to the T nucleotide at the first position of amino acid P287 in the labeled AmpR gene.
7. Ensure Golden Gate overhangs do not repeat between fragments (this is of particular concern when two gRNAs in the array have similar sequences). More information about Golden Gate overhang optimization can be found online from various sources [37].
8. For short fragments (such as each gRNA-tRNA module from hCtRNA_FT), at least 30 cycles of the [95°C, 20 s; at 60°C, 30 s; at 70°C, 30–60 s/kb] step are recommended. Longer fragments (such as the backbone) will amplify more effectively (i.e., with fewer errors) using fewer cycles (25 cycles is recommended here).
9. The cycle number during the Golden Gate reaction can be increased (to 20–25 cycles) in cases of more complicated (6+ fragment) assemblies. We also recommend, in the 8 uL total DNA fragment mixture, using roughly equimolar amounts of each fragment for the Golden Gate assembly for the best reaction efficiency.
10. When testing a DAP array in a mammalian cell context, it is recommended that, in parallel with the DAP array construction, singleplex gRNA-containing plasmids (such as with Addgene hU6-promoted gRNA expression vector #47511) should also be cloned. During the experiment, these

constructs can serve as a positive control to verify the editing efficiency of the DAP array, to ensure that the latter is editing at a desired efficiency level. Moreover, although the described DAP array construction utilizes the more compact tRNA promoter design, the DAP array itself can alternatively be assembled on a backbone where its expression is driven by a hU6 promoter (or other promoter) if desired.

11. Due to its similar architecture to the PE DAP array described here, we also include an outline for constructing the DAP array for use in the recently developed Minimal Versatile Genetic Perturbation Technology ("mvGPT") system (where prime editing is combined with MPH-mediated gene activation and RNAi-mediated repression) [19].

    Sect. 3.1, Step 1 is modified as follows: In a combined strategy where multiple orthogonally-operating genetic-perturbation systems (such as SpCas9-based prime editing and Cas9-based gene activation [38]) are used, multiple different guide RNA scaffolds will be necessary.

    In this protocol's previous examples, the Addgene plasmid hCtRNA_FT serves as a convenient template for each fragment of the DAP array, as using unique junction regions between hCtRNA_FT-derived fragments eliminates the issue of using primers that may hybridize nonspecifically to the tRNA sequence repeated multiple times in the array. hCtRNA_FT contains a SpCas9-specific gRNA scaffold; however, with additional cloning, it can be modified to contain the SAM activation system's MS2-modified chimeric guide RNA scaffold (which can be obtained from Addgene #118650, nucleotides 437-572 inclusive [39]). Such a plasmid will be referred to below as "hCtRNA_FT(SAM)." This initial cloning is highly recommended prior to DAP array assembly, and reduces the number of fragments that must be included in the final DAP construct.

    Sect. 3.1, Steps 2–3 remain the same. The result of Sect. 3.1, Step 3 will be a combination of components, such as a prime editing protospacer and extension, an activation sgRNA (agRNA) compatible with the SAM system, a shRNA designed for gene repression, etc. agRNAs or shRNAs can be generated manually, or designed using online tools such as the Broad Institute's CRISPick [40] or GPP RNAi design tool [41], respectively.

    Sect. 3.2.1, Step 1 is modified as follows: Using a Golden Gate assembly tool, select the following components for assembly:

    1. Backbone: hCtRNA_VT nucleotides 747-650.

2. Fragment 1: hCtRNA_FT nucleotides 854-1004 (this encompasses the Cas9 sgRNA scaffold and the hCtRNA).
3. Gene repression module:
   1. Junction 1: ~40–50 bp shRNA designed in Sect. 3.1, Step 1.
   2. Fragment 2: hCtRNA_FT nucleotides 854-1004.
4. Gene activation module:
   1. Junction 2: ~10 bp agRNA guide sequence designed in Sect. 3.1, Step 1.
   2. Fragment 3: hCtRNA_FT(SAM) sequence, encompassing the SAM chimeric guide RNA scaffold and following tRNA.
5. Prime editing module:
   1. Junction 3: ~20 bp nicking gRNA sequence designed in Sect. 3.1, Step 1.
   2. Fragment 4: hCtRNA_FT nucleotides 854-1004.
   3. Junction 4: ~20 bp prime editing protospacer.
   4. Fragment 5: hCtRNA_FT nucleotides 854-929 (this encompasses just the Cas9 sgRNA scaffold).
   5. Junction 5: the extension sequence of the prime editing guide RNA.

For as many repeats of any components as one desires to include in the DAP array, include additional fragments and primer-templated junction regions in the desired order (*see* **Note 9**). Sect. 3.2.1, Step 2 onward remains the same.

Acknowledgments

This work was supported by National Institutes of Health grants R01HL173243 and R01HL157714, and National Science Foundation Award 2445285 to X.G. D.A.G. was supported by the National Science Foundation Graduate Research Fellowship (award DGE-2236662). T.C.D. was supported by the National Institute of Neurological Disorders and Stroke (award 2T32NS091006-11A1).

References

1. Komor AC, Kim YB, Packer MS et al (2016) Programmable editing of a target base in genomic DNA without double-stranded DNA cleavage. Nature 533:420–424. https://doi.org/10.1038/nature17946
2. Anzalone AV, Randolph PB, Davis JR et al (2019) Search-and-replace genome editing without double-strand breaks or donor DNA. Nature 576:149–157. https://doi.org/10.1038/s41586-019-1711-4
3. Xue C, Greene EC (2021) DNA repair pathway choices in CRISPR-Cas9-mediated genome editing. TIG 37:639–656. https://doi.org/10.1016/j.tig.2021.02.008

4. Chen L, Zhang C, Xue R et al (2024) Deep whole-genome analysis of 494 hepatocellular carcinomas. Nature 627:586–593. https://doi.org/10.1038/s41586-024-07054-3
5. Musunuru K, Grandinette SA, Wang X et al (2025) Patient-specific in vivo gene editing to treat a rare genetic disease. N Engl J Med 392:2235–2243. https://doi.org/10.1056/NEJMoa2504747
6. Li B, Sun C, Li J, Gao C (2024) Targeted genome-modification tools and their advanced applications in crop breeding. Nat Rev Genet 25:603–622. https://doi.org/10.1038/s41576-024-00720-2
7. Neugebauer ME, Hsu A, Arbab M et al (2023) Evolution of an adenine base editor into a small, efficient cytosine base editor with low off-target activity. Nat Biotechnol 41:673–685. https://doi.org/10.1038/s41587-022-01533-6
8. Richter MF, Zhao KT, Eton E et al (2020) Phage-assisted evolution of an adenine base editor with improved cas domain compatibility and activity. Nat Biotechnol 38:883–891. https://doi.org/10.1038/s41587-020-0453-z
9. Doman JL, Pandey S, Neugebauer ME et al (2023) Phage-assisted evolution and protein engineering yield compact, efficient prime editors. Cell 186:3983–4002.e26. https://doi.org/10.1016/j.cell.2023.07.039
10. Yan J, Oyler-Castrillo P, Ravisankar P et al (2024) Improving prime editing with an endogenous small RNA-binding protein. Nature 628:639–647. https://doi.org/10.1038/s41586-024-07259-6
11. Uranga M, Aragonés V, Selma S et al (2021) Efficient Cas9 multiplex editing using unspaced sgRNA arrays engineering in a potato virus X vector. Plant J 106:555–565. https://doi.org/10.1111/tpj.15164
12. Wei T, Cheng Q, Min Y-L et al (2020) Systemic nanoparticle delivery of CRISPR-Cas9 ribonucleoproteins for effective tissue specific genome editing. Nat Commun 11:3232. https://doi.org/10.1038/s41467-020-17029-3
13. Yuan Q, Gao X (2022) Multiplex base- and prime-editing with drive-and-process CRISPR arrays. Nat Commun 13:2771. https://doi.org/10.1038/s41467-022-30514-1
14. Knapp DJHF, Michaels YS, Jamilly M et al (2019) Decoupling tRNA promoter and processing activities enables specific Pol-II Cas9 guide RNA expression. Nat Commun 10:1490 https://doi.org/10.1038/s41467-019-09148-3
15. Xie K, Minkenberg B, Yang Y (2015) Boosting CRISPR/Cas9 multiplex editing capability with the endogenous tRNA-processing system. Proc Natl Acad Sci U S A 112:3570–3575. https://doi.org/10.1073/pnas.1420294112
16. Ni P, Zhao Y, Zhou X et al (2023) Efficient and versatile multiplex prime editing in hexaploid wheat. Genome Biol 24:156. https://doi.org/10.1186/s13059-023-02990-1
17. Howard MJ, Liu X, Lim WH et al (2013) RNase P enzymes: divergent scaffolds for a conserved biological reaction. RNA Biol 10:909–914. https://doi.org/10.4161/rna.24513
18. Schiffer S (2002) Assigning a function to a conserved group of proteins: the tRNA 3′-processing enzymes. EMBO J 21:2769–2777. https://doi.org/10.1093/emboj/21.11.2769
19. Yuan Q, Zeng H, Daniel TC et al (2024) Orthogonal and multiplexable genetic perturbations with an engineered prime editor and a diverse RNA array. Nat Commun 15:10868 https://doi.org/10.1038/s41467-024-55134-9
20. Jinek M, Chylinski K, Fonfara I et al (2012) A programmable dual-RNA–guided DNA endonuclease in adaptive bacterial immunity. Science 337:816–821. https://doi.org/10.1126/science.1225829
21. (2025) Benchling [Biology Software]
22. Hwang G-H, Jeong YK, Habib O et al (2021) PE-designer and PE-analyzer: web-based design and analysis tools for CRISPR prime editing. Nucleic Acids Res 49:W499–W504. https://doi.org/10.1093/nar/gkab319
24. Marillonnet S, Werner S (2019) Assembly of complex pathways using type IIs restriction enzymes. In: Santos CNS, Ajikumar PK (Eds) Microbial metabolic engineering. Springer, New York, NY, pp 93–109
23. Anderson MV, Haldrup J, Thomsen EA et al (2021) pegIT - a web-based design tool for prime editing. Nucleic Acids Res 49:W505–W509. https://doi.org/10.1093/nar/gkab427
25. New England Biolabs (2025) NEBridge Golden Gate Assembly Tool [Biology Software]
26. Kluesner MG, Nedveck DA, Lahr WS et al (2018) EditR: a method to quantify base editing from sanger sequencing. CRISPR J 1:239–250. https://doi.org/10.1089/crispr.2018.0014
27. Hao Y, Xu B, Zong W et al (2025) Directional improvement of agronomic traits in salt-tolerant rice by multiplex-genome-editing. JIPB 67:2480–2490. https://doi.org/10.1111/jipb.13926
28. Nie Q, Sun C, Liu S et al (2025) Enzymatic ring contraction for the biosynthesis of sulfur-containing cyclopentachromone. J Am Chem Soc 147:548–556. https://doi.org/10.1021/jacs.4c11906
29. Guo X, Geng L, Jiang C et al (2023) Multiplexed genome engineering for porcine fetal fibroblasts with gRNA–tRNA arrays based on CRISPR/Cas9. Anim Biotechnol

34:4703–4712. https://doi.org/10.1080/10495398.2023.2187402
30. Nguyen Tran MT, Mohd Khalid MKN, Wang Q et al (2020) Engineering domain-inlaid SaCas9 adenine base editors with reduced RNA off-targets and increased on-target DNA editing. Nat Commun 11:4871 https://doi.org/10.1038/s41467-020-18715-y
31. Chen PJ, Hussmann JA, Yan J et al (2021) Enhanced prime editing systems by manipulating cellular determinants of editing outcomes. Cell 184:5635–5652.e29. https://doi.org/10.1016/j.cell.2021.09.018
32. Han D, Xiao Q, Wang Y et al (2023) Development of miniature base editors using engineered IscB nickase. Nat Methods 20:1029–1036. https://doi.org/10.1038/s41592-023-01898-9
33. Xiao Q, Li G, Han D et al (2025) Engineered IscB–ωRNA system with expanded target range for base editing. Nat Chem Biol 21:100–108. https://doi.org/10.1038/s41589-024-01706-1
34. Xue N, Hong D, Zhang D et al (2024) Engineering IscB to develop highly efficient miniature editing tools in mammalian cells and embryos. Mol Cell 84:3128–3140.e4. https://doi.org/10.1016/j.molcel.2024.07.007
35. Jeong YK, Song B, Bae S (2020) Current status and challenges of DNA base editing tools. Mol Ther 28:1938–1952. https://doi.org/10.1016/j.ymthe.2020.07.021
36. Sousa AA, Hemez C, Lei L et al (2024) Systematic optimization of prime editing for the efficient functional correction of CFTR F508del in human airway epithelial cells. Nat Biomed Eng 9:7–21. https://doi.org/10.1038/s41551-024-01233-3
37. Leeson R (2023) High complexity golden gate assembly with NEB's DAD. In: Addgene Blog. https://blog.addgene.org/high-complexity-golden-gate-assembly-with-nebs-dad
38. Konermann S, Brigham MD, Trevino AE et al (2015) Genome-scale transcriptional activation by an engineered CRISPR-Cas9 complex. Nature 517:583–588. https://doi.org/10.1038/nature14136
39. Lu B, Klingbeil O, Tarumoto Y et al (2018) A transcription factor addiction in leukemia imposed by the MLL promoter sequence. Cancer Cell 34:970–981.e8. https://doi.org/10.1016/j.ccell.2018.10.015
40. Sanson KR, Hanna RE, Hegde M et al (2018) Optimized libraries for CRISPR-Cas9 genetic screens with multiple modalities. Nat Commun 9:5416. https://doi.org/10.1038/s41467-018-07901-8
41. Broad Institute (2025) GPP Web Portal [Biology Software]

Check for updates

Chapter 12

Production of Cell Models with Differential Zygosity Using CRISPR Base Editors

Shun Zhang and Basil P. Hubbard

Abstract

CRISPR base editors have revolutionized the ease with which single-nucleotide gene editing can be performed in cell and organismal models. Unlike conventional CRISPR/Cas9 gene editing, which induces a double-strand DNA break and relies on endogenous homology-directed repair (HDR) or non-homologous end joining (NHEJ) repair, base editors only induce a single-strand nick and exploit the mismatch repair (MMR) system for repair. In many instances, this results in fewer off-target effects and less cytotoxicity. Moreover, because base editors use an alternative repair mechanism, the zygosity of repair is often distinct from that of traditional HDR-based editing. Here, we provide a detailed protocol for performing gene editing using an adenine base editor (ABE) to induce a single A-G transition mutation in a gene of therapeutic relevance. We also illustrate how to sort and isolate cells with differential zygosity state, enabling the production of cell models suited to the study of both dominant and recessive mutations. This workflow can easily be adapted to the use of other variants of base editors (e.g., cytosine base editors) and can be employed in diverse cell lines.

Keywords Genome engineering, Gene editing, CRISPR/Cas9, Base editors, Adenine base editors (ABEs), Cytosine base editors (CBEs), Heterozygous cells, Homozygous cells, Selection of single cells, Cell models

1 Introduction

Next-generation CRISPR/Cas9 DNA base editors have expanded the variety of mutations and genetic sequences that can be efficiently incorporated into a cell line or organismal model, revolutionizing the study of genotype-phenotype in both targeted and screening contexts [1–3]. Rather than inducing a double-strand DNA break and integrating repair information via homology directed repair (HDR) in the presence of a single-strand oligodeoxynucleotide (ssODN) template like the original CRISPR system [4], base editors fuse nickase variants of Cas9 (nCas9) to enzymes that carry out deamination of adenine (adenine base editors; ABEs) or cytosine (cytosine base editors; CBEs) bases in the vicinity of the target site, which ultimately

Huan Peng and S. Brian Kim (eds.), *Biological Macromolecule Engineering Technologies*,
Methods in Molecular Biology, vol. 3032, https://doi.org/10.1007/978-1-0716-5257-2_12,

results in gene conversion [1–3]. Thus, base editors are ideally suited to the correction of one or a few adjacent bases, such as disease causing single-nucleotide variants (SNVs) [5]. Key benefits of the system include high efficiency and comparably low off-target effects, although modifications to bases outside of the target window can occur, an effect termed bystander editing [6]. Further, modification of cellular RNAs and Cas9-independent off-target effects have been reported but addressed with newer variants [7, 8]. Because base editors do not induce double-strand breaks, p53 activation is attenuated [9], and the chance of deletion or recombination events at the target locus is minimized [10].

DNA base editors may be divided into two main classes, CBEs and ABEs [11]. Typically, CBEs use nickase Cas9 (nCas9) tethered to cytosine deaminase enzymes such as APOBEC1 to carry out C•G→T•A editing in cells [11]. Briefly, nCas9 is recruited by its guide RNA to a target DNA site, and recruits the attached cytidine deaminase with it [11]. Next, the deaminating enzyme targets cytosine bases within a window of ~5 nucleotides, approximately 15 bp upstream of the CRISPR protospacer adjacent motif (PAM), converting them to uracil [1, 11]. Several versions of CBEs have a uracil glycosylase inhibitor (UGI) that is added to the base editor architecture to inhibit the base excision repair enzyme uracil DNA glycosylase (UDG), which induces reversion of the edit during repair [11]. After deamination, the nick generated by nCas9 stimulates mismatch repair (MMR) using the target strand as the repair template, leading to gene conversion [1, 12]. Alternatively, gene conversion can occur via DNA replication [1]. Similarly, as shown in Fig. 1a, ABEs use an nCas9 fused to an adenine deaminase enzyme variant such as TadA to carry out A•T→G•C replacements within the window of positions 4–9, relative to the PAM-distal base [2]. Following deamination of adenine bases to inosine, which is much less frequently excised than uracil, MMR-mediated repair or subsequent DNA replication results in nucleotide editing efficiencies of up to 53%, with favorable specificity [2]. Aside from these founding base editor classes, other architectures have been designed to carry out transversion mutations such as C•G→G•C editing, further expanding the available toolkit [13].

In the case of ABEs, a number of improvements have subsequently been made to improve efficiency, expand the scope of the sequences that can be edited, and especially, fine-tune the editing window to reduce bystander effects. For example, the deaminase component of ABE8e was evolved using phage-assisted evolution to contain eight mutations that impart an >500-fold improvement in activity over its predecessor, ABE7.10 [14]. Furthermore, ABEs have been designed to circumvent SpCas9 PAM-restrictions by fusing the deaminating enzyme to variants with alternative or non-existent PAMs [15], or even Cas12 [16]. Finally, a number of protein engineering and evolution efforts have been undertaken to restrict the

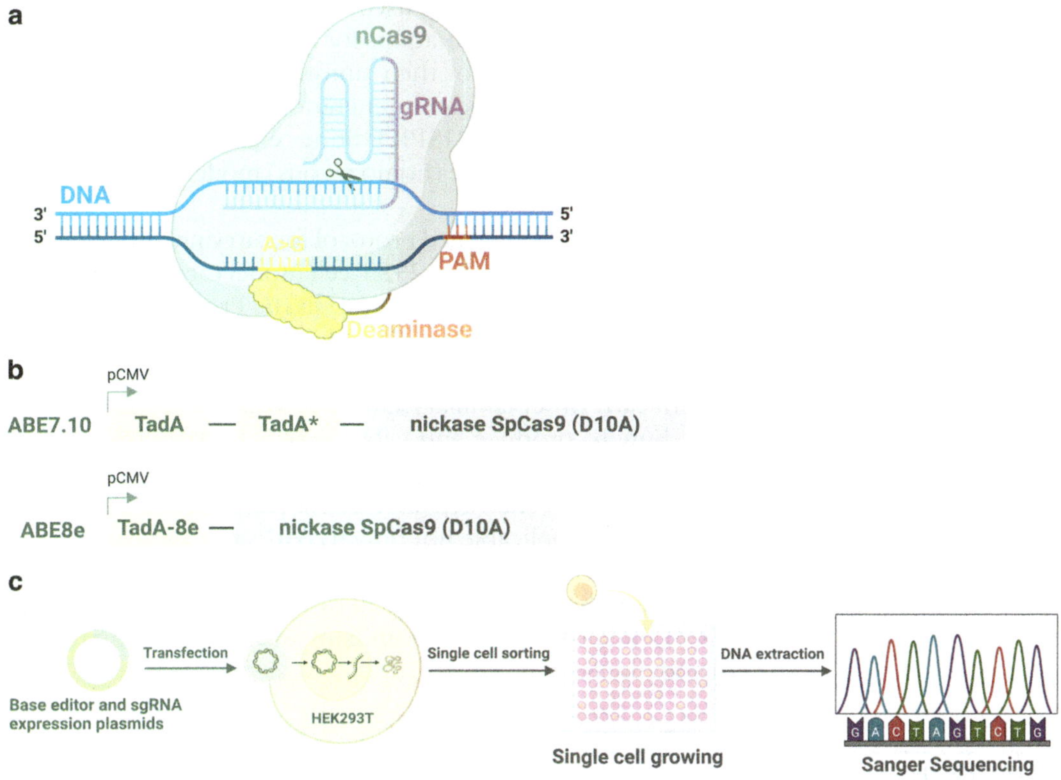

Fig. 1 Diagram of Adenine base editors and schematic of the workflow. (**a**) Diagram outlining the adenine base editing mechanism. A guide RNA (gRNA) directs the Cas9 nickase (nCas9) to a specific DNA target site, resulting in the formation of a single-stranded R-loop. The deaminase enzyme acts on the exposed single-stranded DNA, converting adenine (A) to inosine (I), which is read as guanine (G) during replication or repair. (**b**) Architectures of ABE7.10 and ABE8e. ABE7.10 consists of a wild-type *E. coli* TadA enzyme fused to an evolved TadA* (7.10), which is linked to nCas9. ABE8e contains a single, highly evolved monomeric TadA-8e directly fused to nCas9. (**c**) Overview of the workflow for isolating and genotyping single-cell clones following base editing in HEK293T cells

editing window and thereby reduce bystander edits [17, 18]. These have included changing linker sequences between the protein components [6] and introducing mutations in the deaminating enzyme [19, 20].

Base editors have demonstrated pre-clinical promise in the treatment of both rare monogenic diseases [21] and complex polygenic diseases such as cancer [22]. Moreover, they have been instrumental in developing cell culture and organismal models to study these and many other diseases in the lab, especially in the case of pathogenic SNVs [23]. For example, they have been used to generate mouse models with clinically relevant mutations in the *Ar* and *Hoxd13* genes, with microinjection conversion frequencies up to 100% [24], as well as to generate a monkey model of Hutchinson-Gilford progeria [25].

Further, base editors have been used to rapidly develop cells bearing human cancer-associated mutations for high-throughput functional evaluation [26]. Because of their unique repair mechanism, base editors allow for more efficient generation of heterozygous mutations than conventional CRISPR editing, enabling both dominant (heterozygous) and recessive (homozygous) models to be generated [26, 27].

Here, we provide a detailed protocol for carrying out ABE-mediated base editing to generate a 293T cell model carrying an M146V substitution (from an A→G mutation) in the Presenilin 1 (PSEN1) gene [27] using both ABE7.10 [2] and ABE8e [14, 28]. This mutation is naturally heterozygous in the human population and is associated with early onset familial Alzheimer's disease (FAD) [27]. We illustrate how to produce and validate bulk edited cells and how to isolate both individual homozygous and heterozygous mutant clones using fluorescence-activated cell sorting (FACS). The workflow presented is applicable not only to ABEs but also to CBEs and other base editor variants, and is compatible with the use of diverse cell lines. Together, these approaches can be used to engineer cell models to study the effects of dominant and recessive mutations in health and disease.

2 Materials

2.1 sgRNA Plasmid Design and Cloning

1. An appropriate GFP-tagged base editor plasmid: ABE7.10-GFP (Addgene, #162509), ABE8e-GFP (Addgene, #185910), or BE4max-GFP (Addgene, #112099), etc.
2. Human single guide RNA (sgRNA) expression plasmid backbone BPK1520 (Addgene, #65777). This sgRNA mammalian cell expression backbone can be used with both ABEs and CBEs.
3. BsmBI restriction enzyme (NEB).
4. T4 DNA ligase (NEB).
5. DH5Alpha Competent Cells (NEB).
6. sgRNA target cloning oligonucleotides, e.g., PSEN1 oligonucleotides forward: 5′-CACCGTGTCATGACTATCCTCCTGG-'3 (IDT), PSEN1 oligonucleotides reverse: 5′-AAACCCAGGAGGATAGTCACGACAC-'3 (IDT). The target sequence should be selected taking into account CRISPR PAM requirements.
7. TE buffer (IDT).
8. P10 sterile pipette tips.
9. P200 sterile pipette tips.
10. P1000 sterile pipette tips.

11. UltraPure Agarose (Invitrogen).
12. 50X TAE electrophoresis buffer (2 M Tris, 1 M glacial acetic acid, 50 mM EDTA).
13. QIAquick Gel Extraction Kit (Qiagen).
14. Liquid LB medium.
15. Carbenicillin powder.
16. 100 × 15 mm Petri Dishes (Fisher brand).
17. Nuclease-free water (Invitrogen).
18. Wizard Plus SV Minipreps DNA Purification System (Promega).
19. NanoDrop instrument or similar.
20. 1.5 mL microcentrifuge tubes.
21. Tabletop centrifuge.

2.2 Mammalian Cell Culture and Transfection

1. Mammalian cell lines: HEK293T cells (ATCC) or similar.
2. Cell culture incubator.
3. Dulbecco's Modified Eagle Medium (DMEM) (1X) (Gibco).
4. Fetal Bovine Serum (Gibco).
5. Penicillin-Streptomycin (Gibco).
6. GlutaMAX™-I Supplement (Gibco).
7. TrypLE Express (Gibco).
8. Hemocytometer.
9. 100 × 20 mm TC-Treated Culture Dish (Corning).
10. 24-Well cell culture plate (Falcon).
11. 96-Well cell culture plate (Falcon).
12. PBS, pH 7.4 (1X) (Gibco).
13. Lipofectamine™ 3000 Reagent (Thermo Fisher Scientific).
14. Opti-MEM I Reduced-Serum Medium (1X) (Gibco).

2.3 Cell Sorting and Isolation

1. 40 μm cell strainer (Fisher brand).
2. DRAQ7™ Dye (Invitrogen).
3. 12 × 75 mm Centrifuge and Test Tubes (5 mL) (Falcon).
4. Cell sorter, e.g., BD FACSymphony™ S6 SE.

2.4 Genomic DNA Extraction, PCR, and Analysis

1. QuickExtract DNA Extraction Solution (Biosearch Technologies).
2. Q5 Hot Start High-Fidelity 2X Master Mix (NEB).
3. 8-strip PCR tubes.
4. PCR thermocycler.

3 Methods

This protocol was adapted from previous studies that employed ABE7.10 [2] and ABE8e [14] (Fig. 1b) for gene editing. While we illustrate base editing of a single A-G transition in the *PSEN1* gene in 293T cells, this workflow, depicted in the scheme in Fig. 1c, can be adapted (*see* **Notes**) for the use of other base editors against different genes and in different cells.

3.1 sgRNA Plasmid Design and Cloning

1. Analyze the genomic region of the target gene (e.g., *PSEN1*). Target genomic DNA sequence retrieval can be accomplished using the NCBI Nucleotide database tool (https://www.ncbi.nlm.nih.gov/nuccore/) by searching for the gene name. Once the sequence has been retrieved, make sure to verify the appropriate genome build (e.g., hg38 for human in this case) and gene coordinates to prevent mismatches. Example *PSEN1* nucleotide information can be found here: https://www.ncbi.nlm.nih.gov/nuccore/NC_000014.9?report=genbank&from=73136417&to=73223691 (*see* **Note 1**).
2. Find potential PAM sites (e.g., 5′-NGG-'3 for SpCas9) within ~30 nucleotides upstream or downstream from the target base(s) to be edited by searching both the forward and reverse strands for the appropriate PAM motifs. For each PAM sequence identified, take note of the upstream 20-nt sequence on the 5′ side. These are potential spacer sequences that will constitute the targeting portion of the single guide RNAs (sgRNAs).
3. Record genomic positions, strand orientations, and spacer sequences for each candidate site.
4. Examine each candidate spacer sequence to see where the target base to be edited stands in relation to the PAM. For effective ABE or CBE editing, the target base(s) must occur within the editing window, which typically consists of the PAM-distal positions 4–9 in the spacer sequence (position 1 = the most PAM-distal base of the spacer sequence). Ideally, sites should be selected such that the target adenine/cytosine base resides centrally within this window (e.g., base positions 4–6 usually yield maximum activity). For *PSEN1*, the targeted adenine is located in position 5 (5′-TGTC**A**$_5$TGACTATCCTCCTGG-'3), followed by the TGG PAM motif for SpCas9. The same criteria can be followed for choosing sites for cytosine editing.
5. If multiple candidate sgRNAs are identified, prioritize those that only have a single target base or fewer nearby untargeted adenines within the editing window to reduce bystander edits. Bystander edits can be circumvented in the initial design by

selecting spacer sequences with individual C or A target bases in the editing window (*see* **Note 2**).

6. After selecting a spacer sequence, clone it into the sgRNA expression construct. To do this, single-strand DNA (ssDNA) oligonucleotides must be ordered as follows (examples for *PSEN1*): PSEN1 forward: 5′-**CACC**G*TGTCATGACTATCCTCCTGG*-'3 and PSEN1 reverse: 5′-**AAAC***CCAGGAGGATAGTCACGACA*C-'3. Letters in bold correspond to sequences that are complementary to the enzyme-digested sgRNA cloning vector, which are necessary for restriction enzyme cloning, while the italicized sequence corresponds to the 20-nt spacer sequence (*see* **Note 3**).
7. Prepare annealed oligonucleotides for T4 ligation by first diluting the above forward and reverse oligonucleotides (e.g., *PSEN1*) to 10 μM with TE buffer. Mix 5 μL of forward and 5 μL of reverse oligos with 10 μL nuclease-free water in a 0.2 mL PCR tube. Move the mixture to a thermocycler and heat at 95°C for 5 min and then slowly cool to 25°C over a 1 h period. The annealed DNA can be stored at −20°C until use.
8. Digest 2 μg of human sgRNA expression plasmid backbone BPK1520 with BsmBI enzyme according to the manufacturer's instructions for 1 h. Load the reactions onto a 1.5% TAE agarose gel (with ethidium bromide stain or similar) and carry out electrophoresis of the product at 110 V for 1 h, alongside a DNA ladder. This will serve to separate out the cleaved fragment of ~25 bp from the main fragment. Using a UV imager and cutting tool (e.g., razor blade), collect the digested plasmid, which should run at ~2.2 kb in size.
9. Extract the digested plasmid from the gel using the QIAquick Gel Extraction Kit, following the instructions in the associated product manual. Elute the DNA using 25 μL of elution buffer. Quantify the plasmid concentration using a Nanodrop spectrophotometer or similar (Absorbance at 260 nm). Typically, 2 μg of BPK1520 input plasmid will yield a product concentration of around 30–80 ng/μL. Increase the amount of digested plasmid to 4 μg if the product yield is below this concentration.
10. Prepare T4 DNA ligation reactions with the purified BPK1520 vector: 2 μL of 10X T4 DNA Ligase Buffer, 2 μL of T4 DNA Ligase, 100 ng of purified BPK1520 vector, and 1 μL of 2.5 μM annealed spacer oligonucleotides (e.g., *PSEN1*) in a final volume of 20 μL of nuclease-free water. Incubate the reactions at room temperature for 1 h. A control reaction can be prepared as above but without the annealed oligos, to monitor the intact empty vector.

11. Transform the T4 DNA ligation mixture into DH5Alpha Competent Cells or similar cells, according to the manufacturer's instructions. Briefly, add 5 μL of the T4 DNA ligation reaction to 50–100 μL of chemically competent DH5α *E. coli* cells on ice, flick to combine/mix, and incubate on ice (typically 20–30 min). Next, move the reaction to a heated water bath (or similar) and heat shock at 42°C for approximately 45 s to allow for DNA uptake to occur; subsequently, immediately return to ice for 2 min. Finally, add LB medium and culture at 37°C while shaking for 1 h to permit expression of antibiotic resistance genes prior to plating cells on selective agar plates.
12. After transformation, inoculate 100 μL of the recovered culture on LB agar plates containing 50 μg/mL carbenicillin, invert the plate, and incubate at 37°C overnight. The empty vector T4 ligation control should have few to zero colonies growing on it the next day.
13. In parallel, use sterile P200 pipette tips to inoculate base editor plasmids (e.g., ABE7.10, ABE8e, or other CBE plasmids) from agar stabs onto LB agar plates containing 50 μg/mL carbenicillin or another corresponding antibiotic, invert the plates, and incubate at 37°C overnight.
14. For each plasmid preparation, pick two bacterial colonies and inoculate 15-mL culture tubes with 5 mL of liquid LB medium supplemented with 50 μg/mL carbenicillin. Grow the cultures overnight at 37°C in a bacterial shaker at 200 rpm.
15. Extract plasmid DNA using the Wizard Plus SV Miniprep DNA Purification System, following the corresponding kit instructions. Purified DNA concentration should be quantified by NanoDrop or similar (Absorbance at 260 nm) (*see* **Note 4**).
16. Use Sanger sequencing to confirm the correct sequence of the base editor and sgRNA expression plasmids.

3.2 Mammalian Cell Culture and Transfection

1. 24 h prior to transfection, seed HEK293T cells (or desired cell line) in 24-well plates at ~50% confluency and incubate at 37°C with 5% CO_2. Cells are expected to have 70%–80% confluency on the day of transfection. The DMEM media for transfection should be supplemented with 10% FBS and 1% GlutaMAX (*see* **Notes 5 and 6**).
2. The next day, transfect cells simultaneously with 750 ng of ABE/CBE plasmid and 250 ng of the sgRNA expression plasmid using Lipofectamine 3000, according to the manufacturer's protocol. 24 h post-transfection, cells should display strong GFP-positive signals resulting from expression of ABE7.10-GFP or ABE8e-GFP (*See* **Note 7** for tips on ABE and CBE

transfection as well as **Note 8** for alternative approaches for delivering gene editing systems).

3. Monitor cells for viability and GFP expression after 48 h. Typically for HEK293T cells, ~10%–20% cell death is to be expected following Lipofectamine 3000 transfection. However, extensive cell death (>50%) could indicate excessive reagent, poor cell viability/contamination, or over-confluency. Adjusting the seeding confluency and optimizing the Lipofectamine 3000 transfection reagent ratio and plasmid amounts are necessary for specific experiments when observing overwhelming cell death or weak/low GFP signals.

3.3 Cell Sorting and Isolation

1. 72 h post-transfection, remove media, rinse with PBS (without calcium and magnesium), and digest cells with 80 μL TrypLE reagent, according to the manufacturer's instructions. Dilute and neutralize the reaction by the addition of 200 μL of PBS.
2. Pass the cell suspension through a 40 μm cell strainer (Fisher brand) to remove clumps and dead cells, and transfer into 5 mL Centrifuge and Test Tubes (Falcon) to ready them for cell sorting (*see* **Note 9**).
3. Apart from these sample tubes, prepare a separate batch of test tubes by adding 500 μL of PBS to each for bulk population cell collection.
4. In parallel, add 100 μL of fully supplemented DMEM media (10% FBS, 1% GlutaMAX, and 1% Penicillin-Streptomycin) into wells of a 96-well plate using a multi-channel pipette. This plate will be used for single-cell collection (*see* **Note 10**).
5. As shown in Fig. 2, a number of fluorescence-activated cell sorting (FACS) strategies should be implemented to eliminate dead and un-transfected cells. These may include forward/side scatter gates or the use of live/dead cell dyes such as DRAQ7™ Dye (Invitrogen) to remove dead cells and debris, and a GFP gate to select cells that have been transfected and are expressing the GFP-base editor fusion proteins (*see* **Note 11** for details on gating strategies).
6. Before starting single-cell sorting, use a test tube to calibrate the sorter to make sure the cell/liquid can be sorted into the center of each well. Next, sort single live GFP-positive cell clones into single wells of a 96-well plate. Transfer the plate back into an incubator and incubate at 37°C with 5% CO_2 (*see* **Note 12**).
7. Following single-cell sorting, perform bulk population sorting. In this case, sort 15,000–20,000 GFP-positive cells from ABE7.10 or ABE8e-treated samples. This will be used to assess the bulk population editing efficiency.

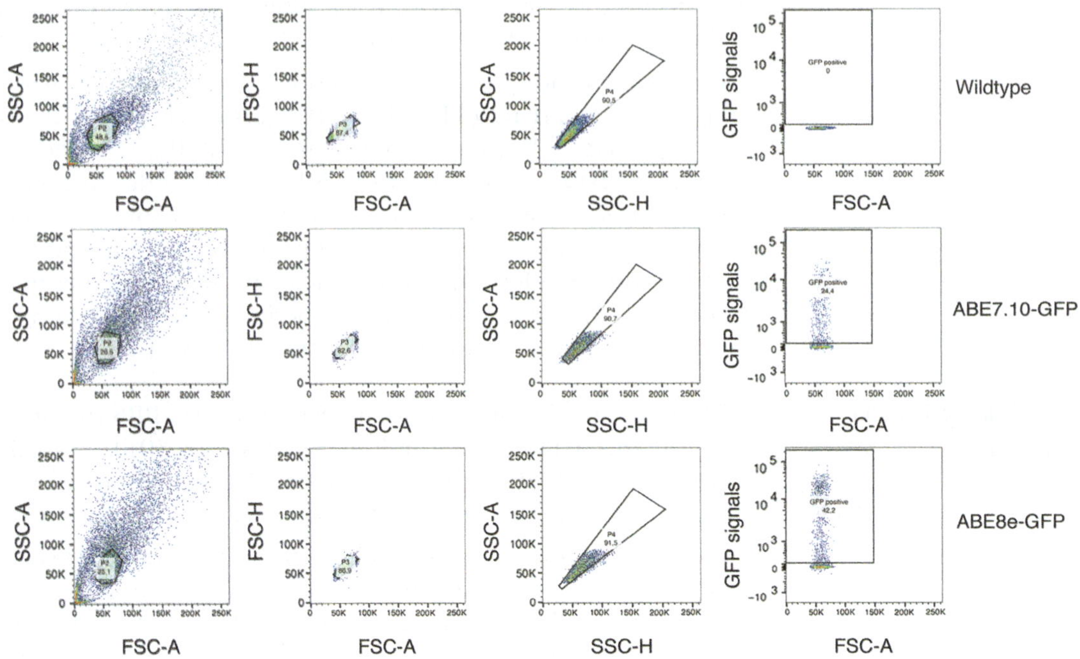

Fig. 2 Example plots outlining FACS gating strategies. FACS plots showing selection of GFP+ cells from the subset of live cells. The top panels correspond to Wild-type (untreated) cells, while the middle panels correspond to ABE7.10-GFP-treated cells, and the bottom panels correspond to ABE8e-GFP-treated HEK293T cells. The gating criteria are consistent throughout all groups and respective replicates

3.4 Genomic DNA Extraction, PCR, and Analysis

1. Extract genomic DNA from the bulk cell population and the individual single cell clones using 50 μL of QuickExtract DNA Extraction Solution per sample, according to the manufacturer's instructions. Quantify the genomic DNA (gDNA) using a Nanodrop or similar instrument (Absorbance at 260 nm).
2. Design primers for amplification of the genomic region of interest from gDNA. To do this, design forward and reverse primers with standard parameters such that the total length of the corresponding product is ~500–1000 bp in length. Ensure that the reference sequence used for primer design matches the genomic sequence of the cells being used. Ideally, the target base(s) being edited should fall within the middle of this product. To assist with primer design, the Primer-BLAST tool from NCBI may be used (https://www.ncbi.nlm.nih.gov/tools/primer-blast/). To ensure that annealing temperature can be optimized and non-specific PCR products can be reduced/eliminated, designing 2–3 pairs of primers is suggested. Example primers for amplifying the region of interest in the *PSEN1* gene are as follows: PSEN1Fwd - 5′-GGAGGTGGTAATGTGGTTGGT-′3, and PSEN1Rev - 5′-GGCTACCAGTTTAGGCAAGGT-′3.

3. Perform PCR to amplify the extracted gDNA. For each PCR reaction, mix 10 μL of Q5 Hot Start High-Fidelity 2X Master Mix, 1 μL of forward and reverse primers (10 μM), 100 ng of gDNA, and complete the volume to 20 μL with Nuclease-free water. Run the PCR program according to the Q5 Hot Start High-Fidelity Polymerase protocol.
4. Perform Sanger sequencing of each sample. To do this, either the forward or reverse primer from the amplification reaction above can be used, or a new primer can be designed. If a new primer is made, ensure that it is positioned at least 50–100 bp upstream of the targeted edit to avoid background noise. For the sequencing of *PSEN1* in HEK293T cells, the results of which are shown in Fig. 3, the PSEN1Fwd primer was used.
5. Analyze the data from the Sanger sequencing using the EditR [29] algorithm to obtain an estimate of bulk population editing efficiency, as well as single clone zygosity distribution (wild-type, homozygous mutant, heterozygous mutant) as shown in Fig. 4. EditR [29] is a tool for quantifying base editing efficiency for target sites, which requires uploading the target reference sequence and the Sanger Sequencing file to the software website. It can be used for quantifying both ABE and CBE editing results. *See* **Note 13** for tips on how to improve low editing efficiency.

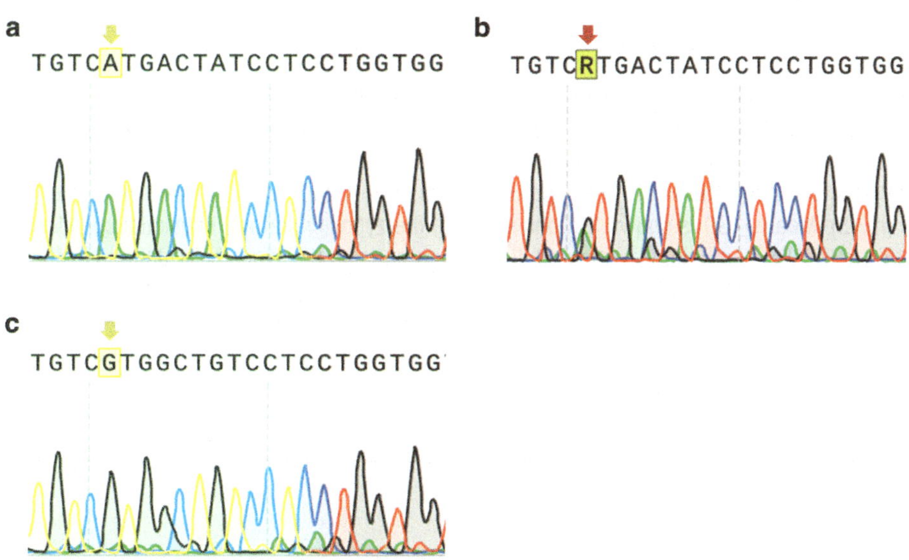

Fig. 3 Example Sanger sequencing traces following base editing of *PSEN1* in HEK293T cells. Sanger traces showing bases surrounding the target editing position in (**a**), wild-type (untreated) cells, (**b**), heterozygous or partially edited cells, and (**c**), homozygous edited cells. The red arrows indicate the target adenine in position five. "R" denotes a mix of bases at the corresponding position in the trace

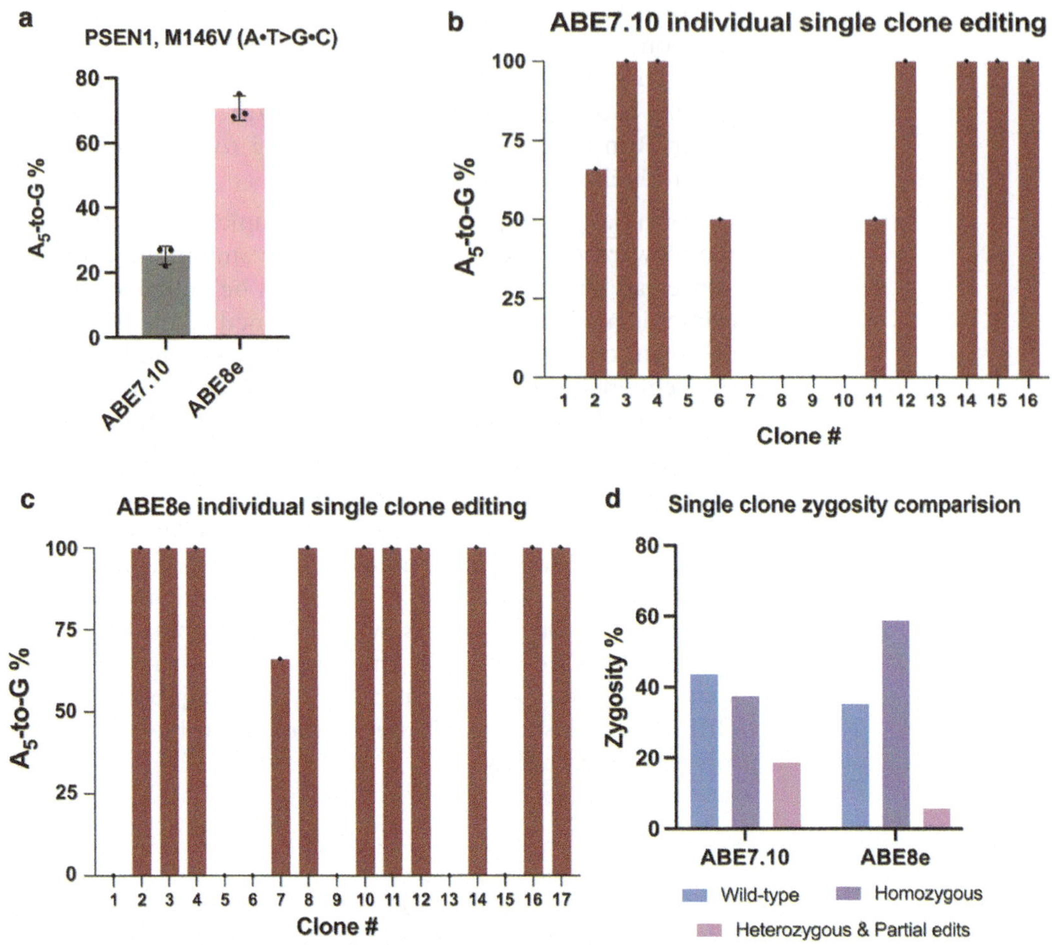

Fig. 4 *PSEN1* (M146V) zygosity comparison following base editing with ABE7.10 or ABE8e in HEK293T cells. (**a**) Bulk population editing efficiency at the target site (*PSEN1* gene - M146V) by ABE7.10 or ABE8e. Genomic DNA was collected and amplified 48 h post-transfection. Efficiency was determined by Sanger sequencing followed by analysis using the EditR [29] algorithm. Results are derived from three independent biological replicates ($n = 3$). Mean ± SD is shown. Graphs showing the zygosity state of individual cell clones following (**b**) ABE7.10-, or (**c**) ABE8e-mediated editing. (**d**) Graph summarizing the zygosity %, defined as the percentage of cells in each category following editing, using data from (**b**) and (**c**). Unedited (wild-type) or homozygous cells are defined as ~0% or ~100% editing, respectively, while heterozygous cells have ~50% allelic editing frequency, and other mosaic allelic edits are categorized as partial edits. For ABE7.10, data is derived from $n = 16$ clones, while for ABE8e, data is derived from $n = 17$ clones

4 Notes

1. In order to ensure that there are no variants or single-nucleotide polymorphisms (SNPs) in the vicinity of the spacer sequence that could disrupt Cas9 binding in the cell line of interest, be sure to sequence the genomic region of interest first and compare it to the reference sequence. To do this,

design primers flanking the spacer sequence (e.g., sgRNA targeting portion) such that the final product is ~500–1000 bp in length. Use Sanger sequencing to read the product sequence and compare it to the reference sequence obtained using the NCBI Nucleotide tool.

2. Low or absent gene editing efficiency can occur if the target region contains methylated DNA or other epigenetic modifications (e.g., histone methylation) [30].
3. The underlined G base is added since the U6 RNA polymerase III promoter used to express the sgRNA prefers a 5′ terminal guanine base as the first base of its transcript (improves efficiency of sgRNA spacers starting with a non-G base) [31].
4. Ensure that the concentration of the ABE plasmid is no lower than 200 ng/μL, as lower concentrations may indicate a sub-optimal prep.
5. It is important not to add Penicillin-Streptomycin to the cell culture media during transfection, as this can negatively interact with transfection reagents.
6. For optimal transfection efficiency, try to optimize the seeding confluency from 40% to 70% based on the growth metrics of the particular cell line being used. Cultures that are either too dense or too sparse on the day of transfection will yield low transfection efficiency or low viability.
7. For HEK293T cells, GFP+ cells should comprise 50%+ of the total cell population following GFP-tagged ABE or CBE transfection. Factors that might influence this metric include cell type, cell viability and density, transfection efficiency, codon optimization of the plasmid, and self-cleaving peptide efficiency (if applicable).
8. There are a number of alternative delivery methods (e.g., viruses, viral-like particles (VLPs), and formats (e.g., ribonucleoprotein (RNP) complexes) that can be used instead of lipid-based transfection of DNA plasmids. A review of these is included in Cavazza et al. [32]. For many hard-to-transfect cell lines, nucleofection can serve as an effective alternative way to deliver gene editing systems. For example, to use this method in HEK293T cells, seed cells 2–3 days before nucleofection and use the TrypLE dissociation reagent to aid in obtaining a cell count. Collect ~0.3×10^6 cells for each reaction, wash twice with PBS, and resuspend the cells in 20 μL of Opti-MEM media or recommended nucleofection media in the provided kit. Each reaction should contain a total of 500 ng–1 μg of ABE8e plasmid and sgRNA. Nucleofection can be carried out in a 16-well Nucleocuvette strip (Lonza) using the program CM-138 (refer to Lonza guidelines) alongside a GFP plasmid control to monitor

efficiency. Subsequently, transfer cells to 12-well plates with pre-warmed media and incubate at 37°C and 5% CO_2 for 48 h before genomic DNA isolation. Cell numbers per reaction, amount of plasmid, and the nucleofection program used will vary for each cell line.

9. DRAQ7™ Dye (Thermo) is a far-red fluorescent dye that can be used to distinguish between live and dead/permeabilized cells. It is compatible with GFP and can be used in conjugation with FACS to quickly exclude non-viable cells from the population.

10. If low clonal survival is anticipated for the cell type being used, a higher percentage of FBS (e.g., 30%–50%) can be used to support cell viability and growth. Other additives that may improve survival include Rat Tail Collagen Coating Solution (Sigma-Aldrich), which can be applied to coat the bottom of 96-well plates the night before sorting. To do this, 30–40 μL of the 50 μg/mL collagen solution can be added to each well. Ensure that the surface is evenly covered (tilt plate gently) and incubate overnight at 37°C. Rinse with PBS once before use.

11. Forward and side scatter (FSC/SSC) can be used to eliminate dead cells and debris from the entire sample. Plotting forward scatter area versus height can be used to perform FSC doublet discrimination, which eliminates events that indicate cell aggregates. To further guarantee that only single cells are included, apply a second SSC doublet discrimination step using side scatter area versus height. Lastly, define the gate for GFP-positive cells using the wild-type control sample as a guide. Determine the GFP fluorescence threshold to differentiate positive from negative cells, and modify the gate stringency based on your desire to capture a more exclusive or broader population of GFP-positive cells.

12. Colonies are expected to grow from single cells in ~2–3 weeks. Exclude samples where more than one colony is present in a well, or those that don't grow in this timeframe.

13. If the editing efficiency is low, a number of troubleshooting steps can be taken to improve it. First, ensure that the PCR amplification step is working correctly and producing on-target bands that are being analyzed accurately. If this isn't the case, optimize primers and the PCR reaction conditions. If the analysis pipeline is working, consider designing and testing additional sgRNAs in different locations to avoid potential epigenetic issues (e.g., DNA methylation), optimizing the transfection process (e.g., confluency) and method of delivery, increasing the amount of plasmid input or amount of material being delivered, and switching to other cell types as an alternative model.

Acknowledgments

This work was supported by a CIHR grant (PJT-189974) to BPH.

References

1. Komor AC, Kim YB, Packer MS, Zuris JA, Liu DR (2016) Programmable editing of a target base in genomic DNA without double-stranded DNA cleavage. Nature 533:420–424
2. Gaudelli NM et al (2017) Programmable base editing of A*T to G*C in genomic DNA without DNA cleavage. Nature 551:464–471
3. Kerek EM, Cromwell CR, Hubbard BP (2021) Identification of drug resistance genes using a pooled lentiviral CRISPR/Cas9 screening approach. Methods Mol Biol 2381:227–242
4. Cromwell CR, Jovel J, Hubbard BP (2021) Methods for measuring CRISPR/Cas9 DNA cleavage in cells. Methods Mol Biol 2162:197–213
5. Kar B, Castillo SR, Sabharwal A, Clark KJ, Ekker SC (2023) Mitochondrial base editing: recent advances towards therapeutic opportunities. Int J Mol Sci 24(6):5798. https://doi.org/10.3390/ijms24065798
6. Valdez I et al (2025) A streamlined base editor engineering strategy to reduce bystander editing. Nat Commun 16:8115
7. Rees HA, Wilson C, Doman JL, Liu DR (2019) Analysis and minimization of cellular RNA editing by DNA adenine base editors. Sci Adv 5:eaax5717
8. Doman JL, Raguram A, Newby GA, Liu DR (2020) Evaluation and minimization of Cas9-independent off-target DNA editing by cytosine base editors. Nat Biotechnol 38:620–628
9. Haapaniemi E, Botla S, Persson J, Schmierer B, Taipale J (2018) CRISPR-Cas9 genome editing induces a p53-mediated DNA damage response. Nat Med 24:927–930
10. Kosicki M, Tomberg K, Bradley A (2018) Repair of double-strand breaks induced by CRISPR-Cas9 leads to large deletions and complex rearrangements. Nat Biotechnol 36:765–771
11. Porto EM, Komor AC, Slaymaker IM, Yeo GW (2020) Base editing: advances and therapeutic opportunities. Nat Rev Drug Discov 19:839–859
12. Rees HA, Liu DR (2018) Base editing: precision chemistry on the genome and transcriptome of living cells. Nat Rev Genet 19:770–788
13. Kurt IC et al (2021) CRISPR C-to-G base editors for inducing targeted DNA transversions in human cells. Nat Biotechnol 39:41–46
14. Richter MF et al (2020) Phage-assisted evolution of an adenine base editor with improved Cas domain compatibility and activity. Nat Biotechnol 38:883–891
15. Cao X et al (2022) Engineering of near-PAM-less adenine base editor with enhanced editing activity and reduced off-target. Mol Ther Nucleic Acids 28:732–742
16. Chen F et al (2022) Multiplexed base editing through Cas12a variant-mediated cytosine and adenine base editors. Commun Biol 5:1163
17. Neugebauer ME et al (2023) Evolution of an adenine base editor into a small, efficient cytosine base editor with low off-target activity. Nat Biotechnol 41:673–685
18. Chen L et al (2024) Author correction: engineering a precise adenine base editor with minimal bystander editing. Nat Chem Biol 20:1094
19. Shang M et al (2025) A motif preferred adenine base editor with minimal bystander and off-targets editing. Nat Commun 16:9153.
20. Chen L et al (2023) Engineering a precise adenine base editor with minimal bystander editing. Nat Chem Biol 19:101–110
21. Musunuru K et al (2025) Patient-specific in vivo gene editing to treat a rare genetic disease. N Engl J Med 392:2235–2243
22. Wang KC, Zheng T, Hubbard BP (2025) CRISPR/Cas technologies for cancer drug discovery and treatment. Trends Pharmacol Sci 46:437–452
23. Anzalone AV, Koblan LW, Liu DR (2020) Genome editing with CRISPR-Cas nucleases, base editors, transposases and prime editors. Nat Biotechnol 38:824–844
24. Liu Z et al (2018) Efficient generation of mouse models of human diseases via ABE- and BE-mediated base editing. Nat Commun 9:2338
25. Wang F et al (2020) Generation of a Hutchinson-Gilford progeria syndrome monkey model by base editing. Protein Cell 11:809–824
26. Kim Y et al (2022) High-throughput functional evaluation of human cancer-associated mutations using base editors. Nat Biotechnol 40:874–884
27. Paquet D et al (2016) Efficient introduction of specific homozygous and heterozygous mutations using CRISPR/Cas9. Nature 533:125–129
28. Gaudelli NM et al (2020) Directed evolution of adenine base editors with increased activity

and therapeutic application. Nat Biotechnol 38:892–900
29. Kluesner MG et al (2018) EditR: a method to quantify base editing from sanger sequencing. CRISPR J 1:239–250
30. Weiss T et al (2022) Epigenetic features drastically impact CRISPR-Cas9 efficacy in plants. Plant Physiol 190:1153–1164
31. Cong L, Zhang F (2015) Genome engineering using CRISPR-Cas9 system. Methods Mol Biol 1239:197–217
32. Cavazza A et al (2025) Advanced delivery systems for gene editing: a comprehensive review from the GenE-HumDi COST action working group. Mol Ther Nucleic Acids 36:102457

Part IV

Directed Evolution and Selection

Chapter 13

Phage-Assisted Continuous Selection of Bioactive Cyclic Peptides

Linwei Yang and Tina Wang

Abstract

Cyclic peptides (CPs) are a class of molecules that can display various biological activities and have been utilized extensively as a source for therapeutic leads and chemical probes. Recently, a strategy inspired by phage-assisted continuous evolution (PACE) was established for high-throughput discovery of CP protein aggregation inhibitors. This method leverages customizable genetic circuits to link desired activities of CP to expression of M13 phage minor coat protein pIII, allowing enrichment of active CP sequences by serial phage propagations through *E. coli* host cells. Compared with existing strategies, this approach enables rapid cellular selection of bioactive CPs while requiring only standard molecular biology techniques and equipment. In this chapter, we provide a comprehensive protocol for applying this strategy to select CP inhibitors against aggregation-prone proteins of interest. Details and procedures covered in this protocol can also provide a foundation for experimental design to discover CPs possessing other bioactivities.

Keywords Cyclic Peptides, Peptide library, Peptide inhibitor, Protein aggregation, Phage-assisted selection

1 Introduction

Cyclic peptides (CPs) have emerged as an attractive molecular scaffold for therapeutic development [1, 2]. CPs are versatile ligands for different targets, including proteins that lack well-defined binding pockets [3–5]. Because they are composed of amino acids, CPs possess a genetically encodable composition, enabling the application of directed evolution methods to their discovery. In turn, this has inspired numerous selection methods for high-throughput *de novo* discovery of CPs active on disease-associated targets [6–8]. For example, phage display and mRNA display have revolutionized CP discovery by enabling the generation and selection of enormous CP libraries for binders of a wide range of targets [9, 10].

Supplementary Information The online version contains supplementary material available at https://doi.org/ 10.1007/978-1-0716-5257-2_13

Huan Peng and S. Brian Kim (eds.), *Biological Macromolecule Engineering Technologies*,
Methods in Molecular Biology, vol. 3032, https://doi.org/10.1007/978-1-0716-5257-2_13,

Both display-based methods have yielded numerous CP ligands with high affinity to many proteins of interest [11, 12]. Despite their success in finding potent CP binders, display methods heavily rely on binding assays that require purification and immobilization of the target protein in a non-cellular environment [13]. This not only limits the scope of targets available for selection but also makes it a challenge to directly select for biological activities of CPs outside of target binding.

Genetically encoded CP libraries can also be selected in cells [7, 14]. In contrast to display strategies, cellular selections can select for more sophisticated biological activities against challenging targets [13]. For example, CP libraries have been selected for the inhibition of protein-protein interactions, protease activity, and protein aggregation using either genetic or phenotypic selection strategies in cells [15–20]. However, cellular selection strategies reported to date have generally relied upon laborious procedures and/or expensive instrumentation (e.g., cell sorters) [18, 20].

We recently developed a strategy for the selection of CP inhibitors using features of phage-assisted continuous evolution (PACE), which can streamline the procedure and timeline for cellular CP selections [21]. In this system, the gene for CPs generated through the split-intein circular ligation of peptides and proteins (SICLOPPS) method [22] is encoded on an engineered M13 phage in place of *gIII*, which encodes an essential phage protein pIII. Upon phage infection of host *E. coli* cells, the CP-encoding gene is delivered into the *E. coli* cytosol and used to produce CPs. Critically, the host *E. coli* carries plasmids encoding a user-defined genetic circuit that links the desired biological activity of CP to expression of *gIII*. Therefore, phage encoding active CP sequences can obtain pIII after host cell infection, allowing them to produce progeny phage. This strategy allows rapid enrichment of active CP sequences over inactive ones and enables identification of CP inhibitors from large libraries ($>10^6$ variants) within days.

This approach, compared with existing cellular selection strategies, is more accessible to general research laboratories. As it is derived from PACE, this platform also has the potential to couple with in situ mutagenesis to allow continuous evolution of bioactive CPs [23]. In the original report of this platform, head-to-tail CPs composed of canonical amino acids were selected. Nevertheless, incorporation of ncAAs [24] or other genetically encodable CP scaffolds for greater chemical diversity is in principle compatible with the approach. Moreover, due to its modularity, this system can be equipped with different genetic circuits to select for a broad spectrum of biological activities.

In this chapter, we provide a roadmap for phage-assisted continuous selection of bioactive CPs, based on a recent proof-of-concept report presenting successful selection of CP inhibitors against protein aggregation (Fig. 1). This protocol covers the generation of phage-encoded CP libraries, the selection procedure, and characterization of candidate CPs.

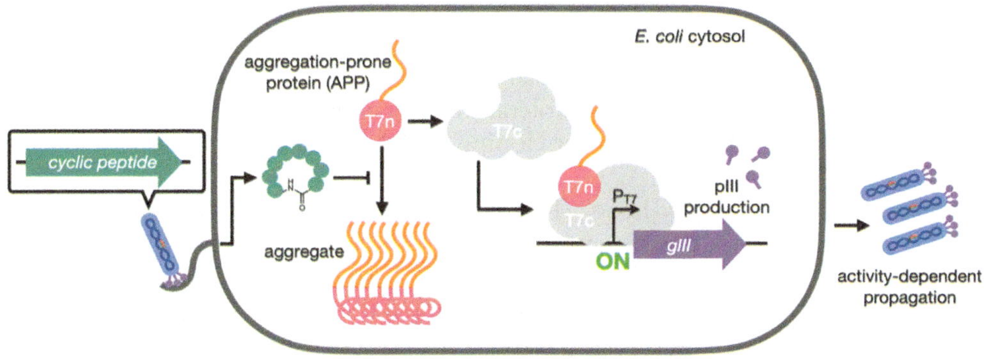

Fig. 1 Overview of the selection strategy. In this selection, the N-terminal half of a split T7 RNA polymerase (T7n) is fused to a target aggregation-prone protein (APP), which aggregates and sequesters T7n in the absence of active CP inhibitors, preventing interaction with the C-terminal half (T7c) to reconstitute a functional T7 RNA polymerase. Because the T7 promoter is used to control the expression of *gIII*, only phages carrying aggregation inhibitors can propagate through the host cells encoding this selection. All components are supplemented *in trans* by plasmids in *E. coli* host cells except the cyclic peptide, which is encoded on and delivered by phage via infection

2 Materials

2.1 E. coli *Strains*

1. *E. coli* strain S2060 (*F' proA+B+ Δ(lacIZY) zzf::Tn10 lacIQ1 PN25-tetR luxCDE Psp(AR2) lacZ luxR Plux groESL / endA1 recA1 galE15 galK16 nupG rpsL ΔlacIZYA araD139 Δ(ara,leu)7697 mcrA Δ(mrr-hsdRMS-mcrBC) proBA::pir116 araE201 ΔrpoZ Δflu ΔcsgABCDEFG ΔpgaC λ–*) [25]. It is used for phage propagation assays and selections.
2. *E. coli* strain S2208 (strain S2060 transformed with plasmid pJC175e) [26]. It is used for plaque assays to determine phage titers and for propagating phage without selection pressure.
3. *E. coli* strain SS320. It is used for cloning phage libraries.

2.2 *Engineered M13 Bacteriophage*

1. spLY014 [21]. It is an evolved *ΔgIII* M13 bacteriophage carrying a gene encoding for the SICLOPPS construct for CP biosynthesis. This phage was evolved to be more compatible with the selection for identifying CP aggregation inhibitors that are used in this protocol. It is used as a template for cloning phage-encoded CP libraries. The DNA sequence of this and the following phage (spLY006) is provided in the associated Electronic Supplementary Material file.
2. spLY006 [21]. It is an evolved *ΔgIII* M13 bacteriophage carrying a gene encoding for the SICLOPPS construct for CP

biosynthesis. This phage was evolved to be more compatible with the selection for identifying CP aggregation inhibitors that are used in this protocol. Compared with spLY014, this phage can tolerate higher selection pressures. It is used as a template for cloning phage-encoded CP libraries.

2.3 Plasmids

1. pJC175e. This SC101-origin plasmid contains *gIII* under the psp promoter (P_{psp}). It is used to supplement *gIII* for *ΔgIII* phage, enabling them to propagate independently of their activity. The Addgene catalog numbers of this and the following plasmids are provided in the associated Electronic Supplementary Material file.
2. pTW357b. This ColE1-origin plasmid is a plasmid containing *gIII* under a weakened T7 promoter and deactivated C-terminal half of T7 RNAP (T7c R632S) under a constitutive promoter P_{pro1}[27]. It is used for lower-stringency selection of protein aggregation inhibitors.
3. pTW357d5. This ColE1-origin plasmid is an accessory plasmid containing *gIII* under a weakened T7 promoter and further deactivated T7c (R632S/Q649S) under a constitutive promoter P_{pro5}[27]. It is used for higher-stringency selection of protein aggregation inhibitors.
4. pTW357bN. This ColE1-origin plasmid has the same architecture as pTW357b, except that *gIII is* replaced with a negative selection marker *gIII-neg*. It is used for negative selection to remove CPs sequences that have activity on an off-target APP.
5. pLY034b. This SC101-origin plasmid has the same architecture as pJC175e, except that a weaker Ribosome-binding site sequence is used for *gIII*. It is used to supplement *gIII* for activity-independent phage propagation during negative selection.
6. pTW358b. This p15A-origin plasmid is a plasmid containing an *Aβ42 (amyloid-β 42)-N-terminal half of T7 RNAP (T7n)* fusion under the psp promoter (P_{psp}). In this protocol, it is used in negative selection to provide the off-target APP. This plasmid can be used as the vector for cloning other target APP-T7n fusions.
7. pTW358f. This p15A-origin plasmid is a complementary plasmid containing *hIAPP (human islet amyloid polypeptide)-T7n* fusion under Ppsp promoter. In this protocol, it is used in positive selection to provide the target APP. This plasmid can be used as the vector for cloning other target APP-T7n fusions.

2.4 Primers

1. LY0003 (5′-ACAGTTGUGGGCGATCGC-3′). This primer is used together with a degenerate primer (*see* **Note 1**) for cloning CP libraries and for re-cloning candidate cyclic peptide sequences after completion of selections through USER cloning.
2. TW2027 (5′-TAATGGAAACTTCCTCATGAAAAAGTCTT TAG-3′) and DM0027 (5′-GGTAATTGAGCACTAATATC AGAGAG-3′). These primers are used to PCR amplify the SICLOPPS gene that contains the CP-encoding sequence to produce a 1230-bp DNA product for Sanger sequencing.
3. TW2065 (5′-GACTCCCTGCAAGCCTCAG-3′). This primer anneals to an upstream region outside the SICLOPPS gene and is used for Sanger sequencing.

2.5 Media

1. 2xYT media (US Biologicals T9200). It is used for routine bacterial culturing and non-selective propagation of M13 bacteriophage in *E. coli*.
2. OG media (2xYT media supplemented with 25 mM glucose). It is used for cell recovery after electroporation or heat shock during transformation.
3. Davis rich media (DRM) (US Biologicals CS050H-001 and CS050H-003). It is used for phage propagation assays and phage selections.
4. LB agar (US Biologicals L1515).

2.6 Consumables and Kits

1. E.Z.N.A. Cycle Pure Kit (Omega Bio-tek D6492).
2. E.Z.N.A. Gel Extraction Kit (Omega Bio-tek D2500).
3. PCR strips.
4. 1.5 mL microcentrifuge tubes.
5. 15 mL and 50 mL conical tubes.
6. 1.2 mL library tube strips.
7. Syringes and hypodermic needles.
8. 0.22 μm syringe filters.
9. Plastic petri dishes and X-plate petri dishes.
10. 12 mL culture tubes.
11. 1 mm gap electroporation cuvettes.

2.7 Equipment

1. Agarose gel electrophoresis system.
2. Thermocycler.
3. Gel imaging system.
4. Gel illuminator for gel extractions.
5. Refrigerated centrifuge.
6. Bacterial incubator.

7. Incubated shaker.
8. Water bath.
9. Baffled Erlenmeyer flasks.
10. Electroporator (Eppendorf 4309000027).

2.8 Reagents and Buffers

1. Maintenance antibiotics, including streptomycin, tetracycline, ampicillin, kanamycin, and spectinomycin (Gold Biotechnology). Their stock (1000X) concentrations are 50 mg/mL, 10 mg/mL, 50 mg/mL, 50 mg/mL, 100 mg/mL, respectively.
2. HyClone nuclease-free water (Omega Bio-tek 63275DY85).
3. Milli-Q (MQ) water.
4. 5-bromo-4-chloro-3-indolyl β-D-galactopyranoside (X-gal) (Gold Biotechnology X4281C). It is used as a 2% w/v solution in DMSO. Store the solution in the dark.
5. Agarose (Gold Biotechnology A-201-100).
6. Ethidium bromide (EtBr) (bio-world L25022004EPEP).
7. 50X Tris-acetate-EDTA (TAE) buffer (Omega Bio-tek AC10089).
8. Glycerol
9. Q5U Hot Start High-Fidelity DNA polymerase (NEB M0515). It is used for PCRs with uracil-containing primers to generate fragments for USER cloning.
10. Q5 High-Fidelity DNA polymerase (NEB M0491). It is used for PCRs to generate templates for Sanger Sequencing.
11. 10X rCutSmart buffer (NEB B6004).
12. USER enzyme (NEB M5505). It is used for USER cloning.
13. DpnI (NEB R0176).

3 Methods

3.1 Generating Phage-Encoded CP Libraries by USER Cloning

3.1.1 Generation of CP Library DNA

A) *PCR*

1. Perform PCR with Q5U DNA polymerase using the reverse primer LY0003 and a degenerate forward primer (*see* **Note 1**) with an evolved *ΔgIII* M13 phage (spLY014 or spLY006) as the template (*see* **Note 2**). The 10^5 pfu phage template is recommended for a 100 μL PCR reaction to reduce the amount of template present in the final

library. Optimize the PCR conditions (especially annealing temperature and extension time) to maximize yield and minimize off-target products. Scale up the reaction to produce sufficient PCR product for the following steps.

2. Prepare an EtBr-containing 0.8% agarose TAE gel for gel extraction. Assemble the electrophoresis system and fill the tank with TAE buffer to cover the agarose gel. Load all the PCR products onto the agarose gel. Perform gel electrophoresis at 140 V for 30 min.
3. Image the gel to confirm the size of the desired PCR product.

B) *Gel Extraction*

1. Use a gel extraction kit and follow the manufacturer's instructions to purify the desired PCR product from the agarose gel.
2. Use Nanodrop to determine the concentration of purified DNA. If the concentration is too high, dilute the DNA with nuclease-free water to a final concentration >= 230 ng/μL.

C) *USER Digestion*

1. In a 0.2 mL PCR tube, add 1 μL 10X rCutSmart buffer, 1 μL USER enzyme, 0.5 pmol DNA, and nuclease-free water to a final volume of 10 μL (*see* **Note 3**). Mix well by pipetting up and down.
2. Transfer the PCR tube onto a thermocycler and perform the following cycling protocol:
Incubate at 37°C for 60 min → Incubate at 80°C for 3 min → Ramp down to 12°C at the rate of 0.1°C/s.
3. Take out the PCR tube upon completion of digestion for the next step.

D) *DNA clean up* (*see* **Note 4**)

1. Use a DNA cleanup kit and follow the manufacturer's instructions to purify the product. Pre-warmed nuclease-free water (45–55°C) is recommended to facilitate efficient elution. This eluent is the final CP library DNA ready for transformation.
2. Use Nanodrop to determine the concentration of CP library DNA. A final concentration higher than 100 ng/μL is ideal for the following steps (*see* **Note 5**).

3.1.2 Preparation of Electrocompetent Cells

Day 1

1. Transform pJC175e into *E. coli* strain SS320. Plate the transformed cells on an LB agar plate with 50 μg/mL ampicillin. Incubate at 37°C overnight (*see* **Note 6**).

Day 2

2. In the evening, pick a single colony of *E. coli* SS320 transformed with pJC175e from the agar plate and inoculate it into 2 mL 2xYT media with 50 µg/mL ampicillin in a culture tube. Let grow at 37°C overnight with shaking.
3. Prepare and chill in advance 15 mL MQ water and 2 mL 10% glycerol (in MQ water) at 4°C. Additionally, pre-warm 1 mL OG media to 37°C. Those reagents will be used in the following steps.

Day 3

4. In the morning, back dilute 8 µL E. coli SS320/pJC174e saturated overnight culture into 8 mL 2xYT media with 50 µg/mL ampicillin. Let grow at 37°C with shaking until the OD_{600} reaches 0.6 (*see* **Notes 7 and 8**).
5. Transfer all 8 mL of culture into a 15 mL conical tube. Centrifuge at 4000 × *g* at 4°C in a pre-chilled centrifuge for 15 min. Discard the supernatant completely (*see* **Note 9**).
6. Add 8 mL cold MQ water into the 15 mL conical tube and gently resuspend the cell pellet by pipetting up and down.
7. Centrifuge the conical tube at 4000 × *g* and 4°C for 15 min. Discard the supernatant completely.
8. Add 1 mL cold MQ water into the 15 mL conical tube and gently resuspend the cell pellet by pipetting up and down. Then transfer the 1 mL resuspended cell into a pre-chilled 1.5 mL microcentrifuge tube.
9. Centrifuge the microcentrifuge tube at 8000 × *g* and 4°C for 3 min. Discard the supernatant. Resuspend the cell pellet in 1 mL cold MQ water.
10. Repeat the above step (Step 9) once more.
11. Centrifuge the microcentrifuge tube at 8,000 × *g* and 4°C for 3 min. Discard the supernatant. Resuspend the cell pellet in 1 mL cold 10% glycerol.
12. Centrifuge the microcentrifuge tube at 8000 × *g* and 4°C for 5 min. Discard the supernatant. Resuspend the cell pellet in 100 µL cold 10% glycerol. This 100 µL competent cell preparation is now ready for electroporation.

3.1.3 Electroporation and Outgrowth

1. Pre-chill a 1 mm electroporation cuvette on ice.
2. On ice, add 100 ng pre-chilled CP library DNA into the 100 µL competent cell preparation. Mix well by gently swirling the pipette tip.
3. Carefully pipette the 100 µL library DNA-and-competent cell mixture into the electroporation cuvette. Check to ensure that

the mixture has completely settled into the gap between the two metal plates and that there are no air bubbles.

4. Quickly wipe excess moisture from the outside of the cuvette and place the cuvette into an electroporator. Electroporate the cells with a voltage setting of 1800 V (this may need to be adjusted depending on the cuvette path length used). For our electroporator, a time constant between 5.4 and 5.8 ms indicates successful electroporation (*see* **Note 10**).
5. After electroporation, immediately add 1 mL pre-warmed OG media (37°C) into the cuvette, pipette up and down twice to mix well, and transfer all liquid into a culture tube.
6. Let the cells recover at 37°C with shaking for 1 h.

3.1.4 Phage Harvesting

1. Upon the conclusion of the 1-h outgrowth, transfer the culture into a microcentrifuge tube.
2. Centrifuge at 8,000 × *g* for 2 min. Discard the cell pellet and keep the supernatant, which contains phage. The phage supernatant should be stored at 4°C.

3.1.5 Plaque Assay

Plaque assay is used to determine the titer of phage. This protocol describes the full procedure for the plaque assay. Steps here regarding the preparation of mid-log *E. coli* S2208 culture can be done simultaneously with previous steps to accelerate the library cloning process.

Day 1

1. Streak *E. coli* strain S2208 from glycerol stock onto an LB agar plate with 50 μg/mL ampicillin. Incubate at 37°C overnight.

Day 2

2. In the evening, pick a single colony of *E. coli* S2208 and use it to inoculate 2 mL of 2xYT media with 50 μg/mL ampicillin in a culture tube. Let grow at 37°C overnight with shaking.

Day 3

3. In the morning, dilute 80 μL overnight *E. coli* S2208 culture into 8 mL 2xYT media with 50 μg/mL ampicillin.
4. Incubate at 37°C with shaking until the OD_{600} reaches 0.4.
5. Remove the mid-log culture from the incubator. This culture can be stored at 4°C for up to a week for use in plaque assays.
6. Microwave a bottle of prepared LB agar to melt it completely (use lower power to prevent the agar from boiling over). Then immediately add 3 mL of the agar to 7 mL of 2xYT media in a 15 mL conical tube. Mix well and keep

the agar/2xYT media mixture in a water bath (55–65°C) to keep it molten for later use. This is the top agar.

7. Add 2 mL of the melted LB agar to cover each quadrant of an X-plate Petri dish. This is the bottom agar.
8. Add 100 μL of the mid-log *E. coli* S2208 culture into each tube of a 4-tube library strip.
9. In another 4-tube library strip, prepare four serial dilutions (normally 10^0, 10^2, 10^4, and 10^6) of the phage supernatant from Sect. 3.1.4. Add 10 μL of each dilution into the library strip containing mid-log *E. coli* S2208 culture.
10. Add 200 μL 2% X-gal (in DMSO) into the 10 mL pre-warmed top agar and mix well.
11. Add 900 μL top agar into each tube of the library strip (from the most diluted phage to the least diluted phage). Mix well by pipetting up and down a few times. Then immediately transfer the top agar onto the bottom agar to evenly cover the quadrant. Repeat until all the top agar in the 4-tube library strip is transferred.
12. Incubate the X-plate petri dish at 37°C overnight.

Day 4

1. Determine the phage titer by counting the number of plaques on a quadrant that has distinct, countable plaques. Calculate the titer using the following formula: titer (in pfu/mL) = (# of plaques in quadrant) × (dilution factor of quadrant) × 100. This titer represents the size of the CP library. This procedure should yield ~10^6–10^7 pfu (plaque-forming units) of phage per transformation (*see* **Note 11**).

3.1.6 Phage Library Expansion

This process increases the volume of the phage library for long-term use and ensures degeneracy of each CP sequence in the library.

Day 1

1. Streak *E. coli* strain S2208 from glycerol stock onto an LB agar plate with 50 μg/mL ampicillin. Incubate at 37°C overnight.

Day 2

2. In the evening, pick a single colony of *E. coli* S2208 and use it to inoculate 2 mL of 2xYT media with 50 μg/mL ampicillin in a culture tube. Let grow at 37°C overnight with shaking.

Day 3

3. In the morning, dilute the overnight *E. coli* S2208 culture 1:100 into 2xYT media with 50 μg/mL ampicillin (*see* **Note 12**).

4. Incubate at 37°C with shaking until the OD_{600} reaches 0.4.
5. Add the harvested phage library to the mid-log *E. coli* S2208 culture. Incubate at 37°C with shaking for 6 h.
6. Centrifuge the *E. coli* culture at 8,000 × *g* for 2 min (or 4000 × *g* for 15 min). Discard the cell pellet and keep the supernatant, which contains phage.
7. Filter the supernatant through a 0.22 µm syringe filter to remove any leftover *E. coli* cells. Store the expanded phage library at 4°C.
8. Determine the titer of the expanded phage library by performing a plaque assay.

3.2 Selection of Phage-Encoded CP Libraries

Different APPs can be used as targets for selection. For demonstration purposes, this protocol focuses on the identification of selective inhibitors for human islet amyloid polypeptide (hIAPP) that exhibit minimal off-target activity for amyloid-β 42 (Aβ42). hIAPP serves as the target in positive selection, whereas Aβ42 serves as the target in negative selection. Since higher selection stringency is required for hIAPP, the accessory plasmid pTW357d5 is used in this protocol for positive selection.

3.2.1 Preparation of Selection Cells

Day 1

1. Transform pTW357d5 and pTW358f into *E. coli* strain S2060. Plate the transformed cells on an LB agar plate with 50 µg/mL spectinomycin and 50 µg/mL kanamycin. Incubate at 37°C overnight. These cells are used for positive selection.
2. Transform pTW357bN, pTW358b, and pLY034b into *E. coli* strain S2060. Plate the transformed cells on an LB agar plate with 50 µg/mL spectinomycin, 50 µg/mL kanamycin, and 50 µg/mL ampicillin. Incubate at 37°C overnight. These cells are used for negative selection.

Day 2

3. Remove agar plates from the incubator. Plates can be stored at 4°C for up to 5 days. Prepare fresh selection cells as needed.

3.2.2 Positive Selection

Day 1

1. Pick a single colony of S2060/pTW375d5/pTW358f cells and inoculate it into 2 mL 2xYT media with 50 µg/mL spectinomycin and 50 µg/mL kanamycin in a culture tube. Let grow at 37°C overnight with shaking.

Day 2

2. In the morning, back dilute the saturated overnight culture 1:100 into DRM media with 50 μg/mL spectinomycin and 50 μg/mL kanamycin. Let grow at 37°C with shaking until the OD_{600} reaches 0.5 (*see* **Note 13**).
3. Infect the mid-log culture with the expanded CP phage library at an input titer of 10^6 pfu/mL. Incubate at 37°C with shaking for 6 hours. This is the first round of selection.
4. Centrifuge the first-round selection culture at 8000×*g* for 2 min (or 4000×*g* for 15 min). Discard the cell pellet and keep the supernatant, which contains phage. Label this as *P1* (passage 1) and store at 4°C.
5. Perform a plaque assay to determine the titer of *P1* phage.
6. Repeat Sect. 3.2.2, Step 1 to prepare an overnight culture for the next round of selection.

Day 3

7. Record the titer of *P1* phage (*see* **Note 14**). The volume of cell culture needed for the second round of selection depends on the titer of *P1* phage (*see* **Note 15**).
8. Back dilute overnight culture into (X+2) mL DRM media with 50 μg/mL spectinomycin and 50 μg/mL kanamycin at a 1:100 dilution factor (where X is the volume determined in Sect. 3.2.2, Step 7). Let grow at 37°C with shaking until the OD_{600} reaches 0.5.
9. Transfer 2 mL of the mid-log culture into a separate culture tube for propagation benchmarking.
10. For the second round of selection, infect the remaining X mL of culture with *P1* phage at a 1:10 dilution factor (e.g., 500 μL *P1* phage into 5 mL culture). Incubate at 37°C with shaking for 6 h.
11. For the benchmark propagation, infect the reserved 2 mL culture with 10^6 pfu/mL *P1* phage. Incubate at 37°C with shaking for 6 h.
12. Centrifuge both the selection culture and benchmark propagation culture at 8000 × *g* for 2 min (or 4,000 × *g* for 15 min). Discard the cell pellet and keep the supernatants containing the phage. Label the supernatant of the selection culture as *P2* (passage 2) and store at 4°C.
13. Perform a plaque assay to determine the titer of both the *P2* phage and the benchmark propagation phage.
14. Repeat Sect. 3.2.2, Step 1 to prepare an overnight culture for the next round of selection.

Day 4

15. Record the phage titer of the benchmark propagation and calculate fold propagation using the equation:

$$\textit{Fold propagation} = \textit{output titer } (\textit{pfu}/\textit{mL})/\textit{input titer } (\textit{pfu}/\textit{mL})$$

where output titer is the titer of benchmark propagation phage and input titer is 10^6 pfu/mL. Record this fold propagation of *P1* phage. This parameter reflects the normalized activity of the phage population and helps evaluate the outcome of each selection passage (Fig. 2c).

16. Record the titer of *P2* phage. Increase the dilution factor for the third round of selection to 1:100 if *P2* has a > 10-fold higher titer than *P1*; otherwise, maintain a dilution factor of 1:10.

17. Repeat Steps 8–13 from Sect. 3.2.2 Day 3, where *P2* is now used for the third round of selection and benchmark propagation.

Day 5+

18. Continue to repeat the procedure described for Sect. 3.2.2 Day 4 until the selected phage pool reaches the desired level of activity (Fig. 2b, c). A steady increase in phage titer and fold propagation activity should be observed during this process if active sequences exist in the CP phage library (Fig. 2b, c).

19. End positive selection when the selected phage no longer shows increased propagation activity (Fig. 2b, c). Purify phage from all selection passages (i.e., *P1*, *P2*,…) according to Sect. 3.1.5, Step 7, and store at 4°C.

3.2.3 Negative Selection

Negative selection follows the same protocol as positive selection, except the selection cells are *E. coli* S2060 transformed with pTW357bN, pTW358b, and pLY034b. Here, a few details regarding negative selection are discussed. Refer to Sect. 3.2.2 for the experimental procedure.

1. Negative selection is usually performed following positive selection to remove false positives or off-target sequences. However, negative selection can be incorporated at any time during continuous selections. It is up to the researcher to decide when negative selection is necessary.

2. Similar to positive selection, it is critical to record the titer and activity of each passage throughout negative selection to help adjust the dilution factor and determine when the selection should be ended.

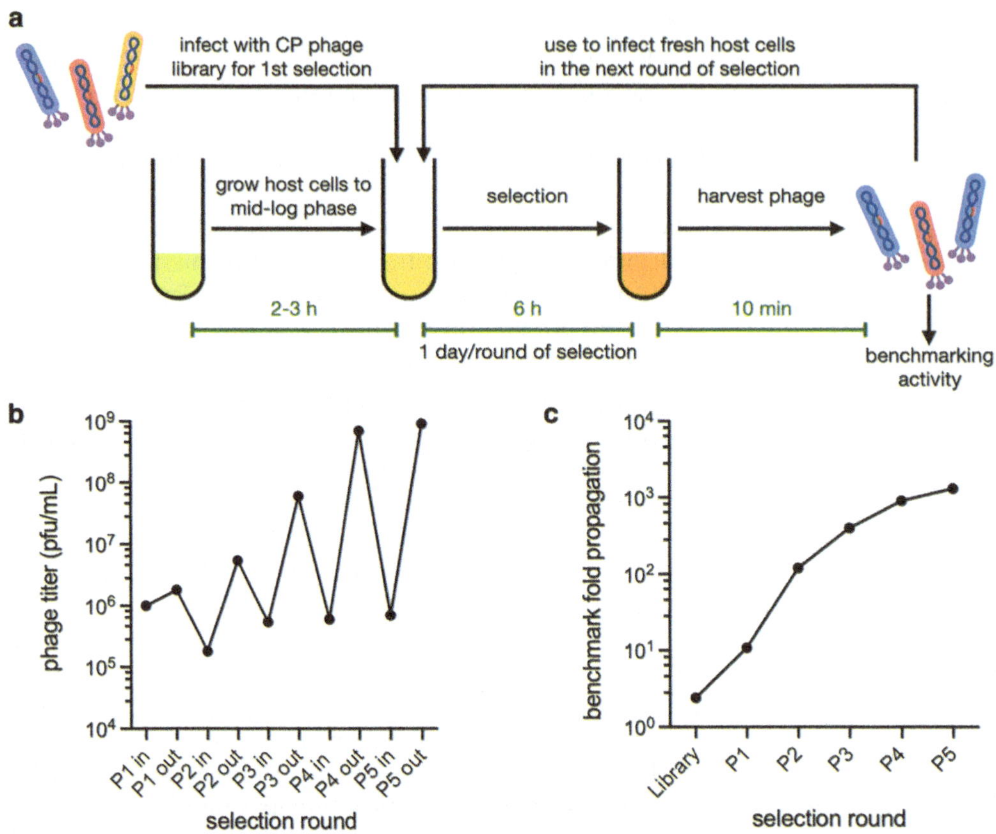

Fig. 2 Workflow of continuous selections and expected outcome. (**a**) An overview of the workflow of phage-assisted continuous selections. As the starting population, the CP-encoding phage library is used to infect the host cells for the first round of selection. Subsequent post-selection phage population is used to infect fresh host cells for the next round of selection. This process is repeated until an optimal activity is reached. (**b**) An example (simulated data) of expected phage titers across the selections. A steady increase of phage titer should be observed with later phage titers reaching the plateau (~$10^{9\text{-}10}$ pfu/mL) regardless of the dilution factor. (**c**) An example (simulated data) of expected fold propagation numbers across the selections. A steady increase of activity should be observed with low initial activity and final activities reaching the plateau (~10^3 fold)

3.3 Characterization of Cyclic Peptide Protein Aggregation Inhibitors

3.3.1 Identification of Candidate Cyclic Peptide Sequences by Sanger Sequencing

1. Perform a plaque assay for the phage stock from the last selection passage.
2. Pick at least 16 clonal plaques from the plate with pipette tips and use them individually as templates to PCR amplify (using Q5 DNA polymerase) the SICLOPPS gene using primers TW0027 and DM0027.

3. Perform gel electrophoresis to confirm the size of the PCR products. ~1200 bp should be expected if spLY014 or spLY006 is used for cloning the phage library.
4. Clean up the PCR products that have the correct size.
5. Mix each cleaned-up PCR product with primer TW2065. Submit the samples for Sanger sequencing.
6. Record the CP sequences identified by Sanger sequencing (*see* **Note 16 and 17**). They are candidate sequences for further activity validation.

3.3.2 Validation of Candidate Cyclic Peptides by Phage Propagation Assay

Since phage carrying mutation(s) outside the CP-encoding sequence can potentially survive the selections via cheating, this procedure helps purge candidate sequences of false positives before downstream in vitro analysis.

A) Reclone Phages Carrying Candidate CP Sequences

Day 1

1. Perform PCR using Q5U DNA polymerase with the same phage template for the library using the reverse primer LY0003 and a forward primer encoding the candidate sequence (*see* **Note 18**).
2. Perform gel electrophoresis to confirm the size and PCR cleanup to purify the DNA.
3. In a 0.2 mL PCR tube, add 1 μL 10X rCutSmart buffer, 0.75 μL USER enzyme, 0.75 μL DpnI, and 7.5 μL DNA. Mix well by pipetting up and down.
4. Transfer the PCR tube onto a thermocycler and run the protocol: Incubate at 37°C for 45 min → Incubate at 80°C for 3 min → Ramp down to 12°C at the rate of 0.1°C/s.
5. Take out the PCR tube upon completion of digestion for the next step.
6. Perform electroporation, outgrowth, and phage harvest sequentially as described previously in this protocol to clone phage carrying candidate CP sequence.
7. Perform a plaque assay using the phage obtained from Step 6.
8. Pick eight clonal plaques from the plate with pipette tips and use them individually as templates to PCR amplify the SICLOPPS gene using primers TW0027 and DM0027.
9. Transfer each pipette tip containing the clonal phage into 2 mL 2xYT in a culture tube. Incubate at 37°C overnight with shaking.
10. Perform gel electrophoresis to confirm the size of the PCR products ~1200 bp should be expected if spLY014 or spLY006 is used for cloning the phage library.

11. Clean up the PCR products that have the correct size.
12. Mix each cleaned-up PCR product with primer TW2065. Submit the samples for Sanger sequencing.

Day 2

1. Check Sanger sequencing results and harvest the clonal phage with the correct sequence from the overnight culture. This is the subcloned phage carrying the candidate CP sequence.

B) Phage Propagation Assay

Day 1

1. Transform corresponding plasmids into *E. coli* strain S2060. Plate the transformed cells on an LB agar plate with maintenance antibiotics. Incubate at 37°C overnight. These cells are used for the phage propagation assay.

Day 2

1. Pick a single colony of the transformed cells and inoculate it into 2 mL 2xYT media with maintenance antibiotics in a culture tube. Let grow at 37°C overnight with shaking.

Day 3

1. In the morning, back dilute the saturated overnight culture 1:100 into 2 mL DRM media with maintenance antibiotics. Let grow at 37°C with shaking until the OD_{600} reaches 0.5.
2. Infect the mid-log culture with the subcloned phage at an input titer of 10^6 pfu/mL. Incubate at 37°C with shaking for 6 h.
3. Centrifuge the culture at 8000 × *g* for 2 min. Discard the cell pellet and keep the supernatant, which contains phage.
4. Perform a plaque assay using the phage.

Day 4

1. Record the titer of the post-propagation phage and calculate fold propagation using the equation:

$$\textit{Fold propagation} = \textit{output titer } (pfu/mL)/\textit{input titer } (pfu/mL)$$

where output titer is the titer of the post-propagation phage and input titer is 10^6 pfu/mL. If the candidate sequence is not a false positive due to off-target mutation(s), this fold propagation should be comparable to the activity of the phage from the last selection passage (*see* **Note 19**).

3.3.3 Synthesis of Cyclic Peptides

Due to the length and precedence of this procedure, cyclic peptide synthesis is not covered in this protocol in detail. Briefly, our procedure consists of five consecutive steps: synthesis of a linear peptide through solid-phase peptide synthesis, purification of the linear peptide using RP-HPLC, cyclization, purification of cyclic peptide using RP-HPLC, and characterization using analytical HPLC and high-resolution mass spectrometry [21]. Users can adopt our procedure or establish their own based on previous studies [28–31]. Critically, extended exposure of the peptide (especially in solution form) to air and repetitive thawing/freezing should be avoided to minimize oxidation of cysteine residues.

3.3.4 Validation of Candidate Cylic Peptides by In Vitro Assays

Depending on the APP of interest, multiple in vitro assays can be employed to validate the activity of candidate CPs and quantify their potency. In this protocol, a brief list of suitable assays for amyloidogenic protein targets (e.g., Aβ42 and hIAPP) is provided in place of detailed procedures.

1. In vitro *ThT fluorescence assays of protein aggregation*. CPs can be tested for their ability to inhibit protein aggregation in the widely-employed ThT fluorescence assay [32]. Thioflavin-T (ThT) can stain amyloid fibrils due to its enhanced fluorescence when bound to fibrils [33]. In these assays, purified APP monomers are mixed with a solution of ThT and varying concentrations of the CP and allowed to aggregate. Amyloid formation can then be measured as a function of ThT fluorescence. Inhibition of aggregation can manifest as a delay in the onset of ThT fluorescence or a change in the slope of ThT fluorescence increase over time. Malmos et al provided a detailed treatment of the ThT fluorescence assay [32].
2. *Assays for binding*. Several assays can be used to investigate the binding of CPs to their target proteins. For instance, surface plasmon resonance (SPR) and biolayer interferometry (BLI) are widely used for measuring the binding affinity of various inhibitors, including CPs [34–37]. Although these assays have been shown to produce reliable binding constants, their requirement of immobilized target protein, preference for ligands with higher molecular weight, and insensitivity to weaker binders can pose challenges when using them to investigate CP binding to APPs. In comparison, label-free methods like isothermal titration calorimetry (ITC) can also be used to measure the binding affinity of CPs to their target protein [38, 39]. However, interactions between APPs themselves can complicate experimental procedures and data analysis of these methods. Native mass spectrometry, on the other hand, can detect both stable binding and weaker protein-protein interactions, albeit it is less frequently used for accurate assessment

of binding affinity [40, 41]. Despite the limitations discussed here, the assays mentioned above can be considered for activity validation of candidate CPs.

3. *Microscopy*. Due to drastic morphological differences between early conformers and late-stage fibrils formed during aggregation of amyloidogenic proteins, methods like transmission electron microscopy (TEM) and atomic force microscopy (AFM) can be used to visually validate CP protein aggregation inhibitors [42]. This can be observed as a delayed appearance or complete suppression of aggregate-related morphology.

4 Notes

1. The degenerate forward primer consists of a 5′ USER junction matching the one in primer LY0003, a middle CP-encoding sequence encoded by degenerate codons, and a 3′ homology sequence matching the SICLOPPS construct. An example is 5′-**ACAACTGU**NNKNNKNNKNNKNNK**TGTCTCAGTTTCGGCACC**- 3′, which encodes a randomized CP sequence CXXXXX (X represents any 20 amino acids).
2. Use spLY014 as template if lower selection stringency is intended, whereas spLY006 should be used as template if higher selection stringency is needed.
3. Multiple reactions can be performed in parallel using PCR strips. Alternatively, scale up the reaction volume while maintaining same concentrations of DNA, buffer, and enzymes.
4. Reagents from Omega bio-tek are used in this protocol. However, other commercially available kits can also be used for DNA cleanup. Follow the manufacturer's instructions if other kits are used.
5. A maximum loss of 50% in DNA can be expected after this DNA preparation procedure, depending on the quality of columns and reagents used.
6. The *E. coli* SS320 transformed with pJC175e used for electroporation must be from a freshly streaked plate. For convenience, a glycerol stock of *E. coli* SS320 transformed with pJC175e can be used to streak fresh plates.
7. This is about 5×10^8 cfu cells in total.
8. Once spun down, the cells need to be kept cold on ice throughout the cell preparation process.
9. Residual supernatant needs to be completely removed by pipetting out to ensure a low salt concentration in competent cells.

10. A low time constant can stem from poor cell preparation, high salt concentration in DNA, dirty electroporation cuvettes, etc. Troubleshoot accordingly to increase the time constant to 5.6–5.8, or whatever value is optimal for the user's electroporator.
11. To cover a larger library size, multiple transformations can be conducted in parallel. Alternatively, the number of cells and DNA can be scaled up for each individual transformation.
12. The volume of 2xYT media is determined by the library size (titer of the phage library). 1 mL media is needed for every 10^7 pfu phage.
13. The volume of DRM media is determined by the library size. 1 mL media is needed for every 10^7 pfu phage.
14. A decrease in phage titer from the library in *P1* is expected as the vast majority of sequences in the initial library should be inactive. A significant increase in phage titer in the initial passage may be an indication of poor phage library quality or contamination with recombinant wild-type phage.
15. 1 mL DRM media is needed for every 10^7 pfu/mL.
16. Generally, CP sequences enriched from continuous selections should dominate the sampled population, with multiple clones carrying the identical CP sequence(s).
17. Based on observations from our studies, CP sequences enriched from selection for protein aggregation inhibitors tend to possess multiple hydrophobic residues (i.e., Ala, Ile, Leu, Val, and Phe). This is likely because hydrophobic interactions play a critical role in protein aggregation.
18. The forward primer consists of a 5′ USER junction matching the one in primer LY0003, a middle CP-encoding sequence, and a 3′ homology sequence matching the SICLOPPS construct. An example is 5′-**ACAACTGU**CATGTTGTTGGTGTTATT<u>TGTCTCAGTTTCGGCACC</u>-3′, which encodes the CP sequence CHVVGVI (USER junction is annotated in bold and homology sequence is underlined).
19. A phage population enriched with active CP sequences should be able to propagate >= 10^2-fold through selective host cells used in this protocol.

References

1. Ji X, Nielsen AL, Heinis C (2024) Cyclic peptides for drug development. Angew Chem Int Ed 63(3):e202308251. https://doi.org/10.1002/anie.202308251
2. Zorzi A, Deyle K, Heinis C (2017) Cyclic peptide therapeutics: past, present and future. Curr Opin Chem Biol 38:24–29
3. Malde AK, Hill TA, Iyer A, Fairlie DP (2019) Crystal structures of protein-bound cyclic peptides. Chem Rev 119:9861–9914
4. Cardote T, Ciulli A (2016) Cyclic and macrocyclic peptides as chemical tools to recognise protein surfaces and probe protein-protein interactions. ChemMedChem 11(8):787–794

5. Qian Z, Dougherty PG, Pei D (2017) Targeting intracellular protein–protein interactions with cell-permeable cyclic peptides. Curr Opin Chem Biol 38:80–86
6. Li X, Craven TW, Levine PM (2022) Cyclic peptide screening methods for preclinical drug discovery. J Med Chem 65:11913–11926
7. Sohrabi C, Foster A, Tavassoli A (2020) Methods for generating and screening libraries of genetically encoded cyclic peptides in drug discovery. Nat Rev Chem 4:90–101
8. Morioka T, Loik ND, Hipolito CJ et al (2015) Selection-based discovery of macrocyclic peptides for the next generation therapeutics. Curr Opin Chem Biol 26:34–41
9. Deyle K, Kong XD, Heinis C (2017) Phage selection of cyclic peptides for application in research and drug development. Acc Chem Res 50:1866–1874. https://doi.org/10.1021/acs.accounts.7b00184
10. Taylor RD, Rey-Carrizo M, Passioura T, Suga H (2017) Identification of nonstandard macrocyclic peptide ligands through display screening. Drug Discov Today Technol 26:17–23
11. Chen FJ, Pinnette N, Gao J (2024) Strategies for the construction of multicyclic phage display libraries. ChemBioChem 25(9):e202400072. https://doi.org/10.1002/cbic.202400072
12. Josephson K, Ricardo A, Szostak JW (2014) MRNA display: from basic principles to macrocycle drug discovery. Drug Discov Today 19:388–399
13. Passioura T, Katoh T, Goto Y, Suga H (2014) Selection-based discovery of druglike macrocyclic peptides. Annu Rev Biochem 83:727–752
14. Iannuzzelli JA, Fasan R (2020) Expanded toolbox for directing the biosynthesis of macrocyclic peptides in bacterial cells. Chem Sci 11:6202–6208. https://doi.org/10.1039/d0sc01699c
15. Young TS, Young DD, Ahmad I et al (2011) Evolution of cyclic peptide protease inhibitors. Proc Natl Acad Sci U S A 108:11052–11056. https://doi.org/10.1073/pnas.1108045108/-/DCSupplemental
16. Yang X, Lennard KR, He C et al (2018) A lanthipeptide library used to identify a protein-protein interaction inhibitor article. Nat Chem Biol 14:375–380. https://doi.org/10.1038/s41589-018-0008-5
17. Kritzer JA, Hamamichi S, McCaffery JM et al (2009) Rapid selection of cyclic peptides that reduce α-synuclein toxicity in yeast and animal models. Nat Chem Biol 5:655–663. https://doi.org/10.1038/nchembio.193
18. Delivoria DC, Chia S, Habchi J et al (2019) Bacterial production and direct functional screening of expanded molecular libraries for discovering inhibitors of protein aggregation. Sci Adv 5(10):eaax5108. https://doi.org/10.1126/sciadv.aax5108
19. King AM, Anderson DA, Glassey E et al (2021) Selection for constrained peptides that bind to a single target protein. Nat Commun 12(1):6343. https://doi.org/10.1038/s41467-021-26350-4
20. Matis I, Delivoria DC, Mavroidi B et al (2017) An integrated bacterial system for the discovery of chemical rescuers of disease-associated protein misfolding. Nat Biomed Eng 1:838–852. https://doi.org/10.1038/s41551-017-0144-3
21. Yang L, Zhang J, Andon JS et al (2025) Rapid discovery of cyclic peptide protein aggregation inhibitors by continuous selection. Nat Chem Biol 21:588–597. https://doi.org/10.1038/s41589-024-01823-x
22. Scott CP, Abel-Santos E, Wall M et al (1999) Production of cyclic peptides and proteins in vivo. Proc Natl Acad Sci U S A 96:13638–13643
23. Badran AH, Liu DR (2015) Development of potent in vivo mutagenesis plasmids with broad mutational spectra. Nat Commun 6:8425. https://doi.org/10.1038/ncomms9425
24. Bryson DI, Fan C, Guo LT et al (2017) Continuous directed evolution of aminoacyl-tRNA synthetases. Nat Chem Biol 13:1253–1260. https://doi.org/10.1038/nchembio.2474
25. Hubbard BP, Badran AH, Zuris JA et al (2015) Continuous directed evolution of DNA-binding proteins to improve TALEN specificity. Nat Methods 12:939–942. https://doi.org/10.1038/nmeth.3515
26. Miller SM, Wang T, Liu DR (2020) Phage-assisted continuous and non-continuous evolution. Nat Protoc 15:4101–4127. https://doi.org/10.1038/s41596-020-00410-3
27. Davis JH, Rubin AJ, Sauer RT (2011) Design, construction and characterization of a set of insulated bacterial promoters. Nucleic Acids Res 39:1131–1141. https://doi.org/10.1093/nar/gkq810
28. Arbour CA, Belavek KJ, Tariq R, et al (2019) Bringing macrolactamization full circle: self-cleaving head-to-tail macrocyclization of unprotected peptides via mild N-Acyl urea activation. J Org Chem 84:1035–1041. https://doi.org/10.1021/acs.joc.8b02418
29. Gless BH, Olsen CA (2022) On-resin peptide cyclization using the 3-Amino-4-(Methylamino) Benzoic Acid MeDbz linker. Methods in molecular biology. Humana Press Inc, pp 101–115
30. Gless BH, Olsen CA (2018) Direct peptide cyclization and one-pot modification using the MeDbz linker. J Org Chem 83:10525–10534. https://doi.org/10.1021/acs.joc.8b01237

31. Terrier VP, Delmas AF, Aucagne V (2017) Efficient synthesis of cysteine-rich cyclic peptides through intramolecular native chemical ligation of N-Hnb-Cys peptide crypto-thioesters. Org Biomol Chem 15:316–319. https://doi.org/10.1039/C6OB02546C
32. Gade Malmos K, Blancas-Mejia LM, Weber B et al (2017) ThT 101: a primer on the use of thioflavin T to investigate amyloid formation. Amyloid 24:1–16
33. Biancalana M, Koide S (2010) Molecular mechanism of Thioflavin-T binding to amyloid fibrils. Biochim Biophys Acta Proteins Proteom 1804:1405–1412
34. Wang S, Faucher FF, Bertolini M et al (2025) Identification of covalent cyclic peptide inhibitors targeting protein-protein interactions using phage display. J Am Chem Soc 147:7461–7475. https://doi.org/10.1021/jacs.4c15843
35. Bruce A, Adebomi V, Czabala P et al (2024) A tag-free platform for synthesis and screening of cyclic peptide libraries. Angew Chem Int Ed 63(21):e202320045. https://doi.org/10.1002/anie.202320045
36. Giebel LB, Cass RT, Milligan DL et al (1995) Screening of cyclic peptide phage libraries identifies ligands that bind streptavidin with high affinities. Biochemistry 34(47):15430–15435. https://doi.org/10.1021/bi00047a006
37. Liu T, Joo SH, Voorhees JL et al (2009) Synthesis and screening of a cyclic peptide library: discovery of small-molecule ligands against human prolactin receptor. Bioorg Med Chem 17:1026–1033. https://doi.org/10.1016/j.bmc.2008.01.015
38. Reille-Seroussi M, Gaucher JF, Desole C et al (2015) Vascular endothelial growth factor peptide ligands explored by competition assay and isothermal titration calorimetry. Biochemistry 54:5147–5156. https://doi.org/10.1021/acs.biochem.5b00722
39. Gaucher JF, Reille-Seroussi M, Broussy S (2022) Structural and ITC characterization of peptide-protein binding: thermodynamic consequences of cyclization constraints, a case study on vascular endothelial growth factor ligands. Chemistry 28(48):e202200465. https://doi.org/10.1002/chem.202200465
40. Bellamy-Carter J, Mohata M, Falcicchio M et al (2021) Discovering protein-protein interaction stabilisers by native mass spectrometry. Chem Sci 12:10724–10731. https://doi.org/10.1039/d1sc01450a
41. Heck AJR (2008) Native mass spectrometry: a bridge between interactomics and structural biology. Nat Methods 5:927–933. https://doi.org/10.1038/nmeth.1265
42. Bartolini M, Naldi M, Fiori J et al (2011) Kinetic characterization of amyloid-beta 1-42 aggregation with a multimethodological approach. Anal Biochem 414:215–225. https://doi.org/10.1016/j.ab.2011.03.020

Check for updates

Chapter 14

Targeted Transition Mutations in Phage-Assisted Evolution

Thandava Vanapilli Nursimulu, Maryam Ali and Jumi A. Shin

Abstract

Phage-assisted evolution is a powerful strategy for directed evolution in which the protein undergoes continuous mutagenesis under selective pressure to rapidly evolve protein variants. However, conventional approaches often introduce mutations across the entire target plasmid, not just the gene-of-interest (GOI), leading to unwanted genomic changes that enable the evasion of selective pressure. To overcome this limitation, we integrated the MutaT7 system, which uses two base deaminases, into phage-assisted evolution to direct all four transition mutations specifically to the GOI. By using these deaminases, our system promotes the accumulation of all transition mutations, providing greater diversification of the GOI. This targeted yet comprehensive mutation strategy accelerates and enhances the evolutionary outcome of proteins to improve their stability, binding affinity, and specificity.

Keywords Directed evolution, Phage-assisted evolution, Targeted mutagenesis, Deaminases, M13 bacteriophage, T7 promoter, T7 terminator, T7 RNAP, Selection

1 Introduction

Directed evolution is a powerful tool in protein engineering that mimics natural selection to improve the stability or desired function of a protein by subjecting it to iterative rounds of mutagenesis and selection. It consists of four steps: (1) random mutagenesis to create a diverse library of mutants, (2) expression of the mutants to display the various phenotypes, (3) screening of the phenotypes and selection of the best mutants, and (4) amplification of the fittest mutants to serve as the template for the next round. Phage-assisted evolution has revolutionized the field of directed evolution by utilizing M13 bacteriophage, which has a much shorter life cycle than *E. coli* and other common organisms (Fig. 1) [1, 2]. In phage-assisted evolution, the gene-of-interest (GOI), which encodes the protein-of-interest

Huan Peng and S. Brian Kim (eds.), *Biological Macromolecule Engineering Technologies*,
Methods in Molecular Biology, vol. 3032, https://doi.org/10.1007/978-1-0716-5257-2_14,

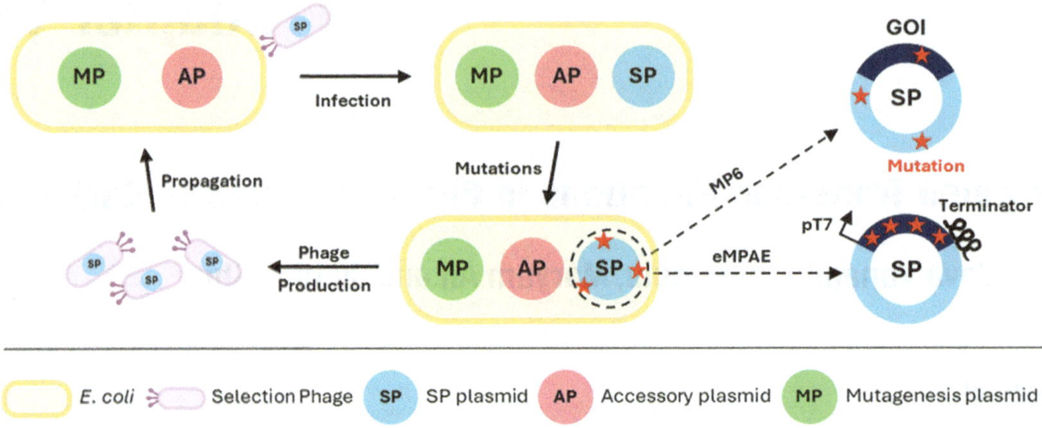

Fig. 1 Schematic of phage-assisted evolution. The selection phage (SP) infects a host *E. coli* cell, thereby inserting the SP plasmid carrying the gene-of-interest (GOI) [1]. Inside the cell, a mutagenesis plasmid (MP) induces mutations (indicated by red stars) into the SP plasmid. These mutations generate variants whose activity is linked to *gIII* expression, which codes for a phage protein essential for infection. Active mutants promote *gIII* expression, leading to the production of infectious phages, whereas less active/inactive mutants lower *gIII* expression levels, leading to non-infectious phages that cannot propagate. The mutagenesis plasmid MP6 has a high rate of undesirable off-target mutations outside the GOI [3], whereas mutations in eMPAE are limited to the GOI, provided the GOI is flanked by a T7 promoter (pT7) and a terminator [4, 5]

(POI), is introduced into the genome of the selection phage (SP), a modified M13 bacteriophage containing a deletion of *gene III* (*gIII*). This essential phage gene is instead expressed from an accessory plasmid (AP), and its production is coupled to the activity of the POI.

To enhance the mutation rate *in cellulo*, host cells are transformed with a mutagenesis plasmid (MP) containing genes encoding proteins that induce mutations. The most common MP is MP6, which is a global mutagen that induces mutations throughout the SP plasmid and the host genome [3]. To avoid unwanted off-target mutations, we developed enhanced mutation phage-assisted evolution (eMPAE) by combining phage-assisted evolution with the MutaT7 system, a targeted mutagenesis tool to ensure only a region-of-interest on the SP plasmid gets mutated [4, 5]. Mutations are restricted to the GOI, as it is flanked by a T7 promoter (pT7) and T7 terminator. The MP expresses a mutation-inducing enzyme, such as a base deaminase, fused to a T7 RNA polymerase (RNAP). To induce mutagenesis, T7 RNAP binds to pT7 upstream of the GOI and begins transcription (Fig. 2). This localizes the deaminase close to the transcription bubble to induce point mutations on the exposed DNA. Upon reaching the terminator, the deaminase-T7 RNAP fusion dissociates. Off-target mutations can be minimized with eMPAE to avoid false positives.

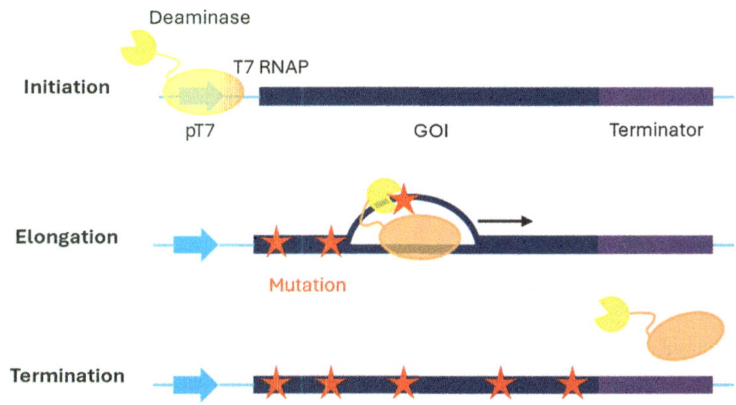

Fig. 2 Mechanism of the MutaT7 system [6]. This system consists of T7 RNA polymerase (RNAP) fused to a deaminase [7–10]. To induce mutagenesis, the deaminase-T7 RNAP fusion binds to a T7 promoter (pT7) and transcribes the GOI, allowing the deaminase to induce point mutations on the exposed DNA in the transcription bubble. When the T7 RNAP reaches the terminator, it dissociates from the DNA, thereby preventing downstream mutations. The native T7 terminator has weak efficiency, which can lead to read-through transcription and unwanted downstream mutations [11]. Several strategies have been implemented to avoid such off-target mutations, such as an array of four native T7 terminators [6], a deactivated Cas9 roadblock [7], and a highly efficient synthetic T7 terminator [5, 11]

Until recently, eMPAE was limited to C:G→T:A mutations that are induced by the cytidine deaminase *pm*CDA1 [5]. By combining both cytidine *and* adenosine deaminases in a deaminase- T7 RNAP fusion, all four possible transition mutations can be generated. We have now introduced the adenosine deaminase TadA8e [9,10,12] to access the complementary A:T→G:C mutations. By expanding our system with more enzymes, eMPAE can generate a more diverse library of mutants to explore more of the evolutionary landscape, which should result in a more successful evolutionary campaign. To ensure that the deaminase is able to mutate the GOI, we recommend increasing the base-pair length of the GOI to at least 800 bp to prevent the premature release of the T7 RNAP when it encounters the terminator [5].

Prior to starting eMPAE, the relevant plasmids (SP, AP, MP) must be acquired and designed with the appropriate GOI and selection circuit (Fig. 3; *see* **Note 1**). Additionally, pT7 and T7 terminator must be inserted in the SP upstream and downstream of the GOI. Phage-assisted evolution is performed by running passages with the AP, SP, and MP transformed in bacterial cells. After each passage, the phage titer is measured via plaque assays to ensure healthy phage propagation. The phage can also be isolated from plaque assays to check for mutations through sequencing analysis.

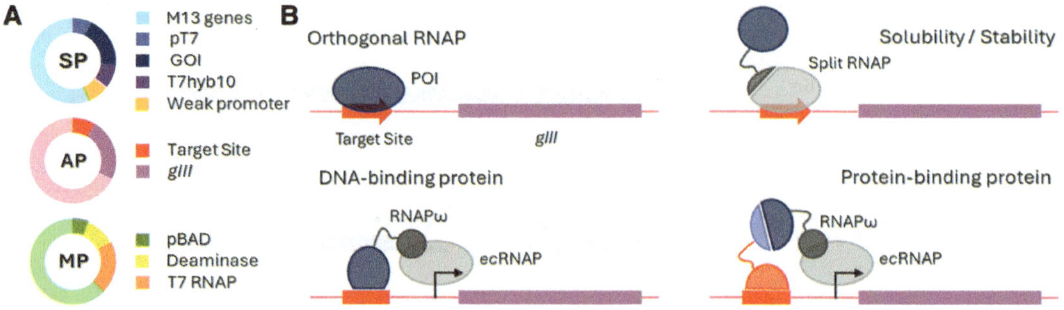

Fig. 3 Design of plasmids and selection circuits. (**A**) Plasmid maps of selection phage (SP) plasmid, accessory plasmid (AP), and mutagenesis plasmid (MP). The SP plasmid carries all the M13 phage genes, except *gIII*, and the GOI is flanked by a T7 promoter (pT7) and terminator (T7hyb10). A weak promoter is also inserted downstream of the terminator to allow expression of downstream phage genes. The AP provides *gIII*, and its expression is controlled by the activity of the protein-of-interest (POI). The MP used in eMPAE is MPT7, which expresses a base deaminase- T7 RNAP fusion under the control of the arabinose promoter (pBAD). (**B**) Common selection circuits used in phage-assisted evolution to improve POI (dark blue): orthogonal RNAP [1], protein solubility and/or stability [13], one-hybrid reporter for DNA-binding proteins [5, 14, 15], and two-hybrid reporter for protein-binding proteins [16]. Orthogonal RNAPs can be evolved to recognize new promoter sequences. Complementation of an inactive, split RNAP can be used to evolve greater POI solubility and/or stability. For POIs that bind to DNA or other proteins, n-hybrid reporters can be used, where binding of the POI to its target recruits *E. coli* RNAP (*ec*RNAP) via a fused RNAP ω subunit

2 Materials

2.1 Bacterial Strains

1. *E. coli* S1030 [17] (Addgene #105063) or S2060 [14] (Addgene #105064) (*see* **Note 2**)
   - Host strain for activity-dependent propagation of phages in selection passages.
   - Carries F-plasmid.
   - Resistant to streptomycin and tetracycline.
2. *E. coli* S1059 or S2208, defined as S1030 or S2060 transformed with pJC175e [17]
   - Permissive host strain for the production and recovery of phages, for plaque assays, and for drift passages (*see* **Note 3**).
   - Resistant to streptomycin, tetracycline, and ampicillin/carbenicillin.
3. *E. coli* DH5α
   - For general molecular biology purposes (e.g., cloning).
   - No antibiotic resistance.

2.2 Plasmids

1. pJC175e [17] (Addgene #79219)
   - Contains *gIII* under the control of a phage-shock promoter for expression upon phage infection.
   - Confers resistance to ampicillin/carbenicillin.
2. Accessory plasmid
   - Expresses *gIII* in an activity-dependent manner.
   - Confers resistance to ampicillin/carbenicillin.
3. Selection phage plasmid
   - Carries GOI and all M13 phage genes, except *gIII*.
   - Confers resistance to kanamycin.
4. MPT7^{CD}[5]
   - Expresses *pm*CDA1- T7 RNAP and uracil glycosylase inhibitor (UGI) for targeted C:G→T:A mutations upon induction with arabinose (*see* **Note 4**).
   - Confers resistance to chloramphenicol.
5. MPT7^{AD}
   - Expresses TadA8e-T7 RNAP for targeted A:T→G:C mutations upon induction with arabinose.
   - Confers resistance to chloramphenicol.

2.3 Reagents

1. **Lysogeny broth (LB)**
   Add LB powder to a final concentration of 25 g/L in water. Autoclave to dissolve and sterilize. Cool to room temperature and add appropriate antibiotics. Store at 4°C.
2. **LB agar plates**
   Add LB powder and agar to a final concentration of 25 g/L and 15 g/L, respectively, in water. Autoclave to dissolve and sterilize. Cool to 55°C and add appropriate antibiotics. Pour into sterile petri dishes. Cool to room temperature to solidify and store at 4°C.
3. **2× YT media**
   Add 2× YT media powder to a final concentration of 31 g/L in water. Autoclave to dissolve and sterilize. Cool to room temperature and add appropriate antibiotics. Store in the dark at 4°C.
4. **2× YT agar plates**
   Add 2× YT media powder and agar to a final concentration of 31 g/L and 15 g/L, respectively, in water. Autoclave to dissolve and sterilize. Cool to 55°C and add appropriate antibiotics. Pour into sterile petri dishes. Cool to room temperature to solidify and store in the dark at 4°C.

5. **2× YT top agar**
   Add 2× YT media powder and agar to a final concentration of 31 g/L and 6 g/L, respectively, in water. Autoclave to dissolve and sterilize. Store at room temperature.
6. **Transformation and storage solution (TSS)**
   Dissolve $MgCl_2$, PEG 3350, and DMSO to a final concentration of 30 mM, 10% w/v, and 5% v/v, respectively, in LB. Filter-sterilize.
7. **Super optimal broth with catabolite repression (SOC)**
   Add super optimal broth powder to a final concentration of 31 g/L in water. Autoclave to dissolve and sterilize. Cool to room temperature and supplement with filter-sterilized 1 M glucose to a final concentration of 20 mM. Store at 4°C for up to 6 months.
8. **Ampicillin stock solution (1000×)**
   Dissolve ampicillin to a final concentration of 50 mg/mL in water. Filter-sterilize with a 0.22 μm sterile filter. Aliquot and store at −20°C for up to 6 months.
9. **Chloramphenicol stock solution (1000×)**
   Dissolve chloramphenicol to a final concentration of 25 mg/mL in ethanol. Filter-sterilize with a 0.22 μm sterile filter. Aliquot and store at −20°C for up to 6 months.
10. **Kanamycin stock solution (1000×)**
    Dissolve kanamycin to a final concentration of 50 mg/mL in water. Filter-sterilize with a 0.22 μm sterile filter. Aliquot and store at −20°C for up to 6 months.
11. **Streptomycin stock solution (1000×)**
    Dissolve streptomycin to a final concentration of 50 mg/mL in water. Filter-sterilize with a 0.22 μm sterile filter. Aliquot and store at −20°C for up to 6 months.
12. **Tetracycline stock solution (1000×)**
    Dissolve tetracycline to a final concentration of 10 mg/mL in 50% v/v ethanol. Filter-sterilize with a 0.22 μm sterile filter. Aliquot and store at −20°C for up to 6 months.
13. **L-arabinose stock solution**
    Dissolve L-arabinose to a final concentration of 10% w/v in water. Filter-sterilize with a 0.22 μm sterile filter. Store at room temperature for up to 3 months.

3 Methods

3.1 Constructing Selection Phage Plasmid

1. Derived from M13 phage genome (*see* **Note 5**).
2. Replace *gIII* with GOI.
3. Insert pT7 and T7hyb10 terminator [11] upstream and downstream of GOI (*see* **Note 6**).

4. For short GOIs, insert a spacer between the GOI and the terminator to increase the length of the target region to at least 800 bp (*see* **Note 7**).
5. Insert weak phage promoter between T7hyb10 terminator and *gVI* to allow expression of essential phage genes downstream of the terminator (*see* **Note 8**).

3.2 Constructing Accessory Plasmid

1. Design an appropriate selection circuit to link *gIII* expression to POI activity.
2. Validate circuit to ensure no autoactivation (*see* **Note 9**).

3.3 Constructing Mutation Plasmids

1. MPT7^{CD} was derived from eMutaT7^{pmCDA1} with optimized RBS [10] (Addgene #187621) with the following modifications:
   Replaced *pm*CDA1 with an *E. coli* codon-optimized variant obtained from pSEVA221*pm*CDA1-T7 RNAP [7] (Addgene #167978; *see* **Note 10**)
   Introduced unique XbaI and XhoI restriction sites upstream and downstream of *pm*CDA1, respectively, for easy substitution of mutators.
2. MPT7^{AD} was derived from MPT7^{CD} with the substitution of *pm*CDA1 by TadA8e obtained from eMutaT7^{transition} [10] (Addgene #187622).

3.4 Preparing TSS Chemically Competent Cells [18]

1. Dilute an overnight culture of the desired cells 50- to 1000-fold in 50 mL LB with appropriate antibiotics and grow at 37°C with shaking at 200 rpm to OD_{600} 0.2–0.5.
2. Chill cells on ice for 10 min.
3. Centrifuge cells at 3000 rpm and 4°C for 10 min.
4. Discard supernatant and resuspend cells gently in 5 mL ice-cold TSS (*see* **Note 11**).
5. Make 100 μL aliquots in pre-chilled, sterile 1.5 mL tubes.
6. Use immediately or store at −80°C.

3.5 TSS Transformation

1. Thaw an aliquot of chemically competent cells on ice for 10–15 min.
2. Add ~1 μL of the desired plasmid (<100 ng) to the cells and stir gently with the pipette tip.
3. Incubate on ice for 30 min.
4. Heat-shock the cell-plasmid mixture at 42°C for 60 s.
5. Incubate cells on ice for 15 min.
6. Transfer cells to 1 mL SOC and recover cells at 37°C with shaking at 250 rpm for 1 h.
7. Plate on LB agar supplemented with appropriate antibiotics to select for the plasmid and incubate at 37°C overnight.

3.6 Assembling Phages

1. Once the SP plasmid has been assembled, transform it into S1059 chemically competent cells.
2. Following heat shock (Step 3.5.4), transfer cells to 10 mL antibiotic-free 2× YT media.
3. Grow cells at 37°C with shaking at 200 rpm for 16 h.
4. Collect phages by centrifuging the mixture at 3000 rpm for 5–10 min and filtering the supernatant containing phages with a sterile 0.22 μm filter into a clean, sterile tube (*see* **Note 12**).

3.7 Activity-Independent Plaque Assays (see Note 13)

1. Dilute an overnight culture of S1059 or S2208 cells 100-fold in 2× YT media supplemented with appropriate antibiotics and grow at 37°C with shaking at 250 rpm to OD_{600} 0.6–0.9. These cells can be stored at 4°C for up to 7 days.
2. While cells are growing, melt 2× YT top agar in a microwave oven and aliquot 4 mL in culture tubes. Incubate in a water bath at 55°C to keep the top agar molten until use for up to 6 h.
3. Before handling phages, plate a negative control by adding 270 μL subcultured cells to one aliquot of molten 2× YT top agar. Invert gently to mix and pour onto pre-warmed 2× YT agar plates (*see* **Note 14**).
4. Serially dilute desired phages in tenfold increments up to 10^8-fold diluted.
5. Add 30 μL of each desired phage dilution to clean, sterile tubes.
6. Add 270 μL of subcultured cells to each phage dilution and incubate at room temperature for 10–15 min.
7. Transfer each cell-phage mixture to one aliquot of molten 2× YT top agar. Invert gently to mix and pour onto pre-warmed 2× YT agar plates. Optional: add Bluo-gal to the top agar (*see* **Note 2**).
8. Incubate at 37°C overnight.
9. Count the number of plaques to determine the phage titer using the formula:

$$Phage\ titer\left(PFU/mL\right)=\frac{\#of\ plaque\ forming\ units\ (PFU)}{Volume\ of\ phage\ (mL)}\times Dilution\ factor$$

3.8 Isolating Clonal Phage

1. Using a sterile P200 pipette tip, poke an isolated plaque (corresponding to cells infected by a single phage) and inoculate 4 mL 2× YT media with no antibiotics.
2. Grow at 37°C with shaking at 200 rpm for 16 h.

3. Centrifuge the mixture at 3000 rpm for 5–10 min to plasmid purify the replicative form of the SP plasmid and/or collect phages by filtering the supernatant.

3.9 Amplification/ Recovery of Phages

1. For titers $<10^6$ PFU/mL, run a recovery to amplify SP (*see* **Note 15**).
2. Dilute an overnight culture of S1059 cells or inoculate a 100 μL aliquot of S1059 competent cells into 25 mL 2× YT media with antibiotics and grow at 37°C with shaking at 200 rpm for 2–3 h.
3. Add 25 mL 2× YT media with antibiotics and up to 1 mL of the desired phage culture.
4. Grow overnight at 37°C with shaking at 200 rpm.
5. Collect phages by centrifugation and filtration (*see* **Note 12**).
6. Quantify new titer with an activity-independent plaque assay.

3.10 Running eMPAE Passages

1. Transform $MPT7^{AD}$ and $MPT7^{CD}$ into S1059 (*see* **Note 3**) and S1030-AP TSS chemically competent cells.
2. Inoculate 4 mL 2× YT media supplemented with appropriate antibiotics with a single colony from the corresponding transformations and grow at 37°C with shaking at 250 rpm for 16 h.
3. Subculture the cells 50- to 100-fold in fresh 2× YT media with appropriate antibiotics and grow the cells to OD_{600} 0.4–0.6. These cells can be stored at 4°C for up to 7 days.
4. For each eMPAE passage, set up a 2 mL culture as follows:

   1.6 mL 2× YT media with antibiotics.
   100 μL S1059/S1030-AP cells harboring $MPT7^{AD}$ (*see* **Note 16**).
   100 μL S1059/S1030-AP cells harboring $MPT7^{CD}$.
   40 μL 10% arabinose.
   Incubate at 37°C for 2–4 h (*see* **Note 17**).
   Add 2–200 μL phages from previous passage and grow for 4–16 h (*see* **Note 18**).
   Collect phages by centrifugation and filtration of supernatant (*see* **Note 12**).
   These phages can be used to start the next passage.

3.11 Activity-Dependent Phage Propagation

1. Dilute an overnight culture of S1030 cells transformed with the AP only in 2× YT media with antibiotics and grow to OD_{600} 0.4–0.6.
2. Add phages to a final titer of 10^4 PFU/mL and grow for 16 h.
3. Collect phages by centrifugation and filtration.
4. Determine titer by running an activity-independent plaque assay and calculate fold propagation by dividing final titer by input titer (*see* **Note 19**).

4 Notes

1. Before starting eMPAE, obtain full plasmid sequencing of each plasmid (SP, AP, MP). Ensure all critical genes in each plasmid are being expressed.
2. *E. coli* S2060 and S2208 have been engineered to express LacZ upon phage infection, thereby allowing the formation of easily identifiable colored plaques in plaque assays when 2× YT top agar is supplemented with 0.04% w/v Bluo-Gal.
3. The Liu lab uses a drift plasmid (DP6), which contains the same mutagenic genes of MP6 as well as *gIII* under the control of a tunable Tet promoter, to induce drift upon addition of anhydrotetracycline. Alternatively, we recommend using the *E. coli* S1059 cells harboring pJC175e for the expression of *gIII* independently of POI activity to allow MPT7 to accumulate mutations during drift passages.
4. Co-expression of UGI is crucial for efficient C:G→T:A mutagenesis. UGI inhibits uracil glycosylase, which excises uracil from the deaminated cytosine, and the abasic site is repaired back to the original cytosine.
5. Although the M13 phage genome is single-stranded, its replicative form inside infected cells is double-stranded and can be handled like normal plasmids. Therefore, the SP plasmid can be extracted directly from infected cells and purified like normal plasmids.
6. We recommend using the T7hyb10 terminator over the wild-type T7 terminator because the T7hyb10 terminator has a 98% termination efficiency compared to the 62% efficiency of the wild-type terminator [11]. Using a more efficient terminator minimizes the risk of readthrough transcription, leading to less undesirable mutations downstream of the terminator and outside of the target region.
7. We have previously found that having a small GOI (~300 bp) causes the premature dissociation of the deaminase-T7 RNAP fusion, while a longer target region (>800 bp) results in more efficient mutagenesis.
8. The highly efficient T7hyb10 terminator prevents transcription of downstream essential phage genes, resulting in the production of few infectious phages. Therefore, we inserted a weak phage promoter to allow transcription of these genes and observed a significant increase in phage propagation.
9. Autoactivation may cause expression of *gIII* independently of the activity of the POI, thereby bypassing the intended selective pressure and leading to cheaters.

10. We noticed an increase in the mutation rate of MPT7^{CD} after *E. coli* codon optimization, suggesting an increase in expression levels inside the cells.
11. Do not vortex the cells. Resuspend the cell pellet by gently stirring with a pipette tip or by shaking on ice.
12. Filtered phages can be directly stored at 4°C or in 50% glycerol at −80°C indefinitely. Phage titers may decrease over time, so determine the phage titer again if needed and run a recovery if titer $<10^6$ PFU/mL.
13. Phages are incredibly infectious. Proper care must be taken to avoid contamination. We have found that sterilizing all tools (e.g., pipettes) and consumables (e.g., pipette tips and tubes) by UV irradiation for 20 min and wiping the bench surface with 6% bleach before and after handling phages help minimize the risk of contamination. Additionally, tubes containing phages should be opened as close to a flame as possible. Preparing fresh reagents (e.g., 2× YT top agar and plates) regularly can help reduce the risk of contamination.
14. We recommend running a negative control during each plaque assay by plating just the cells to check for phage contamination of the cell stock as well as to ensure proper growth of the cells. The control plate also allows for easier determination of plaques on plates that contain phages.
15. Phage titers may drop after storing phages at 4°C for an extended period or after running several drift passages. Before starting selection passages from stored phages, the titer should be determined via activity-independent plaque assays and increased to at least 10^6 PFU/mL by performing an amplification/recovery.
16. Since phages infect cells to transfer their genetic material, MPT7^{AD} and MPT7^{CD} can be transformed into different cells rather than co-transformed into the same cells, thereby reducing stress on the cells. By mixing cells harboring each of the two mutators, eMPAE can access all four transition mutations.
17. We have noticed an increase in the mutation rate when diluting cells and inducing expression of the mutator for a few hours prior to phage infection.
18. In eMPAE, the strength of the selective pressure or stringency can be varied by altering the following parameters:
    Drift: eliminates selective pressure to allow accumulation of mutations.
    Input phage titer: lowering input titer increases stringency by selecting for faster propagating SP.
    Growth time: shorter growth times increase stringency by selecting for faster propagating SP.
19. Activity-dependent phage propagation can serve as a rudimentary test to check for improvements of the POI in a

Table 1
Troubleshooting tips

Step	Problem	Possible explanation	Possible solution
3.5	No colonies observed	Wrong antibiotics used	Check the antibiotic resistance of plasmid transformed and use the correct antibiotic(s)
		Low plasmid quality	Purify the plasmid again; check the sequence by the Sanger or Nanopore method
		Low cell competency	Check competency by transforming a different plasmid. If no colonies, make new competent cells (Sect. 3.4)
3.6	No phages produced	Phages are not propagating properly	After heat shock (Step 3.5.4), transfer cells into molten 2× YT top agar and plate instead of growing transformed cells in 2× YT media. Isolate a clonal phage from the resulting plaques (Sect. 3.8)
		Mutations in SP plasmid backbone	Check the sequence of SP plasmid by the Sanger or Nanopore method
		Mutations in AP plasmid	Check the sequence of pJC175e/AP by the Sanger or Nanopore method
			Transform sequence-verified pJC175e/AP into S1030 to make new competent cells
3.7	Uneven lawn of cells	Cells grow too slowly	Use correct concentration of antibiotics (50 μg/mL ampicillin, 10 μg/mL tetracycline and 50 μg/mL streptomycin)
		Cells not grown to the proper OD_{600}	Grow new cells to late log phase (OD_{600} 0.6–0.9)
	Negative control shows plaques	Host cells have been infected with phages	Grow new cells and take better precautions when handling cells and phages
			Prepare fresh reagents, plates, and media
	No plaques observed	Cells lack F-pilus	Add tetracycline when growing cells to select for F-plasmid
			Add antibiotics to 2×YT agar plates to select for the correct cells
		Phage genes are not being expressed	Sequencing analysis to ensure all phage genes in SP are being expressed
			Ensure the presence of T7 terminator in SP is not preventing expression of downstream phage genes
3.10	No mutations observed	Low arabinose concentration	Prepare fresh 10% arabinose and dilute to 0.2% when running passages
		Mutation inducing enzyme is not expressed	Check the sequence of MP by the Sanger or Nanopore method to ensure mutation inducing enzyme is expressed
	Titer drops significantly when starting selection	Selection stringency is too high	Adjust stringency by running more drift, increasing the strength of the ribosome binding site of *gIII* and/or optimizing the selection circuit

phage population or clonal phages. A high fold propagation indicates faster propagating phages, suggesting an increase in the stability and/or activity of the POI.

20. See Table 1 for more tips.

Acknowledgments

We are grateful for funding from the Cancer Research Society and NSERC (JAS), and PRiME Fellowship PRMUOT2025-001 (MA).

References

1. Esvelt KM, Carlson JC, Liu DR (2011) A system for the continuous directed evolution of biomolecules. Nature 472:499–505
2. Packer MS, Liu DR (2015) Methods for the directed evolution of proteins. Nat Rev Genet 16:379–394
3. Badran AH, Liu DR (2015) Development of potent in vivo mutagenesis plasmids with broad mutational spectra. Nat Commun 6:8425–8434
4. Ali M, Akel R, Botero MJ, Shin JA (2024) Combining a base deaminase mutator with phage-assisted evolution. In: Peng H; Liu J, Chen I (Eds) Phage engineering and analysis. Methods in molecular biology. Springer Nature, Humana Press, New York, pp 55–64
5. Ali M, Khan A, Vanapilli Nursimulu T, Shin JA (2025) Unlocking genetic potential: harnessing phage for targeted mutagenesis in phage-assisted evolution. Nucleic Acids Res 53:gkaf746
6. Moore CL, Papa LJ, Shoulders MD (2018) A processive protein chimera introduces mutations across defined DNA regions in vivo. J Am Chem Soc 140:11560–11564
7. Álvarez B, Mencía M, de Lorenzo V, Fernández LÁ (2020) In vivo diversification of target genomic sites using processive base deaminase fusions blocked by dCas9. Nat Commun 11(1):6436
8. Park H, Kim S (2021) Gene-specific mutagenesis enables rapid continuous evolution of enzymes in vivo. Nucl Acid Res 49:e32
9. Mengiste AA, Wilson RH, Weissman RF, Papa LJ III, Hendel SJ, Moore CL, Butty VL, Shoulders MD (2023) Expanded MutaT7 toolkit efficiently and simultaneously accesses all possible transition mutations in bacteria. Nucleic Acids Res 51(6):e31
10. Seo D, Koh B, Eom G-e, Kim HW, Kim S (2023) A dual gene-specific mutator system installs all transition mutations at similar frequencies in vivo. Nucleic Acids Res 51(10):e59
11. Calvopino-Chavez DG, Gardner MA, Griffitts JS (2022) Engineering efficient termination of bacteriophage T7 RNA polymerase transcription. G3: Genes, Genomes, Genet 12:jkac070
12. Richter MF, Zhao KT, Eton E, Lapinaite A, Newby GA, Thuronyi BW, Wilson C, Koblan LW, Zeng J, Bauer DE, Doudna JA, Liu DR (2020) Phage-assisted evolution of an adenine base editor with improved Cas domain compatibility and activity. Nat Biotechnol 38(7):883–891
13. Wang T, Badran AH, Huang TP, Liu DR (2018) Continuous directed evolution of proteins with improved soluble expression. Nat Chem Biol 14:972–980
14. Hubbard BP, Badran AH, Zuris JA, Guilinger JP, Davis KM, Chen L, Tsai SQ, Sander JD, Joung JK, Liu DR (2015) Continuous directed evolution of DNA-binding proteins to improve TALEN specificity. Nat Methods 12:939–942
15. Popa SC, Inamoto I, Thuronyi BW, Shin JA (2020) Phage assisted continuous evolution (PACE): a guide focused on evolving protein-DNA interactions. ACS Omega 5:26957–26966
16. Badran AH, Guzov VM, Huai Q, Kemp MM, Ishwanath P, Kain W, Nance AM, Evdokimov A, Moshiri F, Turner KH, Wang P, Malvar T, Liu DR (2016) Continuous evolution of Bacillus thuringiensis toxins overcomes insect resistance. Nature 533:58–63
17. Carlson JC, Badran AH, Guggiana-Nilo DA, Liu DR (2014) Negative selection and stringency modulation enable phage-assisted continuous evolution (PACE) of enzymes with altered specificity. Nat Chem Biol 10:216–222
18. Chung CT, Niemela SL, Miller RH (1989) One-step preparation of competent *Escherichia coli*—transformation and storage of bacterial cells in the same solution. Proc Natl Acad Sci U S A 86:2172–2175

Check for updates

Chapter 15

Phage Display In Vivo: Protocols for Identifying Brain-Targeting Peptides in Mouse Models of Neurodegeneration

Roberta Aquino Martins, Luís Felipe Galvão Valdivia, Andressa Godoy Amaral, Daniella Sabino Battagello and Ricardo José Giordano

Abstract

In vivo phage display is a versatile strategy for an unbiased identification of peptide ligands that selectively interact with molecular targets, in vitro, *in cell*, and in vivo, in living organisms. In a previous study, we validated this technology to identify a brain targeting peptide. Here, we detail the protocol for selection using phage display in vivo of peptide targeting the mouse brain, in health and disease. Given the need of alternative models to study degenerative disease, such as Alzheimer's disease, we elected the murine model of neuroinflammation and neurodegeneration as an example. This model utilizes the stereotaxic intracerebroventricular administration of streptozotocin (STZ) to induce neuroinflammation, an event that recapitulates key pathological hallmarks of Alzheimer's disease. Following the establishment of the disease model, a randomized phage display peptide library—a collection of billions of phages, each displaying a different peptide sequence on their surface—is administered intravenously into healthy and diseased mice. This allows selective interaction of individual phage with vascular, tissue-specific receptors under physiological or in a pathological condition with which a compromised blood–brain barrier may alter brain homeostasis. After systemic circulation, animals undergo perfusion and organs are harvested for phage recovery and amplification, followed by iterative rounds of biopanning to enrich for high-affinity ligands. This approach offers a robust platform for the discovery of novel diagnostic probes, therapeutic ligands, and molecular markers relevant to brain biology, in health, neuroinflammation and neurodegeneration. Most importantly, the adaptability of this technology enables its application across diverse murine models of human disease, thereby establishing an easy and systematic approach to facilitate the translation of laboratory discoveries into a wide spectrum of human diseases.

Keywords In vivo phage display, Brain-targeting peptides, Neurodegeneration

1 Introduction

The phage display technology utilizes filamentous bacteriophages, such as M13 or Fd, as platforms for presenting peptide ligands to receptors or other target molecules within a living organism [1–3]. The most common strategy for peptide phage display involves

Huan Peng and S. Brian Kim (eds.), *Biological Macromolecule Engineering Technologies*,
Methods in Molecular Biology, vol. 3032, https://doi.org/10.1007/978-1-0716-5257-2_15,

inserting an exogenous DNA fragment that encodes the peptide of interest into the gene for the minor coat protein pIII, which is located at the tip of the phage and is essential for bacterial infection. During the replication cycle in host bacteria, typically *Escherichia coli*, pIII is expressed as a fusion protein with the peptide, exposing the ligand on the viral surface while maintaining its link to the phage genome [4].

The peptide ligands can be either previously known or unknown, depending on the study's objective. Known sequences can be used to investigate specific interactions with different receptors in vitro, for e.g., in affinity comparison assays between multiple molecules [4, 5]. Conversely, unknown sequences—such as those found in modified phage libraries—can be explored as a tool for mapping target receptors in distinct tissues, particularly in in vivo analyses [3, 6].

When the modified phages interact with compatible receptors—either on solid surfaces (in vitro) or in the tissues of a living animal (in vivo)—only those that display peptides with sufficient affinity bind specifically to the molecular targets, immobilized or present in the tissue. The subsequent recovery of bound phage, followed by isolation, amplification, and iterative cycles of selection (biopanning), allows for the enrichment of high-affinity phages and the identification, through viral DNA analysis, of the sequences responsible for molecular recognition [3, 4]. It is important to highlight that a pioneering study on biopanning was conducted in 1996 by Renata Pasqualini and Erkki Ruoslahti, who demonstrated that when a peptide phage display library was injected into mice, phage particle did not distribute randomly but rather in a tissue-specific manner, guided by the peptide they display [3]. Following this seminal work, they showed that peptides identified by phage display have important theragnostic potential [7–10].

The use of libraries for phage display has been increasingly recognized as a significant technology in the discovery of ligand peptides. High diversity, versatility, and target specificity are some of the benefits provided by the technology created by George P. Smith in 1985 [1]. Subsequently, the technology was expanded to display not only small peptides but also larger proteins and even antibodies, broadening its applicability [11] for which, in 2018, both Smith and *Sir* McCafferty were awarded the Nobel Prize in Chemistry for the invention of phage display and its applications in biotechnology [12].

The phage display in vivo technology serves as a power tool for investigating the heterogeneity of complex and difficult-to-mimic tissues, enabling the mapping of receptors, and the functional characterization of their components under both physiological and pathological conditions. Its applicability is particularly relevant in the study of the blood–brain barrier (BBB), whose structural organization and functional dynamics pose significant challenges in biomedical science [6, 8, 13]. The BBB protects neural functions from blood-borne metabolites and toxins, including fatty acids, which confers a critical metabolic dependence on glucose. Simultaneously, it constitutes a robust barrier to drug delivery to the central nervous system. Furthermore, the barrier exhibits heterogeneity and adaptive plasticity, varying between physiological and pathological states [14, 15]. This characteristic underscore the importance of experimental approaches capable of capturing its multiple functional and molecular features.

Recent studies indicate BBB impairment in neurodegenerative diseases such as Alzheimer's disease [16–18]. It is, in fact, a symptom noted as manifesting early in patients, even before noticeable cognitive decline and neurodegeneration[19]. The phage display in vivo technology can be employed in this context to study differences in vascular permeability between healthy and diseased animals and to evaluate interactions with an intact or compromised BBB.

Despite extensive clinical efforts to prevent the accumulation of amyloid β and phosphorylated tau proteins, the overall results have been largely disappointing, and a disease-modifying therapy has yet to be developed [20]. Perhaps one of the bottlenecks is the lack of alternative animal models to study neurodegenerative diseases such

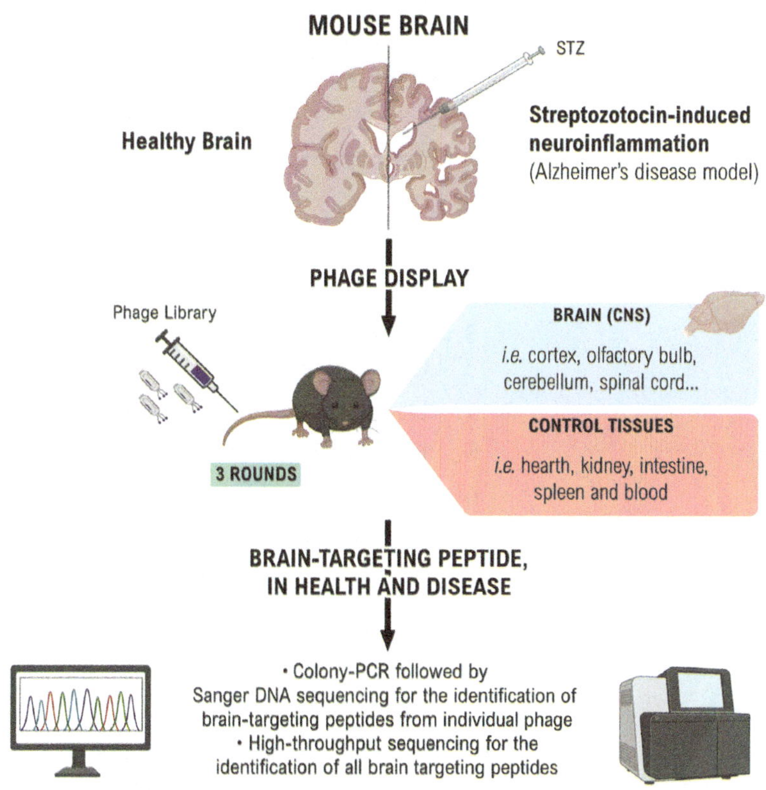

Fig. 1 Representation of an in vivo phage display protocol in animals with neurodegeneration induced by intracerebroventricular injection of streptozotocin (STZ). Following characterization of the neurodegenerative model, animals are injected with the phage library, and tissues are subsequently collected for phage recovery. After three rounds of selection, the sequences can be identified by colony PCR followed by Sanger sequencing, enabling the screening of peptides with affinity for distinct regions of the brain

as Alzheimer [21]. Here, we describe the protocol of the mouse model of neuroinflammation induced by stereotaxic intracerebroventricular administration of streptozotocin (STZ), which mimic the to some extent the early cerebral effects of Alzheimer's disease (Fig. 1) [22, 23].

While classically recognized for its selective toxicity to pancreatic β-cells, the ICV administration of STZ at subdiabetogenic doses reproduces key disease features in the central nervous system, such as β-amyloid aggregation, tau protein hyperphosphorylation, neuroglial inflammation, oxidative stress, and progressive cognitive impairment [22]. These effects are closely related to neuronal insulin resistance, as insulin exerts neuroprotective and anti-inflammatory functions in the brain. The dysfunction of this axis directly impacts the activity of glycogen synthase kinase-3 (GSK3), an enzyme regulated by insulin signaling that, when dysregulated, promotes both tau hyperphosphorylation and amyloid deposition [24]. Thus, the STZ model may provide a robust translational alternative tool for investigating neurodegeneration and how metabolic and cell signaling alterations contribute to the neuropathological mechanisms characteristic of Alzheimer's disease.

2 Materials

2.1 Stereotaxic Surgery Materials

1. 70% Ethanol
2. 0.2% Polyacrylic Acid (Adaptis, Ofta Vision Health)
3. Buprenorphine for Veterinary Use (0.1 mg/kg, Beniv, Ourofino)
4. Disposable Cotton Swabs
5. Disposable Sterile Gauze Pads
6. Hydrogen Peroxide 10%
7. Isoflurane
8. Povidone-iodine 10%
9. Spherica Surgical Burr
10. Sterile Saline Solution 0.9%
11. Sterile Scalpel with Blade #22
12. Sterile Surgical Instruments
13. Sterile Synthetic Suture Materials (Nylon 3-0)
14. Streptozotocin (Sigma #S0130)

2.2 Equipment

1. Anesthesia Induction Chamber with Activated Carbon Filter (Harvard Apparatus)
2. Benchtop centrifuge

3. Dental drill (LB 100, Beltec)
4. Dounce tissue grinder with pestle A and B, 2 mL (Merck, #P1610)
5. Hamilton Syringe, Neuros (Model 75 N, Hamilton)
6. Incubator (37°C)
7. Laminar flow cabinet (Bioseg 09, VECO)
8. Micromotor (LB 100, Beltec)
9. Mouse Ear Bars (Model 922, KOPF)
10. Mouse Gas Anesthesia Head Holder (Model 923-B, KOPF)
11. Peristaltic Pump
12. Small Animal Stereotaxic Apparatus (KOPF)
13. TissueLyser II—Sample Disruption/Homogenizer System (Qiagen)

2.3 Culture Media and Buffers

1. LB Medium (Luria–Bertani): Add tryptone to a final concentration of 10 g/L (1.0%), yeast extract to 5 g/L (0.5%), and NaCl to 10 g/L (171 mM). Add approximately 900 mL of ultrapure water to dissolve all components, adjust the pH to 7.0 with NaOH, bring the volume up to 1 L with ultrapure water, and autoclave.
2. LB Kan/Tet plates: Add 15 g/L (1.5%) bacto-agar to LB medium and autoclave. Allow the bottles to cool to approximately 50°C, then add tetracycline (20 μg/mL) and kanamycin (50 μg/mL). Pour the plates, let them solidify, and store them at 4°C in the dark. Plates containing only LB-agar with kanamycin (50 μg/mL) are also used. It is recommended to prepare stock solutions of tetracycline (20 mg/mL) and kanamycin (50 mg/mL), allowing the addition of 1 μL of each antibiotic per milliliter of medium to reach the target final concentrations, regardless of the total preparation volume.
3. PEG/NaCl solution: Dissolve PEG 8000 to a final concentration of 200 g/L (20%) and NaCl to a final concentration of 150 g/L (2.57 M) in 1000 mL of ultrapure water. Stir until completely homogenized and autoclave.
4. PBS 1×: Dissolve 8.00 g/L NaCl (136.9 mM), 0.20 g/L KCl (2.68 mM), 2.68 g/L $Na_2HPO_4 \cdot 7H_2O$ (10 mM), and 0.24 g/L KH_2PO_4 (1.76 mM) in 990 mL of ultrapure water. Adjust the pH to 7.4, bring the volume to 1.0 L with ultrapure water, and autoclave.
5. DMEM High-Glucose Medium (Dulbecco's Modified Eagle Medium): Add powdered high-glucose DMEM to 990 mL of ultrapure water and stir until completely dissolved. Add 3.7 g/L $NaHCO_3$ (44.04 mM) and adjust the pH to 7.4 using HCl. Filter the solution through a 0.22 μm filter.

6. M9 Minimal Medium: Dissolve 10.5 g/L K_2HPO_4 (60.27 mM), 4.5 g/L KH_2PO4 (33.06 mM), 1 g/L $(NH4)_2SO4$ (7.57 mM), 0.5 g/L sodium citrate dihydrate (1.70 mM), and 15 g/L agar (1.5%) in 990 mL of ultrapure water under stirring. Adjust the volume to 1 L and autoclave. Allow the medium to cool to 45°C, then add 1 mL of 1 M $MgSO4{\cdot}7H_2O$ (1 mM final), 0.5 mL of 1% filter-sterilized thiamine-HCl (5 µg/mL final), 10 mL of 20% filter-sterilized glucose (2 g/L; 111 mM final), and 4 mL of Bacto Casamino Acids (0.04 g/L final, Ref 223050, BD Biosciences). Add 1 mL of kanamycin (50 mg/mL stock solution, 50 µg/mL final). Pour the plates, allow them to solidify, and store at 4°C in the dark.

2.4 E. coli *Strain K91*

The strain is maintained on M9 minimal medium plates supplemented with kanamycin (50 µg/mL final) to prevent mutations. On the day prior to use, it is transferred to LB-agar plates containing kanamycin (50 µg/mL). It is essential that the strain used is F^+ (harboring the sex pilus), as this condition is required for successful phage infection.

3 Methods

3.1 Stereotaxic Surgery for Neurodegeneration Model

3.1.1 Preoperative Procedures

1. Weigh all animals scheduled for surgery.
2. Induce anesthesia by placing the animal in an induction chamber with oxygen flow set to approximately 1.5 L/min and isoflurane vaporizer at 3%. During surgery, adjust the flow according to respiratory pattern and heart rate (1%–2%).
3. Once anesthetized, transfer the animal and place it on the stereotaxic apparatus and position its snout, gently fix the teeth, in the mouse gas anesthesia head holder connected, to continue receiving the mixture of oxygen and isoflurane vaporized from anesthesia.
4. Proceed to the correct animal position in the stereotaxic apparatus: position the ears, adjust lateral pins to avoid injury, and lock the nose clamp. Ensure that the animal's head is firmly stabilized and symmetrically held in the stereotaxic (horizontally aligned). Correct positioning at this stage is critical to avoid measurement errors and to reduce surgical time.
5. Apply eye ointment to animals' eyes bilaterally to prevent corneal drying and damage from exposure during the surgical

procedure. Reapply the eye ointment whenever necessary, throughout the surgical procedure.

6. Shave the cranial region (trichotomy) using a scalpel blade.

3.1.2 Surgical Procedure

Preparation of the injection solution: The STZ dose to be administered is 3 mg/kg of body weight. To this end, prepare STZ (diluted 0.05 M citrate buffer, pH 4.5) at a concentration of 20 mg/mL. Considering that our animals weighed approximately 20 g, we used 3 µL of the 20 mg/mL solution per animal. Citrate buffer is also used as the surgical control. A Hamilton syringe (Neuros 75 RN, 5 µL capacity) should be sterilized by ultraviolet irradiation in a laminar flow hood for 30 min prior to use.

Practical considerations: *STZ is easily degraded; we suggest preparing it for immediate use by diluting it with chilled citrate buffer and keeping it on ice and away from light until needed.*

1. Mount the Hamilton syringe onto the stereotaxic holder.
2. Disinfect the cranial surface with cotton swab in Povidone-iodine 10%, avoiding contact with the eyes.
3. Perform a midline incision of approximately 1 cm, starting a few millimeters above the eyes and extending to the base of the ears. Clean the skull with cotton swab soaked in 10% hydrogen peroxide to remove adherent membranes and expose cranial sutures.
4. Verify dorsoventral (DV) alignment: place the needle tip at the Lambda (posterior intersection of sagittal Suture and the right and left portions of Lambdoid Suture) and record the DV measure. Repeat at the Bregma (anterior intersection of the Sagittal and Coronal Sutures). The difference between the two values must not exceed 0.2 mm. If exceeded, reposition the animal by adjusting the anesthesia head holder and repeat measurements. Once symmetry is confirmed (Bregma and Lambda alignment), record the stereotaxic coordinates (antero–posterior, AP; medio–lateral, ML; dorsoventral, DV) with the needle positioned from Bregma (Fig. 2). Calculate injection coordinates by subtracting the following reference values: AP: −0.6 mm; ML: −1.0 mm (right lateral ventricle); DV: −2.5 mm; intracerebroventricular injection (ICV).
5. Lift the needle, adjust AP and ML coordinates, and slowly lower the needle to mark the entry point. Drill carefully with a spherical burr to avoid brain injury. If bleeding occurs, apply cotton swab until hemostasis is achieved, then rinse with saline.
6. Load 3 µL of STZ solution into the Hamilton syringe.
7. Slowly lower the needle to the predetermined DV coordinate.

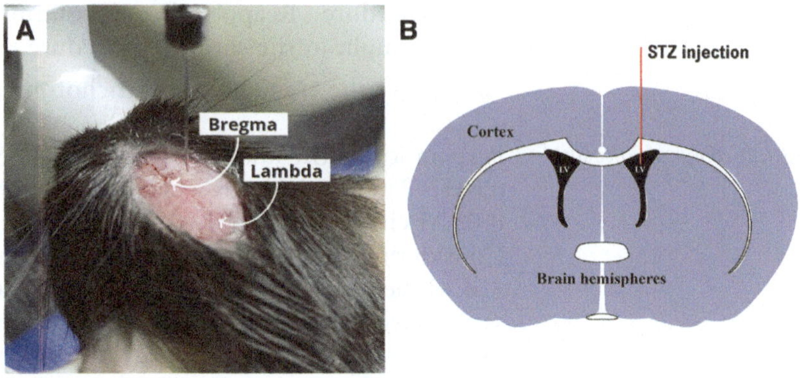

Fig. 2 Cranial landmarks used for stereotaxic positioning. (**A**) Representative image of mouse exposure skull showing the identification of Bregma and Lambda craniometric landmarks; (**B**) Representative coronal section illustrating the target coordinates and trajectory of the local injection. The black-shaded area indicates the space corresponding to the lateral ventricle (LV)

8. Inject 3 μL at a constant rate of 0.5 μL/min. Wait ~2 min before withdrawing the needle slowly over 1 min.
9. Remove the syringe from the stereotaxic frame, cleaning the needle with cotton swab (discard in biological waste).
10. Suture the incision.
11. Disinfect the surgical area with Povidone-iodine 10%, avoiding contact with the animal's eyes.
12. Remove the animal from the stereotaxic apparatus.
13. For analgesic management, administer subcutaneous injection of buprenorphine (0.1 mg/kg). Repeat every 12 h if signs of pain are observed, based on the Grimace scale.
14. Place the animal on its side over a warm surface, keeping the head away from bedding to avoid aspiration and hypothermia during recovery. Monitor until consciousness is restored.
15. House the animal individually in a clean cage for the following days to prevent wound reopening.
16. Between animals, disinfect surgical instruments: immerse in distilled water for 5 min, wash with 70% ethanol, and autoclave afterwards. Alternate instrument sets to ensure sufficient sterilization time. Keep suture materials in 70% ethanol until next use.
17. Monitor animals daily following surgery. The effect of STZ can be assessed by behavioral testing, such as the Barnes Maze, Morris Water Maze, or Novel Object Recognition Test (NOR), as well as by immunohistochemistry using inflammatory markers (Iba1, GFAP) [25].

18. Perform the same procedure in control animals, injecting saline. STZ effects are expected to progressively intensify, while inflammatory responses inherent to surgery diminish, yielding groups with distinct inflammatory profiles.

3.2 In Vivo Panning with Phage Display Library

3.2.1 Procedure

Phage display libraries may be obtained commercially or produced in house [26]. The protocol described here is based on phage display libraries built using the fUSE5 or fUSE55 vector. Please refer to our previous publication regarding more specific details on how to use the phage display system and how to obtain specific reagents such as vectors and bacterial strains [26]. The library used in the assay must undergo two quality assessments before panning: first, the library diversity, which refers to the number of unique peptide sequences present. This number is calculated during library construction and is proportional to the number of transformants obtained during electroporation. If using a commercially available phage display library, please refer to the manufacture for information regarding its diversity. The library we used, for instance, contains sequences of eight random amino acids with a cysteine at each end. To be representative, the library must contain at least one copy of each peptide; therefore, for 20 random amino acids, we have $20^8 = 2.5 \times 10^{10}$ transformants. Keep in mind that a library with a diversity of 10^8–10^9 peptides is fairly diverse; the largest library of chemical compounds is no larger than tens of millions of individual compounds.

The second assessment concerns the titer, which indicates the number of total viral particles present in the library after amplification, expressed in transducing units (TU) or colony-forming units (CFU) per volume unit. For instance, when performing in vivo biopanning, it is essential to ensure that the injected phage population adequately represents the diversity of the library. Considering the practical volume range that can be administered to a mouse (approximately 100 µL), the preparation should contain at least 10^9 transducing units (TU) to maintain a representative sampling of the available clones. In a library with a calculated titer of 5×10^{10} TU/µ L, this corresponds to an injection containing about 5×10^8 to 1×10^9 phage particles, which allows for multiple copies of each unique sequence to be present in the inoculum. Smaller libraries, although less diverse, offer practical advantages in this context: they are easier to construct, yield higher titers, and ensure that each peptide variant is sufficiently represented within the feasible injection volume for in vivo selection.

3.2.2 Phage Display Library Injection

1. Weigh the animal.
2. Check the oxygen level and refill the isoflurane vaporizer if needed. Place the animal in the induction chamber with oxygen flow adjusted to approximately 1.5 L/min and the vaporizer set to 3% isoflurane. During injection, maintain anesthesia between 1% and 2% isoflurane. Adjust the anesthesia flow (isoflurane/oxygen) according to respiratory and heart rate of the animals.
3. Load a syringe with 2×10^{10} TU of the desired library diluted in 1× PBS, for a total volume of 100 μL per animal.
4. Use a red light to promote dilation of the tail vein, taking care to avoid overheating or injury. Clean the tail with gauze soaked in 70% ethanol to enhance vasodilation and the visualization of the vein.
5. Insert the needle approximately 3 mm into the lumen of the lateral tail vein, with the bevel facing upward, preferably in the distal half of the tail (Fig. 3). This vein can be accessed on either side of the mouse tail.
6. Slowly and continuously depress the syringe plunger, monitoring for resistance. Avoid rapid administration, as this may rupture the vessel. Remove the needle after injection.

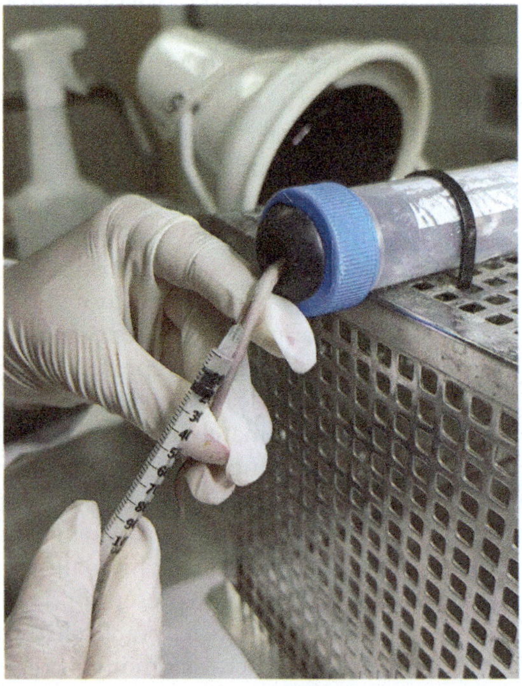

Fig. 3 Intravenous administration of a phage display library in mouse via tail vein injection, using a restraining device to ensure proper immobilization and accurate delivery

7. Apply gentle pressure to the injection site with gauze to achieve hemostasis, and place the animal in a new cage on a heating pad without bedding.
8. 30 min after injection, anesthetize the animals with 4% isoflurane and perform perfusion (10–20 mL/animal) with DMEM. Animals should remain sedated until the diaphragm is opened.

3.2.3 Perfusion to Remove Unbound Phage

1. Place the animal in a supine position, keeping the isoflurane inhaler into the nose. Assess loss of tail and paw responses to confirm the level of sedation. Once verified the loss of reflexes, secure the limbs with tape on a surgical station made of Styrofoam or wood. Next, the animal will be thoracotomized to heart appearance followed by the blood collection and transcardially perfusion via the ascending aorta.
2. Using fine rat tooth tissue forceps, hold the chest skin and make an incision to expose the xiphoid process. Grasp the xiphoid and make lateral incisions under the ribcage to expose the diaphragm and liver.
3. Make cuts on both sides of the thoracic cavity to fully expose the heart and lungs. Carefully tear the pericardial sac and other tissues covering the heart. Proceed to the blood collection.
4. Blood collection: collect at least 20 μL directly from the heart of each animal for injection quality control. For that, gently introduce a small portion of a 1 mL syringe needle positioned into the left ventricle following the aorta angle. Pull lightly until the blood fills the required volume. After that, discard it in a sterile tube and leave it on ice until the infection step. The blood samples should be kept on ice until the Infection Step (3.5).
5. Transcardiac perfusion: stabilize the heart, insert the trimmed needle (infusion needle, 23 gauge) into the left ventricle at an angle approximately parallel to the midline of the heart (aorta artery angle).
6. Before needle introduction into the animal's heart, trim the tip of the infusion needle (23 gauge) to make it slightly blunt, reducing the risk of aortic wall perforation.
7. Apply enough pressure to pierce the ventricular wall, taking care not to advance into the left atrium or right ventricle.
8. Rinse the tubing system with DMEM until all air bubbles are removed. No bubbles should remain in any part of the peristaltic pump, as this is critical for perfusion success.
9. Connect the needle to a 20 mL peristaltic pump system filled with DMEM.
10. Make a small incision in the right atrium; dark venous blood should flow immediately. Start perfusion at a constant rate of

20 mL/min. The liver should gradually blanch. Proceed to the tissue collection and preparation step.

Practical considerations: *If the needle tip enters the right ventricle, perfusion failure may occur, including fluid dripping from the nose or mouth, and incomplete liver blanching, which compromises the efficiency and quality of perfusion.*

3.2.4 Tissue Collection and Homogenization

1. Wash each organ in a 24-well plate containing chilled DMEM. Keep the plate on ice.
2. Weigh each collected organ. This is a necessary step as it allows us to discover the number of infective phage particles recovered per organ mass, which allows us to build a new pool of phages for reinjections in the next rounds.

Collected tissues can be homogenized either manually using a 2 mL Dounce tissue grinder or with the Tissuelyser II (Qiagen). Both options are described below:

(a) Manual Homogenization

- Transfer the tissue into a 2 mL Dounce tissue grinder and add 400 μL of DMEM. It is important to keep the tissue grinder on ice during homogenization to prevent sample heating.
- Perform the initial tissue disruption using pestle A (~10 strokes), followed by pestle B, applying the number of strokes required until the suspension appears fully homogeneous.

Practical considerations: *For fibrous tissues such as muscle and heart, mince the tissue into small pieces (~1 mm³) prior to homogenization with the Dounce tissue grinder.*

(b) Tissuelyser Homogenization

- Preload 2 mL microtubes with ~5 ceramic beads (2.8 mm Ceramic Beads, Omni, catalog no. 19-646).
- Transfer the tissue into the microtube and add 400 μL of DMEM.
- Place the tubes into the Tissuelyser adapter and homogenize at 30 Hz for 1 min.
- Remove the tubes from the instrument and place them on ice to cool the samples.

  The use of the TissueLyser methodology is advantageous, as it allows the simultaneous homogenization of up to 48 samples (in 2 mL tubes)

Posthomogenization processing (applies to both methods):

- Transfer the homogenate to labeled 2 mL tubes and keep on ice.

- Centrifuge at 15,000 × *g*, 4°C, for 5 min.
- Discard the supernatant, add 1 mL of DMEM, and centrifuge again.
- Keep the tubes on ice until infection.

3.2.5 Phage Recovery by Bacterial Infection

1. Grow a streak of *Escherichia coli* K91 colonies in 10 mL of liquid LB with kanamycin (50 μg/mL) in a shaker at 250 rpm and 37°C. Grow until OD 600 nm approximately 0.4–0.6, ~2 h growth.

Practical considerations: *It is important to start the bacterial culture in advance since it may take about 2–3 h to grow and be ready for use. Once the bacterial culture reaches the log-phase stage it should be used with a 15–30 min window).*

2. Add 180 μL of K91 culture (OD 600 nm = 0.4–0.6) to the organ pellet. Incubate for 30 min at 37°C.
3. Grow the bacteria-phage suspension of the tissues in 10 mL LB overnight with kanamycin (50 μg/mL) and tetracycline (20 μg/mL) (16–18 h).

Practical considerations: *In the first round, do not plate recovered phages, as unique sequences may be lost. From the second round onward, plate phages from each organ, as sequences will have been amplified in the first round.*

For verification of injection efficiency, infect 20 μL of blood samples with 180 μL of K91 culture and incubate at 37°C for 30 min. Mix the bacteria–phage suspension thoroughly by pipetting up and down, then plate 1, 10, and 100 μL onto LB agar plates supplemented with tetracycline (20 μg/mL) and kanamycin (50 μg/mL). On the following day, multiple colonies should be observed, indicating successful tail vein injection. In our experience, animals injected with 10^9 TU of the library typically yield titers of approximately 3.75×10^2 TU/μL in blood. However, these values may vary depending on the efficiency of the injection procedure and factors influencing phage binding within the tissues.

3.2.6 Next Day–Phage Recovery

1. Centrifuge the overnight culture at 8000 × *g* for 15 min and transfer the 10 mL of supernatant to a 15 mL tube.
2. Add 1.5 mL of PEG/NaCl solution, close, and mix by inverting the tube several times. Incubate on ice for 30 min.
3. Centrifuge at 10,000 × *g* for 30 min, discard the supernatant, and invert the tube for a few minutes to remove residual liquid.
4. Centrifuge again at 10,000 × *g* for 5 min and discard any remaining liquid.
5. Resuspend the phage pellet in 1 mL of PBS 1×. Incubate for a few minutes at 37°C with rotation at 200 rpm. Centrifuge at 10,000 × *g* for 10 min.

6. Transfer the supernatant to a new 2 mL tube, add 150 μL PEG/NaCl, mix by inverting, and incubate on ice for 30 min.
7. Centrifuge at 10,000 × *g* for 5 min and discard residual liquid.
8. Resuspend the phage pellet in 100 μL sterile PBS 1×, incubate at 37°C with rotation (200 rpm), and centrifuge at 15,000 × *g* for 10 min. Transfer the supernatant to a 1.5 mL tube. Store the phage library at 4°C.

3.2.7 Titration

After amplification and recovery from the second round, perform titration to determine the number of phages recovered from each organ and animal.

1. Inoculate 100 μL of *E. coli* K91 stock or scrape a small amount of bacteria from an agar plate using a loop or P200 tip, into 10 mL of LB or 2xTY medium supplemented with kanamycin (50 μg/mL). Grow in a shaker at 250 rpm, 37°C.

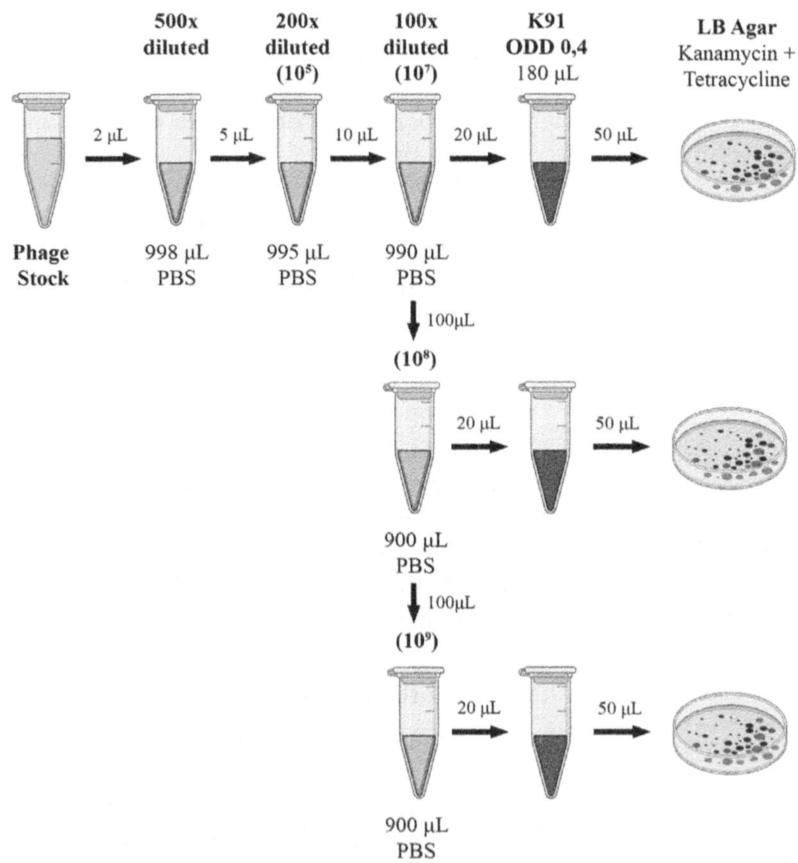

Fig. 4 Representation of the serial dilution of the phage stock for infection of the *E. coli* K91 strain and plating. It is recommended to vortex each dilution for better dilution quality

2. Monitor bacterial growth by measuring OD. (Take 1 mL in a plastic cuvette and read on the spectrophotometer. Record OD every 30 min).
3. When the OD (600 nm) reaches 0.4, inoculate 20 μL of diluted phages into 180 μL of *E. coli* K91 (Fig. 4).
4. Incubate at 37°C for 30 min.
5. Mix the bacteria-phage solution by pipetting up and down, then plate the desired volume onto LB agar plates supplemented with tetracycline (20 μg/mL) and kanamycin (50 μg/mL).

Practical considerations: *(1) If plating 50μL, spread across the entire plate and homogenize the inoculum with glass beads; (2) If plating 10μL, divide the plate into 2–3 sections and homogenize the inoculum using a pipette tip or by gently tilting the plate.*

6. Allow ~5 min for the liquid to be absorbed by the agar. If 50 μL was plated, remove the glass beads afterward.
7. Incubate the plates at 37°C overnight.
8. On the following day, count the colonies.

   Count the colonies on the plates and apply the necessary information to the following formula:
   *TU/μL phage stock = (Number of TU/plated volume) * (total volume of bacteria + phages solution/volume of phages inoculated) * dilution factor of inoculated phages*
   where:
   TU = Transducing Units (phage titer)
   Plated volume = volume plated on agar (μL)
   Total volume of bacteria + phages solution (μL)
   Volume of phages inoculated (μL)
   Dilution factor of inoculated phages = dilution used

For example, if the number of colonies corresponding to the 10^9 dilution is 150, we obtain:

*TU/μL phage stock = (150/50μL) * (200μL/20μL) * 10^9*

The value obtained for the titer per organ should be divided by the mass of the corresponding organ in order to determine the number of phages per gram of tissue. Based on this, a new phage pool must be prepared using phages recovered from all tissues and animals, to be reapplied in the subsequent rounds, considering both the mass of each organ and the amount recovered. It is recommended that, ideally, the new pool must contain approximately 2×10^9 TU diluted in 1× PBS, for a total volume of 100 μL per animal. The same quantity of phages should be used in all rounds in order to avoid the loss of poorly amplified sequences throughout the process.

Furthermore, if phage is recovered from different tissues and organs, for instance, different areas of the brain, a pool representative of each sample should be prepared. We recommend to use the mass of each tissue sample as a reference to prepare the pool of phage. For example, if the cortex weight 100 mg and the olfactory bulb 20 mg, a proportion of 5:1 (v/v or TU/TU) should be used to prepare the pool. Thus, the number of phage particle (in TU) is calculated based on the average weight proportion of each tissue relative to the combined weight of all collected tissues, resulting in a total of 2×10^9 TU per animal. For instance, if the average heart mass corresponds to 8.5% of the total average mass of all collected tissues, then, considering a total of 2×10^9 TU, the reinjection solution should contain 1.7×10^8 TU corresponding to this organ. This calculation is performed for all organs, and the respective volumes are subsequently combined.

3.2.8 Sequences Identification

At the end of the rounds, with all the phages already recovered, the ligand identification step is initiated. For this purpose, the phages must be subjected to PCR for sequence amplification, using primers corresponding to the vector sequence employed in the library construction. These products can then be sequenced and analyzed through bioinformatics to assess specificity by organ or by groups of neurodegenerative versus healthy animals. Here, there are two alternatives. First, individual colonies of phage infected bacteria plated in the last round of selection can be sequence by Phage-PCR (below) followed by Sanger DNA sequencing to identify each individual peptide display by these selected phage clones. However, this method is limited to a few hundred [27] or at most thousands [13] of colonies of individual phage-infected bacteria. In order to sequence all recovered phage present within each sample, next-generation high throughput sequencing must be used. First, a low-amplification PCR is performed followed by the addition of adaptor primer and bar code to amplicons (library preparation) using an appropriate library preparation protocol (i.e., Nextera XT DNA library preparation kit). These protocols are beyond the scope of this book chapter because most sequencing facilities today offer this as a service for a fee. But for reference, a detail protocol can be found in Tang et al. [6].

To perform colony PCR using phage suspensions stored in LB medium supplemented with 10% glycerol, dilute each sample 1:100 in 1× PBS to minimize the inhibitory effect of glycerol on the enzymatic reaction. The PCR mixture should be prepared according to the manufacturer's recommendations for the selected Taq DNA Polymerase, including final concentrations of 0.2 mM for each dNTP, 2 μM for both forward and reverse primers, and approximately 1 unit of Taq DNA Polymerase per reaction.

Thermal cycling conditions are as follows: initial denaturation at 94°C for 5 min, followed by 34 cycles of denaturation at 94°C for 15 s, annealing at 60°C for 20 s, and extension at 72°C for 45 s. A final extension step should be performed at 72°C for 7 min.

If desired, electrophoresis on a 1.5% agarose gel can be used to confirm the presence and size of the amplified fragment.

In our studies employing the fUSE55 phage, the following primers were used:

```
Forward primer: 5'-GATGGTTGTTGTCATTGTCGGCG-3'
Reverse primer: 5'-CAGAGCCACCACCCTCATTTTCAG-3'
```

3.2.9 Evaluating Neurodegenerative Phenotypes

In addition to molecular screening approaches such as in vivo phage display, complementary experimental strategies can be employed to confirm the establishment of neurodegeneration and neuroinflammation in animal models. Suitable evaluation methods include behavioral assessments, which provide indirect evidence of cognitive and motor impairment, as well as biochemical and histological analyses that quantify neuroinflammatory markers and neuronal loss. Such approaches are well established in the literature and contribute to validating the disease-like phenotype induced in the experimental model [25].

4 Conclusion

This protocol demonstrates the integration of stereotaxic neurodegeneration modeling with in vivo phage display as a comprehensive approach for probing blood–brain barrier dynamics and molecular alterations in Alzheimer-like pathology. By combining precise surgical induction of neurodegeneration, iterative ligand selection, and molecular characterization of enriched peptides, the method enables the identification of tissue-specific motifs that distinguish healthy from diseased states. The approach not only advances the understanding of receptor heterogeneity and BBB dysfunction in neurodegeneration but also establishes a robust platform for the discovery of novel diagnostic tools and therapeutic vectors with translational potential in neurological research.

References

1. Smith GP (1985) Filamentous fusion phage: novel expression vectors that display cloned antigens on the virion surface. Science (1979) 228:1315–1317. https://doi.org/10.1126/SCIENCE.4001944
2. Hertveldt K, Beliën T, Volckaert G (2009) General M13 phage display: M13 phage display in identification and characterization of protein-protein interactions. Methods Mol Biol 502:321–339. https://doi.org/10.1007/978-1-60327-565-1_19
3. Pasqualini R, Ruoslahti E (1996) Organ targeting in vivo using phage display peptide libraries. Nature 380:364–366. https://doi.org/10.1038/380364A0
4. Smith GP, Petrenko VA (1997) Phage display. Chem Rev 97:391–410. https://doi.org/10.1021/CR960065D
5. Michon IN, Hauer AD, Von Der Thüsen JH et al (2002) Targeting of peptides to restenotic vascular smooth muscle cells using phage display in vitro and in vivo. BiochimBiophys

Acta Mol Cell Res 1591:87–97. https://doi.org/10.1016/S0167-4889(02)00254-9

6. Tang FHF, Staquicini FI, Teixeira AAR et al (2019) A ligand motif enables differential vascular targeting of endothelial junctions between brain and retina. Proc Natl Acad Sci U S A 116:2300–2305. https://doi.org/10.1073/PNAS.1809483116
7. Arap W, Pasqualini R, Ruoslahti E (1998) Cancer treatment by targeted drug delivery to tumor vasculature in a mouse model. Science 279:377–380. https://doi.org/10.1126/SCIENCE.279.5349.377
8. Ozawa MG, Zurita AJ, Dias-Neto E et al (2008) Beyond receptor expression levels: the relevance of target accessibility in ligand-directed pharmacodelivery systems. Trends Cardiovasc Med 18:126–133. https://doi.org/10.1016/j.tcm.2008.03.001
9. Sidman RL, Li J, Lawrence M et al (2015) The peptidomimetic vasotide targets two retinal VEGF receptors and reduces pathological angiogenesis in murine and nonhuman primate models of retinal disease. Sci Transl Med 7:309ra165. https://doi.org/10.1126/SCITRANSLMED.AAC4882
10. Pasqualini R, Millikan RE, Christianson DR et al (2015) Targeting the interleukin-11 receptor α in metastatic prostate cancer: a first-in-man study. Cancer 121:2411–2421. https://doi.org/10.1002/CNCR.29344
11. McCafferty J, Griffiths AD, Winter G, Chiswell DJ (1990) Phage antibodies: filamentous phage displaying antibody variable domains. Nature 348:552–554. https://doi.org/10.1038/348552A0
12. Barderas R, Benito-Peña E (2019) The 2018 Nobel prize in chemistry: phage display of peptides and antibodies. Anal Bioanal Chem 411:2475–2479. https://doi.org/10.1007/S00216-019-01714-4
13. Arap W, Kolonin MG, Trepel M et al (2002) Steps toward mapping the human vasculature by phage display. Nat Med 8:121–127. https://doi.org/10.1038/NM0202-121
14. Tam SJ, Watts RJ (2010) Connecting vascular and nervous system development: angiogenesis and the blood-brain barrier. Annu Rev Neurosci 33:379–408. https://doi.org/10.1146/ANNUREV-NEURO-060909-152829
15. Park-Windhol C, D'Amore PA (2016) Disorders of vascular permeability. Annu Rev Pathol 11:251–281. https://doi.org/10.1146/ANNUREV-PATHOL-012615-044506
16. Sweeney MD, Zhao Z, Montagne A et al (2019) Blood-brain barrier: from physiology to disease and back. Physiol Rev 99:21–78. https://doi.org/10.1152/PHYSREV.00050.2017
17. Nation DA, Sweeney MD, Montagne A et al (2019) Blood-brain barrier breakdown is an early biomarker of human cognitive dysfunction. Nat Med 25:270–276. https://doi.org/10.1038/S41591-018-0297-Y
18. Montagne A, Barnes SR, Sweeney MD et al (2015) Blood-brain barrier breakdown in the aging human hippocampus. Neuron 85:296–302. https://doi.org/10.1016/j.neuron.2014.12.032
19. Alkhalifa AE, Al-Ghraiybah NF, Odum J et al (2023) Blood–brain barrier breakdown in Alzheimer's disease: mechanisms and targeted strategies. Int J Mol Sci 24(22):16288. https://doi.org/10.3390/IJMS242216288
20. Reiss AB, Glass AD, Wisniewski T et al (2020) Alzheimer's disease: many failed trials, so where do we go from here. J Investig Med 68:1135–1140. https://doi.org/10.1136/JIM-2020-001297
21. King A (2018) The search for better animal models of Alzheimer's disease. Nature 559:S13–S15. https://doi.org/10.1038/D41586-018-05722-9
22. Grieb P (2015) Intracerebroventricular streptozotocin injections as a model of Alzheimer's disease: in search of a relevant mechanism. Mol Neurobiol 53(3):1741–1752. https://doi.org/10.1007/S12035-015-9132-3
23. Kamat PK (2015) Streptozotocin induced Alzheimer's disease like changes and the underlying neural degeneration and regeneration mechanism. Neural Regen Res 10:1050–1052. https://doi.org/10.4103/1673-5374.160076
24. Deng Y, Li B, Liu Y et al (2009) Dysregulation of insulin signaling, glucose transporters, O-GlcNAcylation, and phosphorylation of tau and neurofilaments in the brain: implication for Alzheimer's disease. Am J Pathol 175:2089–2098. https://doi.org/10.2353/ajpath.2009.090157
25. Chen Y, Liang Z, Blanchard J et al (2013) A non-transgenic mouse model (icv-STZ mouse) of Alzheimer's disease: similarities to and differences from the transgenic model (3xTg-AD mouse). Mol Neurobiol 47:711–725. https://doi.org/10.1007/S12035-012-8375-5
26. Giordano RJ, Alecrim LC (2024) Protocols for building and producing high diversity peptide phage display libraries. Methods Mol Biol 2793:3–19. https://doi.org/10.1007/978-1-0716-3798-2_1
27. Giordano RJ, Lahdenranta J, Zhen L et al (2008) Targeted induction of lung endothelial cell apoptosis causes emphysema-like changes in the mouse. J Biol Chem 283:29447–29460. https://doi.org/10.1074/jbc.M804595200

Part V

Nucleic Acid Synthesis and Cellular Recording

Chapter 16

Generation of Diversely Functionalized DNA Libraries Using Ligase-Catalyzed Oligonucleotide Polymerization (LOOPER)

Natalie Khamissi, Luca Rustico and Ryan Hili

Abstract

Ligase-catalyzed oligonucleotide polymerization (LOOPER) enables the synthesis of highly modified nucleic acid libraries, which can be used as pools to evolve functional nucleic acids, such as aptamers. Here, we describe the process to generate a LOOPER-derived ssDNA library comprising 16 different chemical modifications installed in a sequence-defined manner. The preparation of the building blocks, their templated polymerization, and strand separation to generate the modified ssDNA library are outlined.

Keywords Modified nucleic acids, Aptamers, DNA templated, Combinatorial library, Ligation

1 Introduction

Aptamers are single-stranded nucleic acid polymers that can exhibit high affinity and selectivity for their molecular targets. Their development has been a boon to biomedical research as they can be readily evolved in the lab by in vitro selection to bind biological targets ranging from small molecules to proteins. Aptamers demonstrate unique advantages over their protein counterparts—antibodies. Their reversible denaturation/renaturation, thermal and chemical stability, and predictable and scalable synthesis, enables biomedical and diagnostic applications beyond the scope of antibodies [1]. Notwithstanding the benefit of these salient features, canonical aptamer selections have a modest success rate with a reported >70% of selections failing to yield a high-affinity aptamer [2]. This shortcoming can be remedied by addressing the dichotomy between nucleic acids and proteins in chemical diversity. Nucleic acids, particularly DNA, is a highly stable long-term information carrier, which has evolved with a limited set of functional groups to preclude undesired reactivity. Contrast this with the rich chemical diversity displayed in proteinogenic polymers, which use heteromultivalency—the sum of low-affinity molecular interactions—to achieve specific and high-affinity molecular recognition [3].

Huan Peng and S. Brian Kim (eds.), *Biological Macromolecule Engineering Technologies*,
Methods in Molecular Biology, vol. 3032, https://doi.org/10.1007/978-1-0716-5257-2_16,

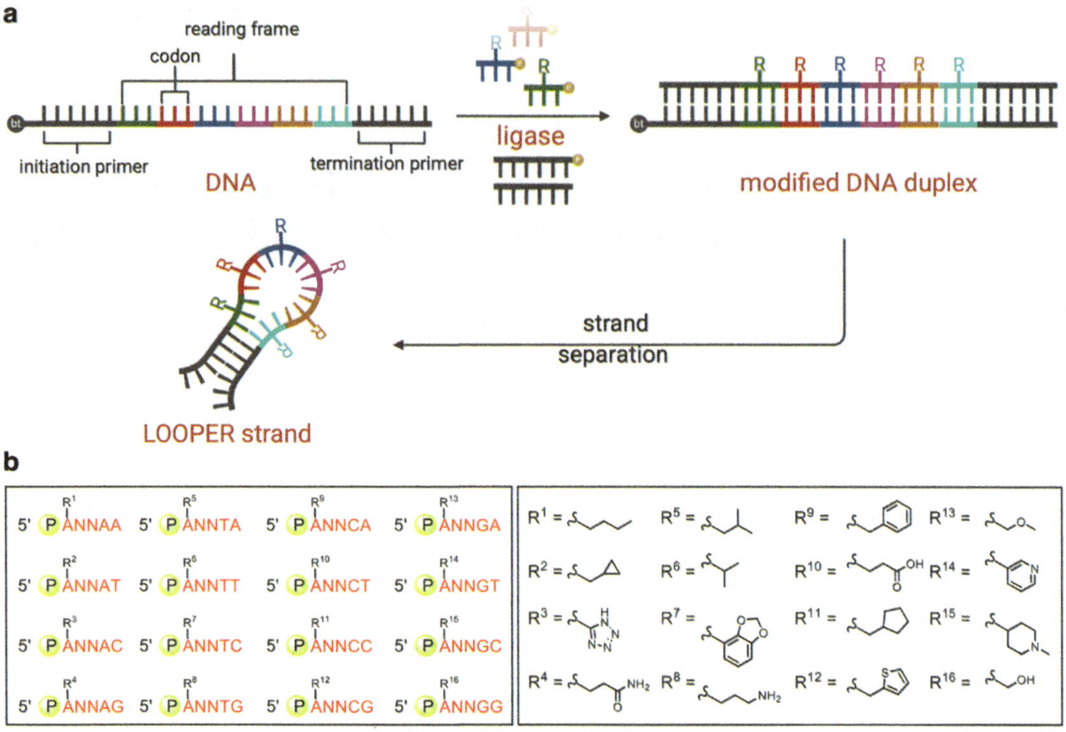

Fig. 1 (**a**) workflow for LOOPER to generate modified single-stranded nucleic acid libraries. (**b**) anticodon modification table highlighting the encoding of modifications by the dinucleotides at the 3'-end of the anticodon. Note that the "N" represents a mixture of A, C, G, and T, and is thus a degenerate base

Chemists have explored ways to expand the chemical repertoire of nucleic acids through modification of the phosphate backbone [4–6], modifications on the nucleobase [7], or by expanding the genetic alphabet beyond canonical A–T/U and C–G base pairing [8]. This strategy has enhanced the functional capacity of nucleic acid polymers for molecular recognition [9], but also enabled new catalytic function that was previously inaccessible with unmodified DNA [10]. These approaches all share one feature in common: they rely upon polymerases to install modifications. This inherently limits the number of unique modifications to a maximum of four when using a standard genetic code [11]. While selections including up to three modifications have been successfully performed and have opened access to new molecular function [12], the scope and diversity of such polymers still falls far below that observed in proteins. With less tractable molecular targets, such as carbohydrates or small molecules, chemical diversity may be the ultimate driving force for a successful selection. A non-polymerase approach is essential to break beyond the limited diversity available through conventional modified aptamer selections.

Our lab has developed ligase-catalyzed oligonucleotide polymerization (LOOPER), which is an enzymatic method that enables the generation of sequence-defined, diversely functionalized DNA libraries with chemical diversity similar to proteins [13–15]. In contrast to polymerase-based methods to generate modified DNA, LOOPER, in the protocol presented here, permits the incorporation of 16 different chemical modifications via the T4 DNA ligase-catalyzed DNA-templated polymerization of a library of modified pentanucleotides anticodons along a library of DNA templates with high efficiency and fidelity (Fig. 1). This protocol describes the synthesis of a validated modified pentanucleotide anticodon library [13] and its polymerization by LOOPER along with the isolation of the single-stranded LOOPER library product. These libraries have been used as inputs for in vitro selection to generate modified aptamers against protein targets [14], the generation of peptide display [16], and can be foreseeably used in other nucleic acids applications.

2 Materials

2.1 Amine-Modified Pentanucleotide Anticodons

The amine site for conjugation of different modifications is positioned at the 5′-end of the pentanucleotide anticodon via a C6-amino dA (Fig. 2) (*see* **Note 1**). Amine-containing pentanucleotide anticodons may be purchased directly from most DNA synthesis companies (*see* **Note 2**); however, they may be synthesized on any DNA synthesizer using standard phosphonamidite chemistry (*see* **Note 3**). The required sequences are below:

1. /5phos//iAmMC6A/NNAA
2. /5phos//iAmMC6A/NNAT
3. /5phos//iAmMC6A/NNAC
4. /5phos//iAmMC6A/NNAG
5. /5phos//iAmMC6A/NNTA
6. /5phos//iAmMC6A/NNTT
7. /5phos//iAmMC6A/NNTC
8. /5phos//iAmMC6A/NNTG
9. /5phos//iAmMC6A/NNCA
10. /5phos//iAmMC6A/NNCT
11. /5phos//iAmMC6A/NNCC
12. /5phos//iAmMC6A/NNCG
13. /5phos//iAmMC6A/NNGA
14. /5phos//iAmMC6A/NNGT
15. /5phos//iAmMC6A/NNGC
16. /5phos//iAmMC6A/NNGG

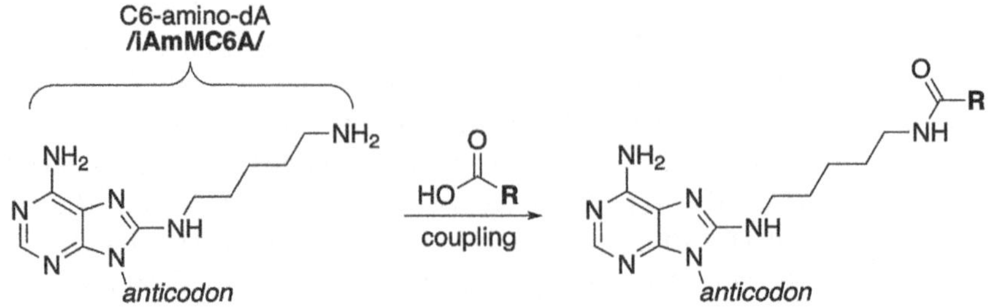

Fig. 2 General process for modification of amine-modified pentanucleotide anticodons to generate functionalized pentanucleotide anticodons. Note that coupling with an anhydride in place of the carboxylic acid is also described

2.2 Functionalization of Pentanucleotide Anticodons

The choice of modifications to present on the anticodon library is critical to the success of in vitro selections. Diverse chemical modifications are compatible with LOOPER (*see* **Note 4**), and a diverse library has been developed explicitly for the selection of high-affinity modified aptamers for protein targets [14, 17]. If anticodon modifications are followed as outlined in this protocol, no fidelity or efficiency assessment will be required; however, customized libraries with significantly different functional group modifications or sizes will require fidelity and efficiency assessment, which has been described in detail elsewhere [18].

1. Valeric acid.
2. Cyclopropylacetic acid.
3. 1*H*-Tetrazole-5-acetic acid.
4. Succinamic acid.
5. Isovaleric acid.
6. Isobutyric acid.
7. 3,4-(Methylenedioxy)phenylacetic acid.
8. Phenylacetic acid.
9. Succinic anhydride.
10. Cyclopentylacetic acid.
11. 3-Thiopheneacetic acid.
12. 1-methyl-4-piperidinecarboxylic acid.
13. Glycolic acid.
14. Nicotinic acid.
15. *N*-Hydroxysulfosuccinimide sodium salt.
16. 1-(3-Dimethylaminopropyl)-3-ethylcarbodiimide (free base form).
17. DMSO, ≥99.9%.
18. Mini tube rotator.

19. 500 mM Na_2CO_3 in H_2O adjusted to pH 9 using conc. NaOH.
20. 500 mM Tris–HCl in H_2O adjusted to pH 8 using conc. HCl.
21. HPLC equipped with a C18 column (Agilent, Eclipse XDB-C18).
22. HPLC solvent A: 0.1 M triethylamine in H_2O adjusted to pH 7 using acetic acid. It can be stored protected from light at 4°C for up to 6 months.
23. HPLC solvent B: HPLC-grade acetonitrile.
24. Vortex.

2.3 Ligase-Catalyzed Oligonucleotide Polymerization (LOOPER)

1. Thermocycler.
2. T4 DNA ligase (New England Biolabs, M0202).
3. Bovine serum albumin.
4. 100 mM ATP: make 250 μM stock in water and store at −20°C.
5. DNA loading dye, 6× SDS dye.
6. MinElute® PCR purification kit.
7. LOOPER template: /5BioTinTEG/GAT TCG CCT GCC GTC GCA NNNNT NNNNT NNNNT NNNNT NNNNT NNNNT NNNNT NNNNT CAC GTG GAG CTC GGA TCC (IDT) (*see* **Note 5**).
8. Primer 1: /5Phos/TGCGACGGCAGGCGAATC (*see* **Note 6**).
9. Primer 2: GGATCCGAGCTCCACGTG (*see* **Note 6**).
10. 4× Ligation buffer: 40 mM $MgCl_2$, 24% w/v PEG6000, 40 mM dithiothreitol, 264 mM Tris pH 7.6. Can be stored at −20°C for up to 12 months.
11. 10% TBE–Urea polyacrylamide gels.
12. Ethidium bromide solution (10 mg/mL).

2.4 Strand Separation of LOOPER Product

1. Magnetic rack.
2. Dynabeads™ MyOne™ Streptavidin C1 magnetic beads (ThermoFisher, 65001).
3. 150 mM NaOH solution.
4. 1.5 M HCl solution.
5. 2× Separation buffer: 80 mM HEPES, pH 7.5, 250 mM NaCl, 10 mM KCl, 2 mM $MgCl_2$, 2 mM $CaCl_2$, and 0.1% TWEEN-20 in H_2O. Can be stored at 4°C for up to 3 months.
6. Centri-Sep Spin Columns (Princeton Separations).

3 Methods

3.1 Coupling of Carboxylic Acid Group Onto Amine-Modified Pentanucleotide Anticodons

1. Freshly prepare a 100 mM solution of carboxylic acid in DMSO (*see* **Note 7**).
2. Freshly prepare a 100 mM solution of 1-(3-Dimethylaminopropyl)-3-ethylcarbodiimide (*see* **Note 8**) in DMSO (*see* **Note 9**).
3. Freshly prepare a 100 mM solution of *N*-Hydroxysulfosuccinimide in 1:1 mixture (*v/v*) of DMSO and water (*see* **Note 9**).
4. In a 1.5 mL microcentrifuge tube, add 7.5 μL DMSO, 25 μL of 100 mM carboxylic acid, 25 μL of 100 mM *N*-Hydroxysulfosuccinimide, and 5 μL of 100 mM 1-(3-Dimethylaminopropyl)-3-ethylcarbodiimide.
5. Incubate at room temperature for 30 min on tube rotator.
6. Add 30 μL 500 mM Na_2CO_3 (pH 9) and 7.5 μL 1.5 mM amine-modified pentanucleotide anticodon.
7. Incubate at room temperature overnight on rotator.
8. Quench reaction with 15 μL Tris–HCl (pH 8) buffer.
9. Lyophilize-quenched reaction mixture.
10. Proceed to HPLC purification (see Sect. 3.3).

3.2 Coupling of Anhydrides Onto Amine-Modified Pentanucleotide Anticodons

1. Freshly prepare a 100 mM solution of anhydride in DMSO.
2. Mix 25 μL 100 mM anhydride, 25 μL 1 M $NaHCO_3$, 163 μL DMSO, 22.4 μL water, and 14.6 μL 1.5 mM amine-modified pentanucleotide anticodon.
3. Incubate at room temperature overnight on vortex.
4. Quench reaction with 50 μL Tris–HCl (pH 8) buffer.
5. Lyophilize-quenched reaction mixture.

3.3 HPLC Purification of Functionalized Pentanucleotide Anticodons

1. Resuspend lyophilized-quenched coupling reaction mixture into 100 μL of HPLC solvent A and transfer to a vial that is compatible with the HPLC instrument used for purification. Run the HPLC purification program shown in Table 1. Collect peak fractions that elute after starting material (Fig. 3).
2. Confirm fraction contains library by negative mode electrospray ionization (ESI) mass spectroscopy (Fig. 4) (*see* **Note 10**).
3. Prepare a 480 μM solution of each functionalized pentanucleotide in water (*see* **Note 11**) and store at ≤−20°C.
4. Once all functionalized pentanucleotides have been synthesized, create a 480 μM anticodon mix library by combining each 480 μM functionalized pentanucleotide at equal volume. Store this mix at ≤ −20°C.

Table 1
HPLC program for purification of functionalized pentanucleotide anticodons

Time (min)	Solvent A (%)	Solvent B (%)	Flow (mL/min)
0	90	10	4
10	80	20	4
23	55	45	4
26	20	80	4
28	20	80	4
29	90	10	4
31	90	10	4

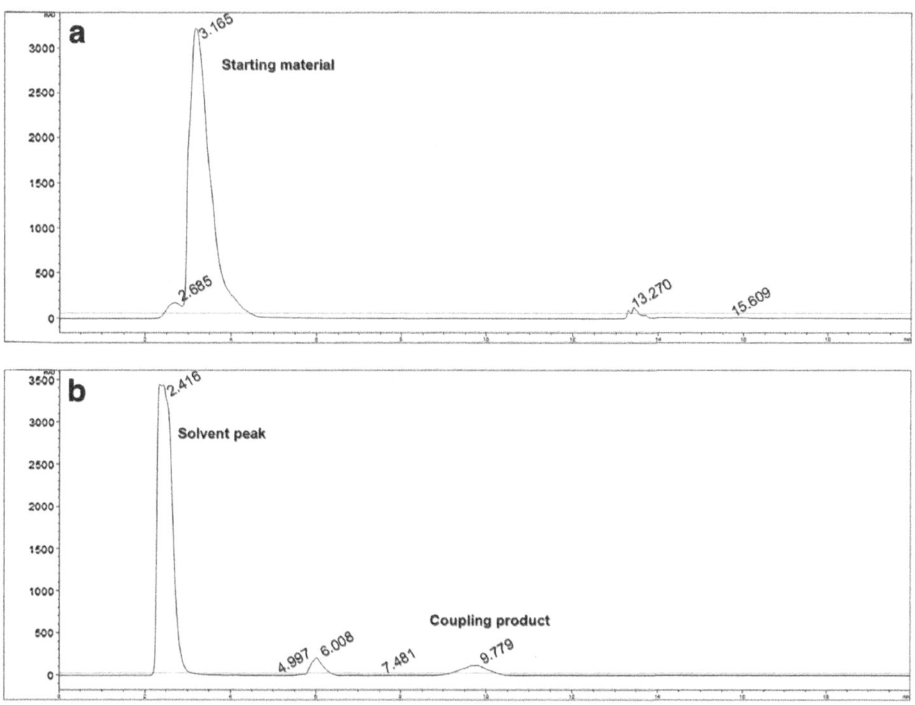

Fig. 3 Example of HPLC chromatograms for (**a**) amine-modified anticodon starting material /5phos//iAm-MC6A/NNCA and (**b**) post-coupling with phenyl acetic acid

3.4 Ligase-Catalyzed Oligonucleotide Polymerization (LOOPER)

1. Mix 1.5 μL of 10 μM DNA template, 2.25 μL of 10 μM primer 1, 2.25 μL of 10 μM primer 2, 2 μL of 250 μM ATP, 5 μL 4× ligation buffer, and 4 μL H_2O in a standard PCR tube (*see* **Note 12**).
2. Heat to 90°C for 2 min on a thermocycler, then cool to 25°C at the rate of 0.1°C/s.

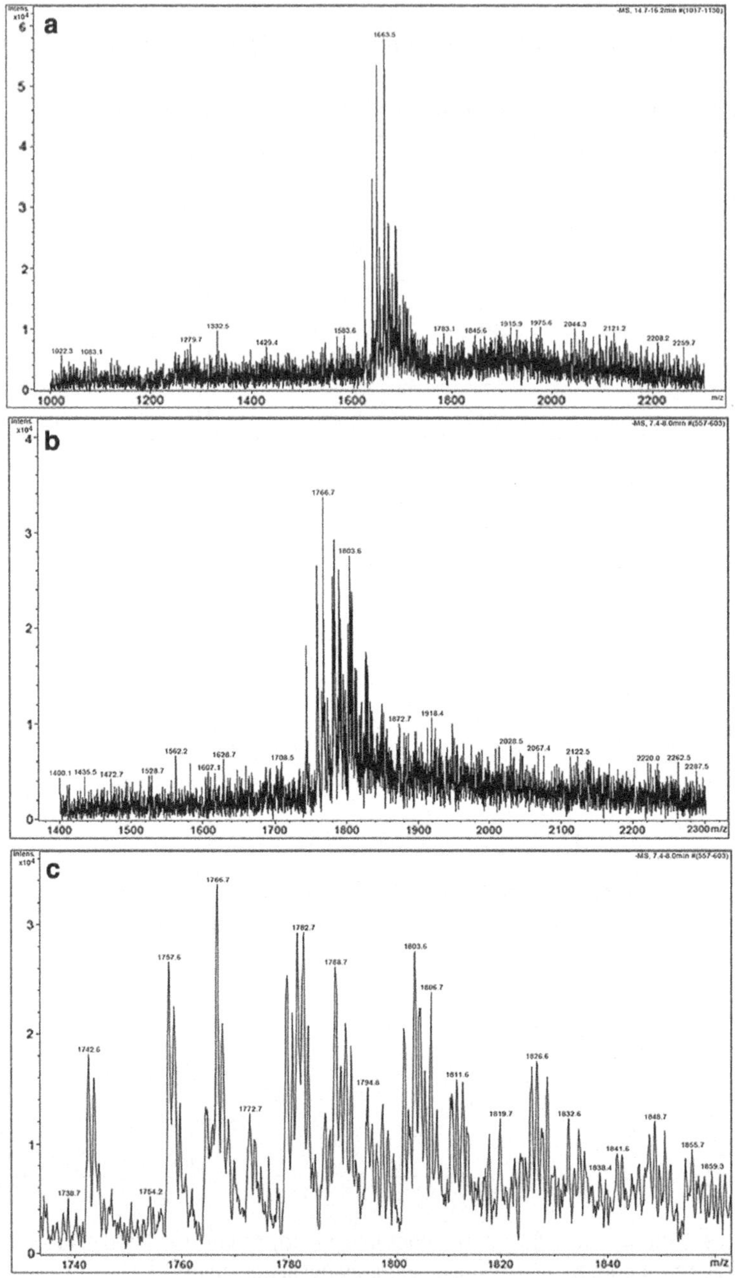

Fig. 4 Example of negative mode electrospray ionization mass spectrometry analysis of (**a**) /5phos//iAmMC6A/NNCA before coupling and (**b**) after coupling to phenyl acetic acid. (**c**) shows a zoomed-in spectra of the post-coupling library mass peaks

3. Add 1 μL 480 μM pentanucleotide mix, 1 μL 2 mg/mL BSA and 1 μL 400 U/μL T4 DNA ligase.
4. Incubate the reaction at 25°C for 24 h.

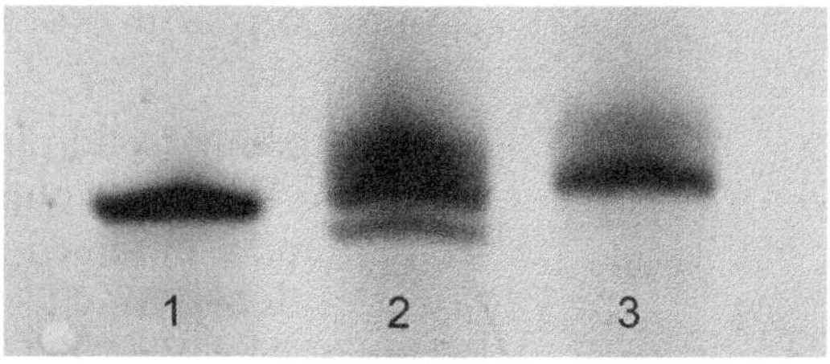

Fig. 5 Denaturing PAGE of LOOPER products. Lane 1: biotinylated 76-mer template; lane 2: LOOPER process with modified anticodon library along LOOPER template. Note that the biotinylated template is the bottom strand; lane 3: strand-separated LOOPER strand

5. Purify the products with MinElute® PCR purification kit (or similar) and elute in water.
6. Evaporate the water to dry the product using a SpeedVac.

3.5 Strand Separation of LOOPER Product

1. Prewash 5 μL streptavidin magnetic beads with 50 μL 1× strand-separation buffer, three times (*see* **Note 13**).
2. Add 25 μL of 2× separation buffer to the purified DNA (25 μL) and add the mixture to the washed beads.
3. Incubate mixture at room temperature for 30 min on rotator. This time can be used to hydrate Princeton Separation columns with water.
4. Wash streptavidin beads with 50 μL 1× strand separation buffer, three times.
5. Add 40 μL 150 mM NaOH to beads to elute non-biotinylated strand. Mix by pipetting; do not vortex.
6. Place mixture on magnetic rack and allow beads to aggregate for 1 min.
7. Transfer supernatant elution to a new microcentrifuge tube.
8. Add 4 μL 1.5 M HCl to neutralize elution.
9. Purify neutralized elution with Princeton Separation.
10. LOOPER and its strand-separated product can be evaluated using 10% TBE–urea polyacrylamide gel electrophoresis at 150 V for 40 min, followed by ethidium bromide (0.5 μg/mL) staining (see Fig. 5).

4 Notes

1. The location and nature of the amine modification within the anticodon greatly impacts the efficiency and fidelity of LOOPER [13, 16, 19].

2. /iAmMC6A/ is an off-catalog order for IDT, but available as a quote.
3. If conducting solid-phase DNA synthesis, we recommend the following phosphoramidite modifiers for synthesis: (1) amine-modified C6 dA (Glen Research, 10-1089); (2) dA+dC+dG+dT−CE Phosphonamidite Mix (Glen Research, Mix N), (3) Chemical phosphorylation reagent II (Glen Research, 10-1901).
4. We have shown that a diverse set of small molecules can be tolerated, including acids, bases, aliphatics, aromatics, and polar groups [13]. We have also demonstrated that LOOPER can tolerate larger fragments, such as oligopeptides, up to eight amino acids in length [16]. However, there is some evidence that aldehydes are poorly tolerated by T4 DNA ligase [20]; thus, we would recommend avoiding aldehydes or other reactive electrophiles as substrates during LOOPER. Since strand separation uses alkaline conditions to generate the modified ssDNA libraries following LOOPER, the modification must not be base labile, such as esters.
5. LOOPER efficiency drops off as the number of codons in the reading frame increases. The system has been optimized to polymerize eight anticodons, resulting in a 40 nt variable aptamer region. This can be increased; however, assessment of polymerization yield would need to be performed by gel analysis as described in this protocol.
6. We have not observed any influence of primer sequence on LOOPER efficiency or fidelity; thus, primer sequences may be changed so long as they are designed to enable efficient PCR.
7. Some carboxylic acids may be less stable than others in solution. Freshly made solutions will ensure highest yields.
8. Reagents stored at ≤4°C should be warmed to room temperature before opening to ensure that no water is condensed inside reagent bottle. Seal the lid of the EDC reagent bottle with parafilm before storing to minimize hydrolysis.
9. *N*-Hydroxysulfosuccinimide and 1-(3-Dimethylaminopropyl)-3-ethylcarbodiimide are carboxylic acid activators. Solutions must be freshly made.
10. The library of pentanucleotides will appear as a multiplet of peaks in ESI-MS. This is because the library comprises 16 anticodons. You should be able to identify each member from lightest to heaviest. In this example of /5phos//Am-C6-dA/NNCA, the lightest peak comes from /5phos//Am-C6-dA/CCCA, while the heaviest peak comes from /5phos//Am-C6-dA/GGCA.
11. Oligonucleotide concentration measurement was performed by UV–vis spectrophotometry using a Nanodrop (Thermo)

monitoring absorbance at 260 nm. Concentrations were calculated using an online calculator (http://oligocalc.eu). Since the measurement is on a library, the average between the lowest and highest calculated concentration was used. Due to lack of known extinction coefficients for diversely modified oligonucleotides, we ignored the modifier contribution to concentration measurements.

12. This will produce a theoretical yield of 15 pmol. If more is required, conduct more polymerizations in parallel. For SELEX, we routinely conduct eight polymerizations in a 8-well PCR strip to yield ~10^{14} library members at 120 pmol scale. For purification, pool all eight LOOPER reactions and purify with one MinElute® PCR Purification Kit, which can bind ~1 nmol of duplex library.
13. The amount scales linearly with the amount of product purified. For instance, in a 120 pmol scale LOOPER, use 40 μL of beads. See manufacturer guidelines.

References

1. Keefe AD, Pai S, Ellington A (2010) Aptamers as therapeutics. Nat Rev Drug Disc 9:537–550
2. Gold L, Ayers D, Bertino J, Bock C, Bock A, Brody EN et al (2010) Aptamer-based multiplexed proteomic technology for biomarker discovery. PLoS ONE 5:e15004
3. Mammen M, Choi S-K, Whitesides GM (1998) Polyvalent interactions in biological systems: implications for design and use of multivalent ligands and inhibitors. Angew Chem Int Ed 37:2754–2794
4. Eschenmoser A (1999) Chemical etiology of nucleic acid structure. Science 284:2118–2124
5. Chaput JC (2021) Redesigning the genetic polymers of life. Acc Chem Res 54:1056–1065
6. Chaput JC, Egli M, Herdewijn P (2025) The XNA alphabet. Nucl Acids Res 53:gkaf635. https://doi.org/10.1093/nar/gkaf635
7. Kong D, Yeung W, Hili R (2016) Generation of synthetic copolymer libraries by combinatorial assembly on nucleic acid templates. ACS Comb Sci 18:355–370
8. Feldman AW, Romesberg FE (2018) Expansion of the genetic alphabet: a chemist's approach to synthetic biology. Acc Chem Res 51:394–403
9. Lipi F, Chen S, Chakravarthy M, Rakesh S, Veedu RN (2016) In vitro evolution of chemically-modified nucleic acid aptamers: pros and cons, and comprehensive selection strategies. RNA Biol 13:1232–1245
10. Huang P-JJ, Liu J (2020) In vitro selection of chemically modified DNAzymes. ChemistryOpen 9:1046–1059
11. Jäger S, Rasched G, Kornreich-Leshem H, Engeser M, Thum O, Famulok M (2005) A versatile toolbox for variable DNA functionalization at high density. J Am Chem Soc 127:15071–15082
12. Wang Y, Liu E, Lam CH, Perrin DM (2018) A densely modified M2+-independent DNAzyme that cleaves RNA efficiently with multiple catalytic turnover. Chem Sci 9:1813–1821
13. Kong D, Lei Y, Yeung W, Hili R (2016) Enzymatic synthesis of sequence-defined synthetic nucleic acid polymers with diverse functional groups. Angew Chem Int Ed 55:13164–13168
14. Kong D, Yeung W, Hili R (2017) In vitro selection of diversely functionalized aptamers. J Am Chem Soc 139:13977–13980
15. Movahedi M, Khamissi N, Hili R (2021) Ligase-catalyzed oligonucleotide polymerization (LOOPER): evolution of chemically diverse aptamer libraries. Aptamers 5:22–30
16. Guo C, Watkins CP, Hili R (2015) Sequence-defined scaffolding of peptides on nucleic acid polymers. J Am Chem Soc 137:11191–11196
17. Kong D, Movahedi M, Mahdavi-Amiri Y, Yeung W, Tiburcio TR, Chen D, Hili R (2020) Evolutionary outcomes of diversely-functionalized aptamers isolated from in vitro evolution. ACS Synth Biol 9:43–52
18. Lei Y, Kong D, Hili R (2015) A high-fidelity codon set for the T4 DNA ligase-catalyzed polymerization of modified oligonucleotides. ACS Comb Sci 17:716–721

19. Guo C, Mahdavi-Amiri Y, Hili R (2019) Influence of linker length on ligase-catalyzed oligonucleotide polymerization. ChemBioChem 20:793–799

20. Guo C, Hili R (2017) Fidelity of the DNA ligase-catalyzed scaffolding of peptide fragments on nucleic acid polymers. Bioconjugate Chem 28:314–318

Check for updates

Chapter 17

Solid Phase Chemical Synthesis of RNA

Joshua O'Grady, Zidi Lyu, Masad J. Damha and David Sabatino

Abstract

Solid phase oligonucleotide synthesis enables the generation of various biologically relevant RNAs. These include RNA sequences 20–80 nucleotides in length, such as microRNA (miRNA), small interfering RNA (siRNA), self-amplifying RNA (saRNA), and RNA aptamers. Modern advances in RNA chemical synthesis have also produced longer RNA sequences (≥100 nucleotides), such as guide RNA (gRNA), messenger RNA (mRNA), riboswitches, and ribozymes. Further, the introduction of orthogonal protecting and reactive groups allows for the chemical synthesis of branch RNA as mimics of intronic RNA lariats. Here, we describe a solid phase chemical synthesis protocol based on *t*-butydimethylsilyl (TBDMS) phosphoramidite chemistry to produce native or branched RNA suitable for structural, functional, or biological studies.

Keywords RNA, Phosphoramidite chemistry, Solid phase oligonucleotide synthesis, Branch and lariat RNA, Therapeutic oligonucleotides

1 Introduction

The evolution of oligonucleotide synthesis began in the 1950–1960s when short block fragments of DNA-based oligonucleotides were generated in solution using the H-phosphonate [1], phosphodiester [2–4], and phosphotriester [5–8] chemistries developed by Todd, Khorana, Letsinger, and others [1–4, 5, 6]. Letsinger and colleagues reported the earliest work on solid-phase oligonucleotide synthesis in 1966, describing the phosphotriester synthesis of dCTTT and dCTA on a polystyrene solid support [9]. This pioneering work laid the foundation for automated solid-phase chemistry as the basis for modern oligonucleotide production [10]. The 1970s produced evolutionary advances in catalysis for phosphotriester synthesis [11–13], in solution [13, 14] and on solid support [15–17], and an increased understanding of the orthogonal deprotection and coupling conditions needed to overcome nucleotide stability issues [13, 14]. The 1970s ended with the introduction of the P(III) phosphite triester method [18], later refined by Caruthers and coworkers [19], which

Huan Peng and S. Brian Kim (eds.), *Biological Macromolecule Engineering Technologies*,
Methods in Molecular Biology, vol. 3032, https://doi.org/10.1007/978-1-0716-5257-2_17,

would dramatically accelerate the chemical synthesis of oligonucleotides, leading to automation [20, 21]. This seminal work served as an entry point into oligonucleotide synthesis on silica beads [22–24] and other solid supports. The development of orthogonal protecting group chemistry was necessary for RNA synthesis to prevent side reactions or degradation due to the presence and reactivity of the 2′-OH group [25]. The *tert*-butyldimethylsilyl (TBDMS) group, a 2′-hydroxyl protecting group developed by Ogilvie and coworkers for RNA synthesis, is widely used as it is stable under many conditions but can be selectively removed with fluoride ions [26]. RNA synthesis using 2′-*O*-TBDMS protection and solid-phase phosphoramidite coupling chemistry has proved successful [27]. The impact of this approach was demonstrated in 1988 when a 77-nt long tRNA analogue was synthesized by stepwise solid-phase synthesis [28]. Synthetic optimization has resulted in the introduction of less sterically hindered 2′-OH protecting groups, such as 2′-*O*-(triisopropylsilyl) oxymethyl (TOM) [29], 2′-*O*-bis(2-acetoxyethoxy)methyl (ACE) [30], 2′-*O*-cyanoethoxymethyl (CEM) [31], 2′-*O*-tert-butyldithiomethyl (DTM) [32], and the 2′-*O*-acetal levunlinic ester (ALE) [33], to boost coupling efficiencies for generating RNA sequences exceeding 100 nucleotides [34–36]. Further advancements have included the development of labile nucleobase protection for rapid removal post-synthesis [37]. Despite these important contributions to the chemical synthesis of RNA, the TBDMS phosphoramidite method remains the gold standard for generating synthetic oligoribonucleotides for biological and therapeutic applications (Fig. 1) [38].

Automated oligonucleotide synthesis initiated a tidal wave of nucleic acids research leading to applications in chemical biology, biosensing, and therapeutics, among many others [39–41]. The field continues to grow with significant contributions and milestones. Notably, the development of DNA and RNA therapeutics has increased significantly, leading to the FDA approval of several oligonucleotide drugs as safe, precise, and effective alternatives to traditional small molecule therapies [41]. There are currently 26 oligonucleotide-based therapeutics approved for clinical use, including RNA-based therapies related to, small interfering RNA (siRNA), self-amplifying RNA (saRNA), mRNA, aptamers, and CRISPR gRNA gene regulatory and editing biotechnologies for a variety of diseases [41]. Notably, the RNA 2′-TBDMS phosphoramidite chemistry was at the cornerstone of Patisiran synthesis, the first siRNA drug approved for clinical use [42].

Despite the success of oligonucleotide therapeutics, several challenges hinder the clinical pipeline advancement of new and improved leads. In a therapeutic context, native oligonucleotide sequences are limited by their susceptibility to nuclease degradation, rapid renal clearance, and immune system recognition [43]. Modern synthetic methods have been used to implement noncanonical modifications to DNA and RNA to address these limitations, thereby enhancing overall bioactivity

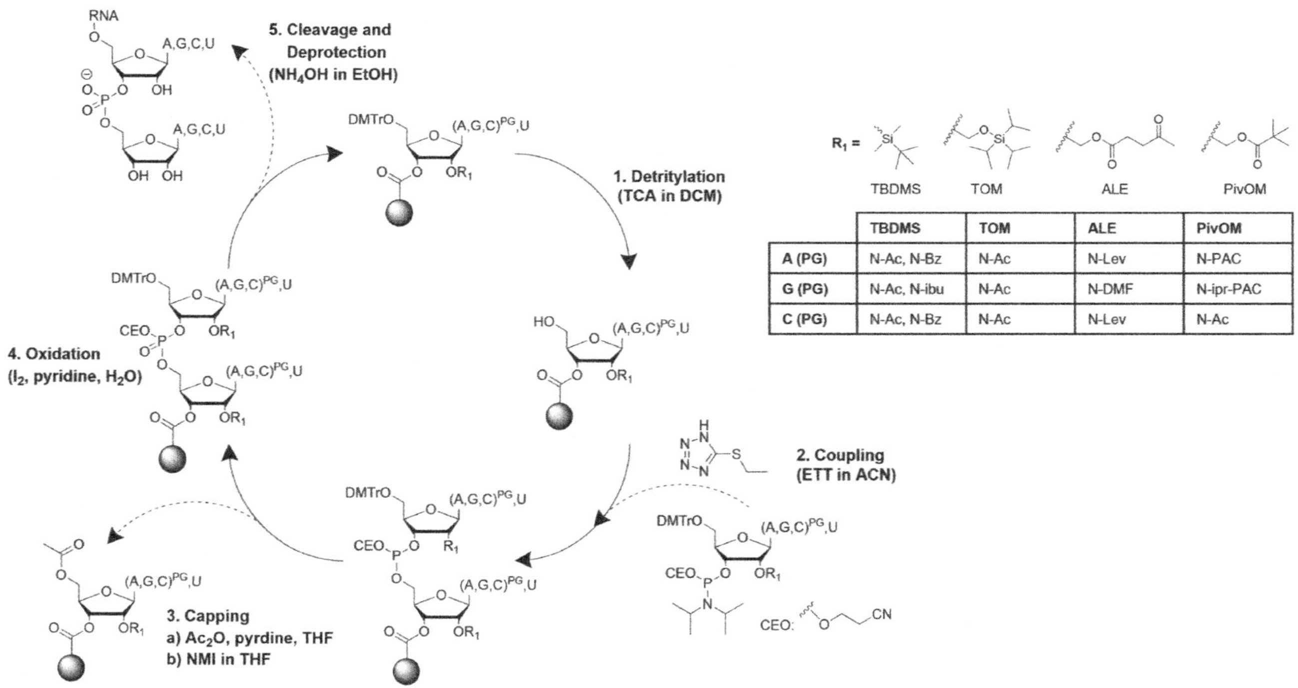

	TBDMS	TOM	ALE	PivOM
A (PG)	N-Ac, N-Bz	N-Ac	N-Lev	N-PAC
G (PG)	N-Ac, N-ibu	N-Ac	N-DMF	N-ipr-PAC
C (PG)	N-Ac, N-Bz	N-Ac	N-Lev	N-Ac

Fig. 1 Phosphoramidite synthesis cycle to manufacture RNA and the associated 2'-OH and nucleobase protecting group chemistries

profiles and increasing therapeutic efficacy [38, 41]. Modifications to the ribose sugar, phosphodiester backbone or the nucleobase have been made to reengineer and improve the structure–function properties of modified RNA in biological systems [41, 44]. Similarly, the use of commercially available phosphoramidite modifiers enables conjugation of oligonucleotides with different substrates or biological probes [45–48]. For example, thiol modifiers allow direct RNA functionalization of gold surfaces for biosensing (surface plasmon resonance chips or gold nanoparticles) applications [49]. Amino modifiers facilitate conjugation to fluorophores, such as fluorescein isothiocyanate (FITC), for visualizing RNA biology in cells [50]. Orthogonally protected branching amidites allow for the generation of branched RNAs (bRNA) that contain vicinal 2′-5′ and 3′-5′ phosphodiester linkages to mimic natural lariat RNA intermediates formed during pre-mRNA splicing (Fig. 2) [51–53]. Branched RNA architectures have served as model systems for studying RNA lariat hydrolysis [54–56], as templates for the assembly of supramolecular RNA nanostructures for nanobiotechnology applications [57, 58], and as potential therapeutic agents [59–61].

In this chapter, we describe a detailed protocol using 2′-*O*-tert-butyldimethylsilyl (TBDMS) RNA phosphoramidites for generating short to medium-length RNA oligonucleotides. The protocol can produce biologically relevant RNAs applicable to therapeutic interventions including siRNAs, antisense oligonucleotides (ASO), microRNAs (miRNA), or CRISPR RNAs (crRNA). Furthermore, solid phase synthesis serves as a robust method for generating structurally complex, biologically active RNA molecules, including synthetic branched RNA (bRNA).

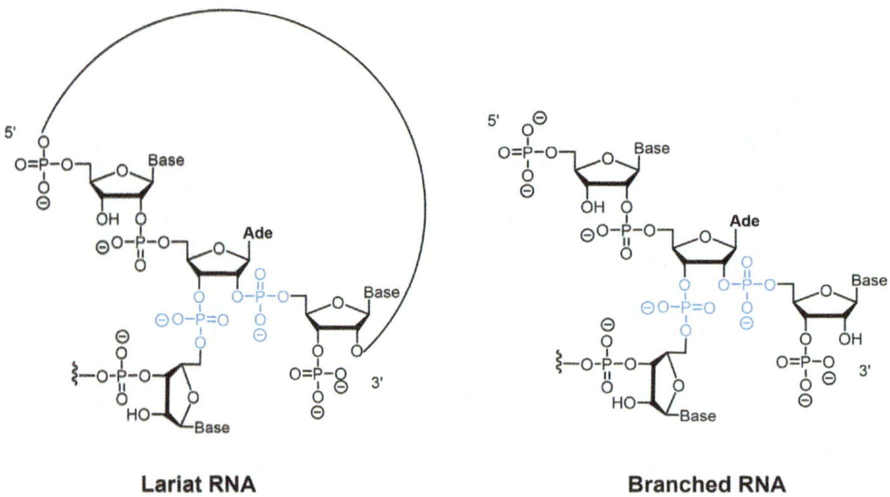

Fig. 2 The structures of lariat RNA (left) and branched RNA (right), featuring vicinal 2′–5′ and 3′–5′ phosphodiester bonds

2 Materials

2.1 Oligonucleotide Synthesis Reagents (see Note 1)

1. Anhydrous acetonitrile (ACN).
2. Detritylation: 3% trichloroacetic acid (TCA) in dichloromethane (DCM).
3. Coupling: 0.25 M ethylthiotetrazole (ETT) in ACN.
4. Capping: acetic anhydride (Ac_2O), pyridine (pyr), tetrahydrofuran (THF) CAP A, 16% *N*-methyl imidazole (NMI) in THF, CAP B.
5. Oxidation: 0.1 M iodine (I_2) in pyr, water (H_2O), and THF (75%, 20%, and 5% v/v/v).
6. Drying trap or molecular sieve.

2.2 RNA Phosphoramidites (see Note 2)

1. 5′-Dimethoxytrityl (DMT) 2′-*tert*-butyldimethylsilyl (TBDMS) Adenosine (N-acetyl, Ac) 3′-cyanoethyl (CED) phosphoramidite, 5′-DMT 2′-TBDMS Guanosine (N-Ac) 3′-CED phosphoramidite, 5′-DMT 2′-TBDMS Cytidine (N-Ac) 3′-CED phosphoramidite, 5′-DMT 2′-TBDMS Uridine 3′-CED phosphoramidite, RNA amidite solutions prepared as 0.15 M final concentration in ACN.
2. 2′-ALE-3′-DMT-Adenosine (N-Bz) 5′-CED phosphoramidite (branch point monomer).
3. 2′-TBDMS-3′-DMT-Adenosine (N-Bz) 5′-CED phosphoramidite, 2′-TBDMS-3′-DMT-Guanosine (N-ibu) 5′-CED phosphoramidite, 2′-TBDMS-3′-DMT-Cytidine (N-acetyl) 5′-CED phosphoramidite, 2′-TBDMS-3′-DMT-Uridine 5′-CED phosphoramidite (reverse RNA phosphoramidites).

2.3 Controlled Pore Glass (CPG) Resin

1. 1000 Å 2′-TBDMS Adenosine (N-Ac) 3′-long chain alkyl amino (lcaa) CPG.
2. 1000 Å 2′-TBDMS Guanosine (N-Ac) 3′-lcaa CPG.
3. 1000 Å 2′-TBDMS Cytidine (N-Ac) 3′-lcaa CPG.
4. 1000 Å 2′-TBDMS Uridine 3′-lcaa CPG.
5. 1000 Å Universal (UnyLinker) CPG.

2.4 Cleavage and Deprotection (see Note 3)

1. 28-30% aqueous ammonium hydroxide ($NH_4OH_{(aq)}$), ethanol (EtOH), 3:1 (v/v).
2. NH_4OH, propylamine, H_2O, 2:1:1 (v/v/v).
3. NH_4OH, methylamine, 1:1 (v/v).
4. Decyanoethylation: Diethylamine:Acetonitrile 1:9 (v/v).
5. 2′-*O*-ALE removal: 0.5 M hydrazine hydrate ($NH_2NH_2{\cdot}H_2O$) in pyridine:acetic acid (pyr:HOAc) 3:2 (v/v).

2.5 Desilylation

1. Dimethyl sulfoxide (DMSO), triethylamine trihydrofluoride (TREAT-HF), 1-butanol (n-BuOH).
2. 3 M sodium acetate ($NaOAc_{(aq)}$) pH 5.5 in H_2O. Add 1.64 g NaOAc to 5 mL H_2O and adjust the pH to 5.5 using acetic acid and/or 1 M $NaOH_{(aq)}$. Dilute to 10 mL using H_2O.

2.6 Reversed Phase Ion Pairing (RPIP) High Performance Liquid Chromatography (HPLC)

1. Buffer A: 0.1 M triethylammonium acetate ($TEAA_{(aq)}$), pH 7.4. Add 13.9 mL of triethylamine (TEA) and 5.7 mL acetic acid (AcOH) to 500 mL H_2O while stirring. Adjust pH to 7.4 with AcOH and/or TEA. Dilute up to 1 L with H_2O (*see* **Note 4**).
2. Buffer B: 0.1 $TEAA_{(aq)}$, 20% ACN, pH 7.4. Add 13.9 mL of triethylamine (TEA) and 5.7 mL acetic acid to 500 mL H_2O while stirring. Adjust pH to 7.4 with AcOH and/or TEA. Dilute to 800 mL with H_2O and add 200 mL of ACN while stirring (*see* **Note 4**).

2.7 Ion-Exchange (IEX) High Performance Liquid Chromatography (HPLC)

1. Buffer A: 15 mM sodium acetate ($NaOAc_{(aq)}$), pH 5.0, 25% $ACN_{(aq)}$.
   Add 5 mL of 3 M $NaOAc_{(aq)}$, pH 5.5 stock to final volume to 750 mL with H_2O. Mix well with 250 mL ACN.
2. Buffer B: 0.5 M lithium perchlorate ($LiClO_{4(aq)}$), 15 mM sodium acetate ($NaOAc_{(aq)}$), pH 5.0, 25% $ACN_{(aq)}$.

Dissolve 80.2 g of lithium perchlorate trihydrate in 500 mL of H_2O while stirring. Add 5 mL of 3 M NaOAc, pH 5.5 stock and bring the final volume to 750 mL with H_2O. Mix well with 250 mL ACN.

2.8 Polyacrylamide Gel Electrophoresis (PAGE)

1. Annealing buffer: 10 mM Tris–HCl, 50 mM NaCl, 1 mM EDTA, pH 7.4. Combine 15.7 mg Tris–HCl, 29.2 mg NaCl, 2.9 mg EDTA in 5 mL H_2O. Adjust pH to 7.4 using 1 M $HCl_{(aq)}$ and/or 1 M $NaOH_{(aq)}$ and dilute to 10 mL with H_2O (*see* **Note 5**).
2. 10× TBE buffer: Combine 108 g Tris base, 55 g boric acid, 7.44 g EDTA disodium salt dihydrate to 500 mL of H_2O and adjust to pH 8.8 using 1 M $HCl_{(aq)}$ and/or 1 M $NaOH_{(aq)}$, as needed. Dilute to 1 L with H_2O.
3. Gel loading buffer (denaturing PAGE): 80% formamide, 20% 10× TBE buffer (2× final concentration). Combine 8 mL formamide with 2 mL of 10× TBE buffer (10 mL total volume) and mix by vortex.
4. Gel loading buffer (nondenaturing PAGE): 50% glycerol, 2.5× TBE. Combine 5 mL of glycerol with 2.5 mL of 10× TBE buffer. Dilute with 2.5 mL of H_2O (10 mL total volume) and mix by vortex.
5. Gel running buffer: 1× TBE

6. Gel tracking dye: bromophenol blue, xylene cyanol FF and orange G (commercially available) in gel loading buffer for denaturing PAGE.
7. DNA/RNA molecular weight ladder (commercially available) in gel loading buffer for denaturing PAGE.
8. Stains All® gel staining solution: 0.005% (w/v) Stains-All® (commercially available), 10% (v/v) formamide, 25% (v/v) isopropanol (IPA), 15 mM Tris–HCl (pH 8.8), and 65% (v/v) H_2O. Add 10 mg of Stains-All® to 10 mL of formamide and dissolve by vortex. Add this stock solution to 50 mL of IPA, 10 mL formamide, 1 mL of 3 M Tris–HCl (pH 8.8), and dilute to 200 mL with H_2O.

2.9 Liquid Chromatography Mass Spectrometry (LC–MS)

1. Buffer A: 100 mM hexafluoroisopropanol (HFIP) and 5 mM triethylamine (TEA) in H_2O.

Add 11.5 mL of HFIP to approximately 950 mL of H_2O and mix thoroughly. Add 0.7 mL of TEA. Mix well and adjust to final volume of 1 L with H_2O.

Buffer B: Methanol (MeOH).

3 Methods

3.1 Synthetic Preparation on the K&A H-8 Oligonucleotide Synthesizer from Sierra BioSystems, Inc.

1. Remove a 100 mL septa-seal bottle and amidite bottles from the oven and allow to cool in a desiccator. Once cooled, fill the 100 mL bottle with ACN, add a drying trap and a septum, then purge with inert nitrogen ($N_{2(g)}$) or argon ($Ar_{(g)}$) gas to prevent water absorption from the atmosphere. Add a septum to the amidite bottles to cool and keep dry as previously described.
2. Using the given sequence(s) to be synthesized, calculate the number of additions for each RNA base (A, G, C, U). Each addition should be 275 μL with an additional milliliter added to account for priming the lines and having a slight excess of reagent. For example, if a given sequences has ten additions of A, then 3.75 mL of amidite solution is required [(10 × 275 μL) + 1 mL]. Based on the volume required, the mass of each amidite can then be calculated using their respective molecular weight. Prepare the respective 2′-*O*-TBDMS RNA phosphoramidites as 0.15 M solutions in ACN accordingly.
3. The synthesizer should have ACN in each of the amidite lines to prevent crystallization. Manually purge each of the lines such that the bottles are empty. Replace these bottles with the respective amidite solutions, ensuring a tight seal with the connector head.

4. Weigh the CPG resin based on the specific loading capacity (g/μmol) and the desired synthesis scale (*see* **Note 6**). Seal the columns well with the screw cap and fit firmly to the desired column position on the synthesizer (*see* **Note 7**).
5. In preparation for initiating the synthetic protocol on the K&A H-8 Oligonucleotide Synthesizer, (a) enter sequence(s) into the "RNASequence" program, (b) add the sequences to the respective column position(s), (c) select the CPG type (*see* **Note 8**), (d) indicate preference for final 5′-DMT removal (*see* **Note 9**), and (e) ensure the correct RNA synthesis method is selected in the "Start" menu of the H8 software. The most common RNA synthesis protocol occurs on a 1 μmol scale (*see* **Note 10**).
6. Using the "Prime DNA" function, each of the reagent and A, G, C, and T amidite lines will be primed automatically. All other amidite positions, including the U line for RNA synthesis, will need to be primed manually using the "manual control" menu. Once the appropriate settings are set and all reagents have been primed, the synthesis can be initiated using the "Start" function.
7. The RNA synthetic cycle (Fig. 1) will follow the standardized steps of the oligonucleotide solid phase synthesis protocol (Table 1). Based on this cycle, each addition takes ~15 min (*see* **Note 10**).
8. Monitor the first five to six couplings, checking periodically to ensure that the orange DMT cation color resulting from the 5′-detritylation step remains prominent and consistent, with no leaks detected from the column positions (*see* **Note 11**).
9. The crude RNA is cleaved from the solid support and deprotected as described in Sects. 3.3 and 3.4. Figure 3 shows representative LCMS characterization data of a synthetic RNA sequence.

3.2 Solid-Phase Synthesis of Branched RNA (bRNA) (Fig. 4)

1. Prepare CPG solid support, ACN diluent, and amidite bottles as described in Sect. 3.1.
2. Prepare the standard 2′-TBDMS-3′-DMT 5′-CED RNA phosphoramidites as 0.15 M solution in dried ACN. Prepare the

Table 1
Steps and corresponding reaction times for automated RNA synthesis

Step	Time (s)
Detritylation	60
Coupling	750
Capping	48
Oxidation	50

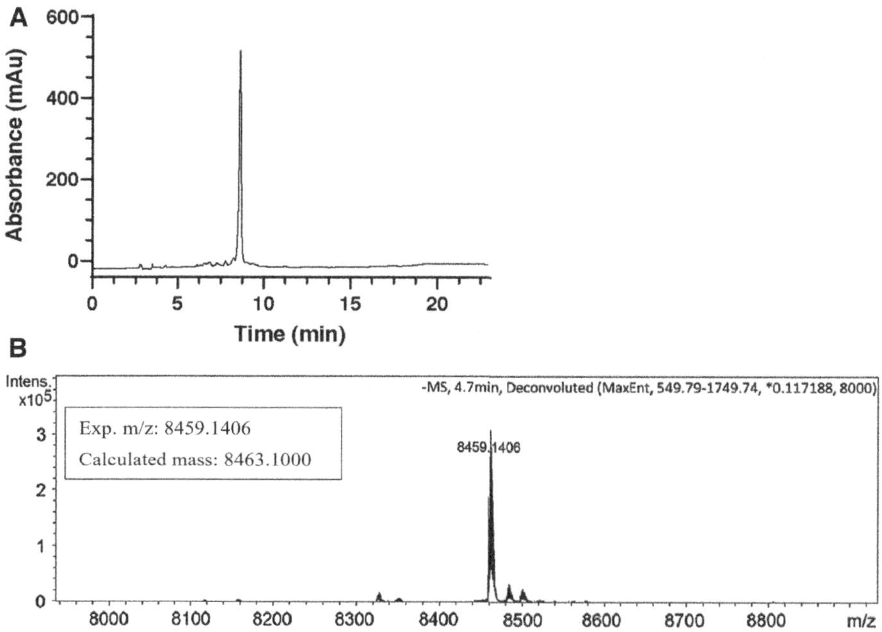

Fig. 3 (**A**) Representative RPIP HPLC chromatogram of a synthetic sense strand siRNA sequence (5'-GCUCCACCAUAUGACUCCCUGUUAGdTdT-3') along with (**B**) the characteristic MS spectrum with inset of the observed experimental and theoretical calculated masses of the synthetic RNA sequence [62].

2′-ALE-3′-DMT-Adenosine (N-Bz) 5′-CED phosphoramidite in 0.1 M solution in dried ACN. The reverse RNA phosphoramidites (2′-TBDMS-5′-DMT-3′-CED phosphoramidites) coupled to the branchpoint are dissolved in 0.3 M in ACN to overcome the steric hindrance at the 2′ branchpoint.

3. Follow the automated synthesis cycle as described in Sect. 3.1. When reaching the branchpoint, incorporate the branchpoint monomer 2′-ALE-3′-DMT-Adenosine (N-Bz) 5′-CED phosphoramidite (0.1 M in ACN) and couple for 10 min, then extend the RNA to the 5′ end until the desired sequence is reached.
4. Remove the last 5′-DMT and perform a capping step to yield the 5′-*O*-acetylated linear fragment, thereby protecting the free 5′-OH from further reactions during the subsequent cycles.
5. Before selectively removing the 2′-*O*-ALE, take off the column from the synthesizer and treat with diethylamine (DEA)/ACN (1:9 v/v) to remove the cyanoethyl phosphotriester protecting group, by pushing the solution through the column using Luer-lock syringe, for a duration of 20 min. Push the solution every 5 min to deliver fresh solution. Then, wash the column with ACN (3 × 10 mL) and dry the support under argon stream (*see* **Note 3**).

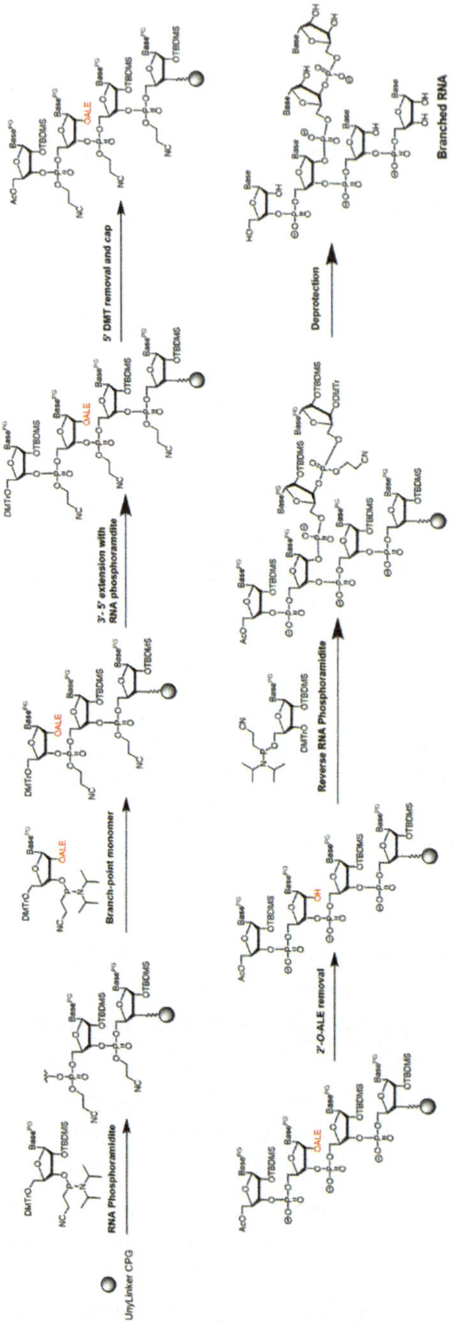

Fig. 4 Solid-phase synthesis scheme of branched RNA using 2′-*O*-acetyl levulinic ester (2′-*O*-ALE) phosphoramidite as the branchpoint monomer and 2′-TBDMS RNA phosphoramidites to produce bRNA

6. Remove the 2′-*O*-ALE group using a freshly prepared solution of 0.5 M hydrazine hydrate in pyridine/acetic acid ($NH_2NH_2H_2O$, pyr:HOAc) (3:2 v/v). Pass through the column using Luer-lock syringe and allow for reaction time of 20 min. Push the solution every 5 min to deliver fresh solution. Wash the column with ACN (3 × 10 mL), and dry the support under argon stream.
7. Return the argon dried column to the synthesizer and grow the branching sequence from the just released 2′-hydroxyl using reverse RNA phosphoramidites (0.3 M in ACN). Couple the first unit of reverse phosphoramidites for 30 min and the remaining ones for 15 min.
8. Complete the synthesis of branch segment until the desired sequence is reached to afford the protected branched RNA.
9. The crude bRNA is cleaved from the solid support and deprotected as described in Sects. 3.3 and 3.4. Figure 5 represents the analytical profile and characterization data of a synthetic 10-nucleotide bRNA.

3.3 Cleavage and Deprotection

1. Upon synthesis completion, remove the columns from the synthesizer and transfer the CPG resin into a screw cap Eppendorf tube. A cannula may be required to gently remove any resin stuck to the side of the column.
2. Based on the reagents provided in Sect. 2.4 above, one of the following cleavage and deprotection conditions may be used depending on the nature of the RNA sequence. The most common RNA cleavage and deprotection conditions involve (a) 3:1 v/v NH_4OH:EtOH (750:250 μL) @ 65°C overnight, with alternate conditions also applicable to native and modified RNA, (b) 2:1:1 v/v/v NH_4OH:propylamine:H_2O (500:250:250 μL) @ 65°C for 1 h, and the most rapid cleavage and deprotection conditions using (c) 1:1 NH_4OH:methylamine (500:500 μL) @ RT for 20 min or 65°C for 10 min.
3. Remove the samples from the heat source and allow it to cool, remove the caps, and place on a centrifugal concentrator to evaporate samples to dryness.
4. Once dry, add 500 μL H_2O to dissolve each sample and vortex. Centrifuge at 9000 × *g* for ~30 s and remove the supernatant, placing into a fresh, sterile screw cap Eppendorf tube. Repeat this process to prepare a 1 mL stock RNA solution. Place the samples back on the centrifugal concentrator to evaporate to dryness. The leftover CPG resin can be stored in a −20°C freezer if additional sample recovery is necessary by repeating Steps 2–4.

3.4 Desilylation

1. Dissolve the final RNA pellet(s) obtained from Sect. 3.3 in 100 μL of DMSO. Add 125 μL of TREAT-HF to each sample and heat for 1.5–2 h at 65°C. Let the tubes cool to room temperature (*see* **Note 12**).

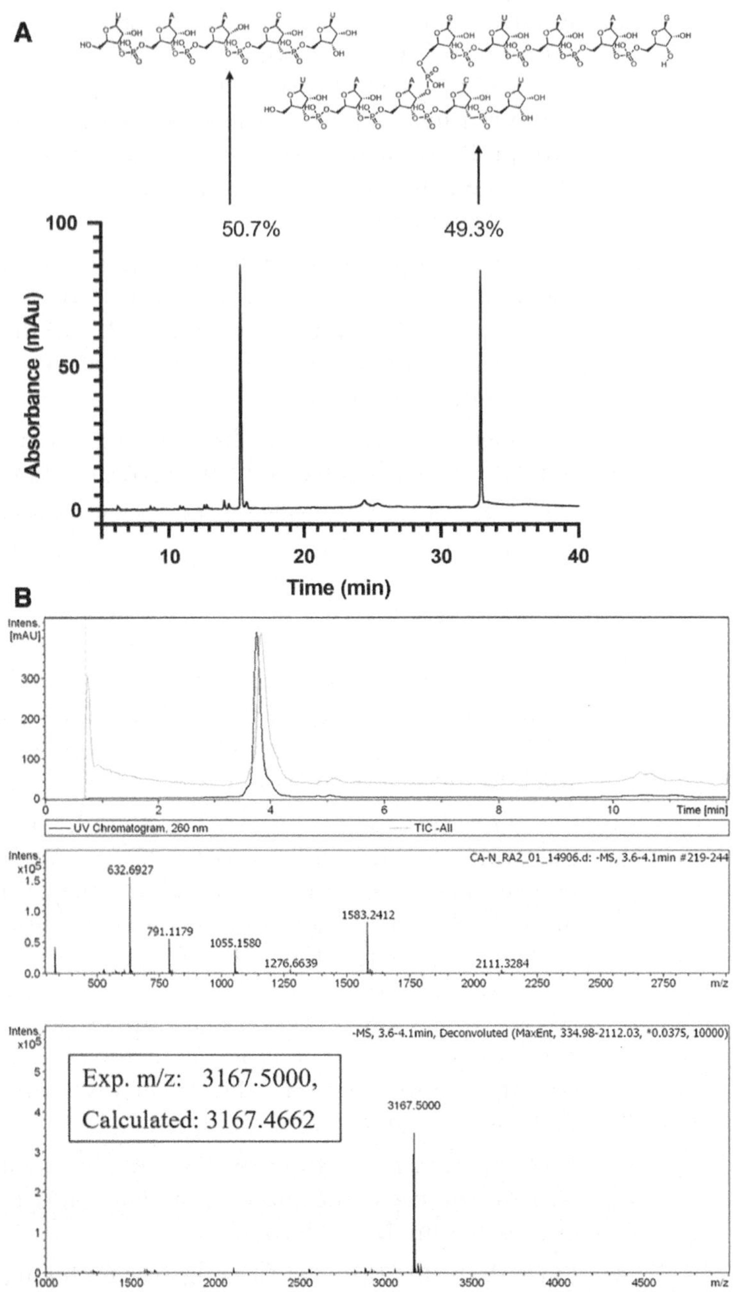

Fig. 5 Characterization of a synthetic bRNA oligonucleotide (5'-UAA(2'-GUAAG)CU-3'). (**A**) Representative IEX-HPLC chromatograms of the crude bRNA. (**B**) LCMS analysis confirming the mass of the HPLC-purified bRNA product

2. Add 50 μL of 3 M $NaOAc_{(aq)}$ (pH 5.5) to each tube to precipitate the RNA out of solution. Add 1 mL of cold n-BuOH and vortex briefly prior to placing the samples in a −20°C freezer for 45 min (or −80°C for 25–30 min).

3. Vortex samples briefly and centrifuge at 9000 × *g* for ~1 min to pellet the RNA. Discard the supernatant and place samples on the centrifugal concentrator for ~30 min to remove the residual solvent. Resuspend samples in 1 mL of H_2O for quantitation.

3.5 Ultraviolet-Visible (UV-Vis) Spectroscopy

1. For standard 1 μmol scale syntheses, prepare a 100× dilution of the RNA stock solution in H_2O for quantitation based on the cuvette volume. Measure absorbance of the diluted sample at 260 nm after blank correction with 1 mL H_2O. Similar measurements can be taken on a smaller volume scale without dilution using a nanodrop spectrophotometer, as needed.
2. Using the Beer–Lambert law, the measured absorbance value(s), and the theoretical molar extinction coefficient(s), calculate the molar concentration of the RNA sample(s) for further use.
3. Alternative to molar concentrations, if working with 1 mL volumes for stock and diluted solutions, optical density (O.D.) values can be used by simply multiplying the absorbance values measured at 260 nm by the dilution factor. This provides quantitative O.D. values of RNA samples dissolved in 1 mL of H_2O.

3.6 HPLC Analysis and Purification

3.6.1 RPIP HPLC Analysis

1. Prepare HPLC buffer solutions A and B according to Sect. 2.6 above. Place these solutions on the HPLC instrument and purge the solvent lines, while allowing the solutions to degas using an in-line degasser.
2. Once purged, flush the buffer solutions through the system to equilibrate the LC column (*see* **Note 4**) to the method starting conditions. For shorter (<80 nucleotides) oligonucleotides, the following gradient conditions are applicable (Table 2).
3. Prepare a 1 O.D. solution for each generated RNA sequence based on the total calculated O.D. value in 1 mL or sterile H_2O.

Table 2
Gradient method conditions for RPIP HPLC (*see* Note 13)

Time (min)	%A	%B	Flow rate (mL/min)	Injection volume (μL)	Column temp (°C)	Detection wavelength (nm)
0	65	35	1	10–100	60	260
14	15	85				
15	5	95				
19	5	95				
20	65	35				
23	65	35				

Syringe filter the solutions through a 0.22 μm filter directly into a clean HPLC vial that is labeled appropriately.

4. Place the vial(s) in the desired positions in the sample tray (autosampler) and set up a sequence or single injection analysis based on the instrument-specific methods.
5. Upon completion of the analysis, integrate the observable peak areas to determine RNA crude purity.
6. Should the purity profile of a given RNA be <90%, purification may be necessary (Fig. 3). If so, prepare a 20 O.D. stock solution of the oligonucleotide in 1 mL of H_2O and conduct repeat injections, with fraction collection of the RNA sample (*see* **Note 14**).

3.6.2 Ion-Exchange (IEX) HPLC Purification and Analysis

1. Prepare IEX-HPLC mobile phase A and B according to Sect. 2.7 above. Place these solutions on the HPLC instrument and purge the solvent lines, while allowing the solutions to degas using an in-line degasser.
2. The IEX method can be used for both purification and analysis by choosing appropriate stationary phase and HPLC system (Tables 3 and 4):
3. Filter the sample as described in Sect. 3.6.1. Inject the filtered RNA sample to the column. For purification purposes, collect the desired peak into appropriate labeled centrifuge tubes depending on the fraction volume.
4. If samples are recovered, dry the collected HPLC fraction in the centrifugal concentrator.
5. Redissolve the sample in maximum 1 mL of H_2O. Desalt the dried RNA sample using commercial desalting columns (e.g., Glen Gel-PAK 2.5 desalting columns, Glen Research) according to the manufacture's protocol.

Table 3
Gradient method conditions for analytical IEX HPLC (*see* Note 15)

Time (min)	%A	%B	Flow rate (mL/min)	Injection volume (μL)	Column temp (°C)	Detection wavelength (nm)
0	100	0				
2	100	0				
30	70	30				
31	0	100	1	10–100	65	260
41	0	100				
42	100	0				
52	100	0				

Table 4
Gradient method conditions for preparative IEX HPLC (*see* Note 15)

Time (min)	%A	%B	Flow rate (mL/min)	Injection volume (μL)	Column temp (°C)	Detection wavelength (nm)
0	100	0				
2	100	0				
50	60	40				
51	0	100	10	200–4000	65	260
61	0	100				
62	100	0				
72	100	0				

3.7 PAGE Analysis (Denaturing and Nondenaturing PAGE)

3.7.1 Denaturing PAGE Analysis

1. Prepare samples by aliquoting the desired amount (*see* **Note 16**) into a separate snap cap Eppendorf tube. Concentrate the solution to dryness using a centrifugal concentrator
2. Prepare the desired number of gel moulds using 18 × 18 cm glass gel plates (Hoefer) using 1.5 mm spacers. Clamp to the casting stand in preparation for gel casting (*see* **Note 17**).
3. To prepare a 16% denaturing PAGE solution, combine 42.04 g of urea in 40 mL of 40% acrylamide/bisacrylamide (19:1 v/v) and 10 mL 10× TBE buffer. Heat to 37°C while stirring to allow the urea to dissolve. Add H_2O as needed to dissolve urea to reach a final solution volume of 100 mL (*see* **Note 17**).
4. Prepare a 10% (w/v) solution of ammonium persulfate (APS) by adding 50 mg to 500 μL H_2O and vortex to dissolve the solution.
5. When ready to begin gel polymerization, add 400 μL APS solution and 40 μL tetramethylethylenediamine (TEMED) to the gel solution. Allow the solution to stir for 8–10 s prior to adding into the prepared gel plates. Place the desired gel comb (1.5 mm) between the gel plates and allow the gel to polymerize for 45–60 min. Replenish the gel cast with extra solution as needed.
6. Remove the comb from the gel and gently rinse the wells with distilled water to remove any excess acrylamide solution. Fill the lower gel running reservoir with 1× TBE buffer, ~1/3 of the way up. Attach the casted gels to the upper reservoir and

place the assembly into the lower gel running reservoir and fill the upper reservoir with 1× TBE buffer.

7. Dissolve samples in 30 μL of gel loading buffer (80% formamide, 20% 10x TBE) and place on a heat block at 90°C for 5 min to fully denature the samples. In addition to test samples, a molecular weight (MW) ladder and/or gel tracking dye can be included. Gel tracking dye and MW ladder 1:1 (20 μL total) in formamide loading buffer (*see* **Note 18**).
8. Load samples into the corresponding wells of the gel. Close the assembly, attach the leads to the power supply and start the gel at 250 V for 30 min to allow the oligonucleotide samples and standards to enter the gel. The voltage can then be adjusted up to 400 V in the presence of a cool water circulator to keep the gel assembly from overheating. Run the gel until the bromophenol tracking dye has migrated ~66%–75% down the gel.
9. Carefully remove the gel(s) from the assembly and separate from the glass plates. Visualize the gel(s) using UV-shadowing (low UV at 250–260 nm) and (if needed) place the gel into the Stains-All® solution overnight. Store in a dark place.
10. The following day, place the gel into the light to destain and visualize/image the colored RNA bands (*see* **Note 19**).

3.7.2 Nondenaturing PAGE Analysis

1. Complete the steps above for gel cassette assembly. Urea will be excluded from the gel mixture, and the nondenaturing gel loading buffer (50% glycerol, 2.5x TBE) will be used. Voltage for nondenaturing PAGE should not exceed 300 V to ensure that the samples do not denature. The use of a cool water circulator is recommended, if available.
2. For annealing double-stranded (ds) RNA samples, combine the desired (equimolar) amount of complimentary RNA strands in a 1.5 mL snap cap Eppendorf tube and evaporate the H_2O. Resuspend the dry RNA pellet in ~15 μL of annealing buffer. Heat the solution at 90°C for 5 min and allow to cool to room temperature for 1 h. Place the sample at 4°C until ready to load onto the gel.
3. Add 15 μL of nondenaturing loading buffer to the RNA samples and load into the gel. Denaturing loading buffer (formamide containing) can still be used for the gel tracking dye and the oligonucleotide MW standard.
4. Voltage for nondenaturing (native) PAGE should not exceed 300 V to ensure that the samples do not denature. Run the gel until the bromophenol tracking dye has migrated ~66%–75% down the gel.

Table 5
Gradient method conditions for LC method used in MS analysis (*see* Note 15)

Time (min)	%A	%B	Flow rate (mL/min)	Injection volume (μL)	Column temp (°C)	Detection wavelength (nm)
0	90	10				
2	90	10				
20	60	40				
21	10	90	0.30	5–10	60	260
31	10	90				
32	90	10				
42	90	10				

5. Disassemble the gel cassette and plates. Visualize and stain the gel as previously described.

3.8 Mass Spectrometry Analysis

1. To prepare oligonucleotides for high resolution mass spectrometry (HRMS) analysis, aliquot 1 nmol of each sample into a snap cap Eppendorf tube and evaporate the solvent using a centrifugal concentrator prior to dissolving samples in H_2O for HRMS analysis (*see* **Note 20**).
2. Prepare the mobile phases according to Sect. 2.9.
3. Analyze oligonucleotides by LC–MS with a Dionex Ultimate 3000 UHPLC coupled to a Bruker Maxis Impact QTOF in negative ESI mode, or equivalent instrumentation (*see* **Note 15**) (Table 5).
4. Process the data and deconvolute the spectra using appropriate software (e.g., Bruker DataAnalysis) to obtain the mass (Da) of the oligonucleotide.

4 Notes

1. The reagents provided for RNA synthesis are considered standard but may vary for DNA synthesis or with modified amidites. For example, dichloroacetic acid (DCA) is typically used for DNA synthesis, 0.02 M oxidizer solution is used for DNA synthesis and the addition of thiol modifiers, etc. Check manufacturer standard operating procedures when working with modified amidites.

2. For RNA synthesis, some commercially available amidites contain different 2′-OH and nucleobase protecting groups, that may require different coupling times and/or reagents for synthesis and post-synthesis deprotection. For example, 2′-TOM protected RNA amidites require 6 min coupling times in comparison to 12 min for traditional 2′-TBDMS amidites. Check manufacturer application guides to ensure the proper coupling times, coupling, and deprotection conditions are used.
3. Items 1–3 represent different cleavage/deprotection conditions depending on differing protecting group chemistries. Items 4 and 5 relate specifically to synthetic protocols for bRNAs. Item 4 is essential to prevent isomerization or cleavage at the branchpoint 3′, 5′-internucleotide linkage.
4. Due to the sensitivity of RNA oligonucleotides, for any instance water is mentioned in this protocol, use deionized (MilliQ) RNase/DNAse free water that has been obtained commercially or autoclaved in-house. The given RPIP HPLC buffer conditions are applicable to C18 reversed phase columns. The column used is the Waters® XBridge C-18 reversed phase column (4.6 × 250 mm, 5 μm). An alternative method to reversed phase ion pairing is ion exchange (IEX) HPLC that utilizes $LiClO_{4(aq)}$ or $NaClO_{4(aq)}$ based buffer solutions. While this method can be effective, issues with salt precipitation, and an additional desalting step is required after sample purification. Alternatively, the buffers used for RPIP HPLC are volatile and can be removed *in vacuo*.
5. The given annealing buffer condition is one example of many that can be used for the hybridization of oligonucleotides. A variety of buffers can be used for the same purpose, based on the intended downstream application.
6. Depending on the sequence composition and availability of materials, different CPG resins may be employed. Universal (Unylinker) CPG resins allow for the addition of any amidite at the 3′ position while nucleoside-tethered CPG resins are sequence-dependent. Both types of CPG resins are commercially available. Follow the manufacturer's instructions for the most appropriate utilization. Each batch will possess a specific loading capacity which will dictate the mass of resin required for a given synthesis scale.
7. Column filter seals will wear after multiple uses and may no longer be viable for oligonucleotide synthesis. To check the column for leaks prior to initiating the synthesis, manually inject ACN through the column position into the waste, followed by $Ar_{(g)}$ purge to dry the column. During this process, visually inspect the column for leaks and/or hold a delicate task wipe to the column. If the column is leaking, attempt to reseal or switch to a new column if needed.

8. If using a nucleoside-tethered CPG, select "ATCG." If using a universal CPG, select "UNIV."
9. Generally, final DMT removal is preferred when taking advantage of RPIP HPLC techniques for analysis and purification of RNA. DMT protecting groups are typically retained when using GlenPak® reversed-phase cartridges to remove failure sequences from the crude product. This step typically requires an additional purification process (HPLC or PAGE) to isolate pure RNA.
10. The K&A H-8 oligonucleotide synthesizer allows for extensive customization of the RNA synthesis cycle, as desired. The provided reaction times (Table 1) are based on optimization of in-house synthetic protocols for high efficiency RNA synthesis using 2′-TBDMS RNA phosphoramidites on a 1 μmol scale. The presence of an additional capping step following oxidation is employed to remove any water present in the reaction column prior to initiating the next detritylation step but is an optional step. Each reaction step can be varied dependent on the nature of the sequence and synthesis scale.
11. If the orange color is no longer visible upon detritylation and the trityl monitoring drops significantly (>50%) in intensity, stop the synthesis and prepare a fresh column with new CPG resin to start a new synthesis.
12. On many occasions, the RNA pellet won't immediately dissolve in DMSO. Heat at 65°C for 5 min then vortex to facilitate dissolution. Repeat as needed prior to adding TREAT-HF. If final DMT retention for purification or chimera DNA/RNA deprotection is required, an alternative desilylation mixture of *N*-methylpyrrolidone (NMP), TEA, and TREAT-HF in a 3:4:6 ratio (300 μL total volume per 1 μmol synthesis) may be used.
13. The injection volumes for HPLC sample analysis may vary based on the given instrument. Pressure limits will vary by column. Gradient conditions may have to be altered depending on column specifications. Typically, columns are heated to denature RNA samples during analysis.
14. Depending on the instrumentation available, purification injections may either be collected manually through the solvent exit line or using an automated fraction collector.
15. Analytical IEX conditions described are conducted using an Agilent 1200 series or equivalent with a DNAPac™ PA200 Oligonucleotide HPLC column, 8 μm, 4 mm × 250 mm. Preparative IEX conditions described are conducted using an Agilent 1260 Infinity II preparative or equivalent with an Agilent PL-SAX anion-exchange HPLC column, 1000 Å pore size,

10 μM. LC–MS analyses were conducted with a Phenomenex Luna C18(2)-HST column (2.5 μM 120 Å, 2.1 × 100 mm) using a gradient of 10%–40% buffer B over 20 min.

16. Molar quantities will vary depending on the sample being used. Some optimization may be required to determine the appropriate amount of oligonucleotide. If using UV shadowing for gel visualization, 0.25–0.5 O.D., will give an observable signal. If lower amounts of material are required (e.g., for an assay) staining or the use of fluorescently tagged oligonucleotides will likely be required.
17. The given preparation for 100 mL of either denaturing or nondenaturing gel solution is sufficient for two gels using the stated gel mould size, with a slight excess remaining. For gels of the specified size, prepare 50 mL of solution per gel as needed. The final concentration of PAGE gels can be adjusted according to the type of RNA samples.
18. The use of a tracking dye is optional for gel PAGE analyses. Different molecular weight ladder options are commercially available depending on the intended application.
19. Denaturing PAGE analysis can also be used for the purification method for RNA. However, in comparison to HPLC methods, PAGE purification is much more tedious, requiring excision of product bands, isolation of RNA from the gel, and an additional desalting step. Therefore, RPIP HPLC is generally the preferred method for RNA purification.
20. Typical LC–HRMS analyses utilize an electrospray ionization (ESI) source with quadrupole time of flight (QTOF) mass analyzer set in negative ion mode. Method optimization may be required based on specific instrumentation.

Acknowledgments

This work is dedicated to the loving memory of Professor Kelvin K. Ogilvie (1942–2025) for his life-long contributions to nucleoside and RNA chemistry. We acknowledge financial support by the Natural Sciences and Engineering Council (NSERC) Discovery Grant to D.S. (RGPIN-2024-05349) and M.J.D. (RGPIN-2022-03372), the Canada Foundation for Innovation John R. Evans Leaders Fund and the Ontario Research Small Infrastructure Fund to D.S. (43604). The coauthors thank Dr. Alexander S. Wahba at the McGill University Mass Spectrometry Lab for oligonucleotide mass spectrometry analysis. The coauthors would also like to thank Marshall Henry and the team at Sierra Biosystems Inc. for technical assistance with automated oligonucleotide synthesis.

References

1. Michelson AM, Todd, Sir AR (1955) Nucleotides part XXXII. Synthesis of a dithymidine dinucleotide containing a 3′: 5′-internucleotidic linkage. J Chem Soc 2632–2638. https://doi.org/10.1039/JR9550002632
2. Khorana HG, Razzell WE, Gilham PT, Tener GM, Pol EH (1957) Synthesis of dideoxyribonucleotides. J Am Chem Soc 79(4):1002–1003
3. Hall RH, Todd Sir AR, Webb RF (1957) Nucleotides part XLI. Mixed anhydrides as intermediates in the synthesis of dinucleoside phosphates. J Chem Soc 3291–3296. https://doi.org/10.1039/JR9570003291
4. Smith M, Moffatt JG, Khorana HG (1958) Carbodiimides. VIII. observations on the reactions of carbodiimides with acids and some new applications in the synthesis of phosphoric acid esters. J Am Chem Soc 80(23):6204–6212
5. Letsinger RL, Mahadevan V (1965) Oligonucleotide synthesis on a polymer support. J Am Chem Soc 87(15):3526–3527
6. Letsinger RL, Ogilvie KK (1967) Convenient method for stepwise synthesis of oligothymidylate derivatives in large-scale quantities. J Am Chem Soc 89(18):4801–4803
7. Eckstein F, Rizk I (1967) Synthesis of oligonucleotides by use of phosphoric triesters. Angew Chem Int Ed Engl 6(8):695–696
8. Reese CB, Saffhill R (1968) Oligonucleotide synthesis via phosphotriester intermediates: the phenyl-protecting group. Chem Commun (London) (13):767–768. https://doi.org/10.1039/C19680000767
9. Letsinger RL, Mahadevan V (1966) Stepwise synthesis of oligodeoxyribonucleotides on an insoluble polymer support. J Am Chem Soc 88(22):5319–5324
10. Egli M (2023) Robert Letsinger and the evolution of oligonucleotide synthesis. ACS Omega 8(36):32222–32230
11. Catlin JC, Cramer F (1973) Deoxy oligonucleotide synthesis via the triester method. J Org Chem 38(2):245–250
12. van Boom JH, Burgers PMJ (1976) Use of levulinic acid in the protection of oligonucleotides via the modified phosphotriester method: synthesis of decaribonucleotide U-A-U-A-U-A-U-A-U-A. Tetrahedron Lett 17(52):4875–4878
13. Stawinski J, Hozumi T, Narang SA (1976) Arylsulfonyltetrazoles as highly efficient condensing reagents for polynucleotide synthesis. Can J Chem 54(4):670–672
14. Rokos H, Myles A, Hutzenlaub W, Pfleiderer W (1975) Nucleotide, II. Über die Verseifung von Dinucleosidmonophosphorsäurephenylester und die Isomerisierung der Internucleotid-Bindung. Chem Ber 108(9):2872–2877
15. Ike Y, Ikuta S, Sato M, Huang T, Itakura K (1983) Solid phase synthesis of polynucleotides VIII. Synthesis of mixed oligodeoxyribonucleotides by the phosphotriester solid phase method. Nucleic Acids Res 11(2):477–488
16. Gait MJ, Matthes HW, Singh M, Sproat BS, Titmas RC (1982) Rapid synthesis of oligodeoxyribonucleotides VII. Solid phase synthesis of oligodeoxyribonucleotides by a continuous flow phosphotriester method on a kieselguhr-polyamide support. Nucleic Acids Res 10(20):6243–6254
17. Ikehara M, Ohtsuka E, Markham AF (1979) The synthesis of polynucleotides. Adv in Carbohydr Chem Biochem 36:135–213
18. Letsinger RL, Lunsford WB (1976) Synthesis of thymidine oligonucleotides by phosphite triester intermediates. J Am Chem Soc 98(12):3655–3661
19. Beaucage SL, Caruthers MH (1981) Deoxynucleoside phosphoramidites—a new class of key intermediates for deoxypolynucleotide synthesis. Tetrahedron Lett 22(20):1859–1862
20. Alvarado-Urbina G, Sathe GM, Liu WC, Gillen MF, Duck PD, Bender R, Ogilvie KK (1981) Automated synthesis of gene fragments. Science 214(4518):270–274
21. Caruthers MH (1985) Gene synthesis machines: DNA chemistry and Its Uses. Science 230(4723):281–285
22. Matteucci MD, Caruthers MH (1980) The synthesis of oligodeoxyprimidines on a polymer support. Tetrahedron Lett 21(8):719–722
23. Ogilvie KK, Nemer MJ (1980) Silica gel as solid support in the synthesis of oligoribonucleotides. Tetrahedron Lett 21(43):4159–4162
24. Brzezinska J, Trzciński S, Strzelec J, Chmielewski MK (2023) From CPG to hybrid support: review on the approaches in nucleic acids synthesis in various media. Bioorg Chem 140:106806
25. Ogilvie KK, Sadana KL, Thompson EA, Quilliam MA, Westmore JB (1974) The use of silyl groups in protecting the hydroxyl functions of ribonucleosides. Tetrahedron Lett 15(33):2861–2863
26. Ogilvie KK, Beaucage SL, Entwistle DW, Thompson EA, Quilliam MA, Westmore JB (1976) Alkylsilyl groups in nucleoside and nucleotide chemistry. J Carb Nucleos- Nucleot 3:197–227
27. Damha MJ, Ogilvie KK (1993) Oligoribonucleotide synthesis—the silyl phosphoramidite method In: Agrawal S (ed) Protocols for

oligonucleotides and analogs. Methods in molecular biology, vol 20. Humana Press, Totowa, NJ.
28. Usman N, Ogilvie KK, Jiang MY, Cedergren RJ (1987) The automated chemical synthesis of long oligoribonucleotides using 2′-O-silylribonucleoside-3′-O-phosphoramidites on a controlled pore glass support: the synthesis of a 43-nucleotide sequence similar to the 3′-half molecule of an *E. coli* formyl-methionine tRNA. J Am Chem Soc 109(25):7845–7854
29. Pitsch S, Weiss PA, Jenny L, Stutz A, Wu X (2001) Reliable chemical synthesis of oligoribonucleotides (RNA) with 2′-O-[(triisopropylsilyl)oxy]methyl(2′-O-tom)-protected phosphoramidites. Helv Chim Acta 84(12):3773–3795
30. Scaringe SA, Wincott FE, Caruthers MH (1998) Novel RNA synthesis method using 5'-O-silyl-2'-O-orthoester protecting groups. J Am Chem Soc 120(45):11820–11821
31. Shiba Y, Masuda H, Watanabe N, Ego T, Takagaki K, Ishiyama K, Ohgi T, Yano J (2007) Chemical synthesis of a very long oligoribonucleotide with 2-cyanoethoxymethyl (CEM) as the 2′-O-protecting group: structural identification and biological activity of a synthetic 110 mer precursor-microRNA candidate. Nucleic Acids Res 35(10):3287–3296
32. Semenyuk A, Földesi A, Johansson T, Estmer-Nilsson C, Blomgren P, Brännvall M, Kirsebom LA, Kwiatkowski M (2006) Synthesis of RNA using 2′-O-DTM protection. J Am Chem Soc 128(38):12356–12357
33. Lyu Z, Katolik A, Yaseen I, Pater AA, Robert F, Huang S, Gagnon KT, Unrau PJ, Damha MJ (2026) Synthesis of long and functionally active RNAs facilitated by acetal levulinic ester chemistry. Nucleic Acids Res 54(2):gkaf1525
34. Beaucage SL, Reese CB (2009) Recent advances in the chemical synthesis of RNA. Curr Protoc Nucleic Acid Chem 38:2.16.1–2.16.31
35. Wunderlich CH, Spitzer R, Santner T, Fauster K, Tollinger M, Kreutz C (2012) Synthesis of (6-^{13}C) pyrimidine nucleotides as spin-labels for RNA dynamics. J Am Chem Soc 134(17):7558–7569
36. Biscans A, Bos M, Martin AR, Ader N, Sczakiel G, Vasseur J, Dupouy C, Debart F (2014) Direct synthesis of partically modified 2′-O-pivaloyloxymethyl RNAs by a base-labile protecting group strategy and their potential for prodrug-based gene-silencing applications. ChemBioChem 15(18):2674–2679
37. Johnsson RA, Bogojeski JJ, Damha MJ (2014) An evaluation of selective deprotection conditions for the synthesis of RNA on a light labile solid support. Bioorg Med Chem Lett 24(9):2146–2149
38. Flemmich L, Bereiter R, Micura R (2024) Chemical synthesis of modified RNA. Angew Chem Int Ed 63(22):1–16
39. Liang JC, Bloom RJ, Smolke CD (2011) Engineering biological systems with synthetic RNA molecules. Mol Cell 43(6):915–926
40. Halloy F, Biscans A, Bujold KE, Debacker A, Hill AC, Lacroix A, Luige O, Strömberg R, Sundstrom L, Vogel J, Ghidini A (2022) Innovative developments and emerging technologies in RNA therapeutics. RNA Biol 19(1):313–332
41. Wang S, Weissman D, Dong Y (2025) RNA chemistry and therapeutics. Nat Rev Drug Discov 24(11):828-851
42. Urits I, Swanson D, Swett MC, Patel A, Berardino K, Amgalan A, Berger AA, Kassem H, Kaye AD, Viswanath O (2020) A review of patisiran (ONPATTRO®) for the treatment of polyneuropathy in people with hereditary transthyretin amyloidosis. Neurol Ther 9(2):301–315
43. Sun X, Setrerrahmane S, Li C, Hu J, Xu H (2024) Nucleic acid drugs: recent progress and future perspectives. Sig Transduct Target Ther 9(316):1–31
44. Glazier DA, Liao J, Roberts BL, Li X, Yang K, Stevens CM, Tang W (2020) Chemical synthesis and biological application of modified oligonucleotides. Bioconjug Chem 31(5):1213–1233
45. George JT, Srivatsan SG (2020) Bioorthogonal chemistry-based RNA labeling technologies: evolution and current state. Chem Commun 56(82):12307–12318
46. Depmeier H, Hoffmann E, Bornewasser L, Kath-Schorr S (2021) Strategies for covalent labeling of long RNAs. ChemBioChem 22(19):2826–2847
47. Klöcker N, Weissenboeck FP, Rentmeister A (2020) Covalent labeling of nucleic acids. Chem Soc Rev 49(23):8749–8773
48. Cultrara CN, Shah S, Cozuch SD, Patel MR, Sabatino D (2018) Solid phase synthesis and self-assembly of higher-order siRNAs and their bioconjugates. Chem Biol Drug Des 93(6):999–1010
49. Wang R, Tombelli S, Minunni M, Spiriti MM, Mascini M (2004) Immobilisation of DNA probes for the development of SPR-based sensing. Biosens Bioelectron 20(5):967–974
50. Kozuch SD, Cultrara CN, Beck AE, Heller CJ, Shah S, Patel MR, Zilberberg J, Sabatino D (2018) Enhanced cancer theranostics with self-assembled, multilabeled siRNAs. ACS Omega 3(10):12975–12984
51. Katolik A, Johnsson R, Montemayor E, Lackey JG, Hart PJ, Damha MJ (2014) Regiospecific solid-phase synthesis of branched oligoribonucleotides that mimic intronic lariat RNA intermediates. J Org Chem 79(3):963–975
52. Damha MJ, Ganeshan K, Hudson RHE, Zabarylo SV (1992) Solid-phase synthesis of branched oligoribonucleotides related to

messenger RNA splicing intermediates. Nucleic acids Res 20(24):6565–6573

53. Nam K, Hudson RH, Chapman KB, Ganeshan K, Damha MJ, Boeke JD (1994) Yeast lariat debranching enzyme. Substrate and sequence specificity. J Biol Chem 269(32):20613–20621
54. Montemayor EJ, Katolik A, Clark NE, Taylor AB, Schuermann JP, Combs DJ, Johnsson R, Holloway SP, Stevens SW, Damha MJ, Hart PJ (2014) Structural basis of lariat RNA recognition by the intron debranching enzyme Dbr1. Nucleic Acids Res 42(16):10845–10855
55. Clark NE, Katolik A, Roberts KM, Taylor AB, Holloway SP, Schuermann JP, Montemayor EJ, Stevens SW, Fitzpatrick PF, Damha MJ, Hart PJ (2016) Metal dependence and branched RNA cocrystal structures of the RNA lariat debranching enzyme Dbr1. Proc Natl Acad Sci 113(51):14727–14732
56. Carriero S, Damha MJ (2003) Inhibition of pre-mRNA splicing by synthetic branched nucleic acids. Nucleic Acids Res 31(21):6157–6167
57. Maina A, Blackman BA, Parronchi CJ, Morozko E, Bender ME, Blake AD, Sabatino D (2013) Solid-phase synthesis, characterization and RNAi activity of branch and hyperbranch siRNAs. Bioorg Med Chem Lett 23(19):5270–5274
58. Patel MR, Kozuch SD, Cultrara CN, Yadav R, Huang S, Samuni U, Koren J, Chiosis G, Sabatino D (2016) RNAi screening of the glucose-regulated chaperones in cancer with self-assembled siRNA nanostructures. Nano Lett 16(10):6099–6108
59. Ye Y, De Leon J, Yokoyama N, Naidu Y, Camerini D (2005) DBR1 siRNA inhibition of HIV-1 replication. Retrovirology 2(1):1–9
60. Armakola M, Higgins MJ, Figley MD, Barmada SJ, Scarborough EA, Diaz Z, Fang X, Shorter J, Krogan NJ, Finkbeiner S, Farese RV, Gitler AD (2012) Inhibition of RNA lariat debranching enzyme suppresses TDP-43 toxicity in ALS disease models. Nat Genet 44(12):1302–1309
61. Galvis AE, Fisher HE, Nitta T, Fan H, Camerini D (2014) Impairment of HIV-1 cDNA synthesis by DBR1 knockdown. J Virol 88(12):7054–7069
62. Yamauchi M, Yoshino I, Yamaguchi R, Shimamura T, Nagasaki M, Imoto S, Niida A, Koizumi F, Kohno T, Yokota J, Miyano S, Gotoh N (2011) N-cadherin expression is a potential survival mechanism of gefitinib-resistant lung cancer cells. Am J Cancer Res 1(7):823-833.

Check for updates

Chapter 18

Fluorescence-Based Assays to Evaluate the Functionality of Phage-Derived Regulatory Elements for Applications in Synthetic Biology

Jorien Poppeliers, Alison Kerremans, Riet Dehaene, Maarten Boon and Rob Lavigne

Abstract

Precise regulation of gene expression is fundamental to microbial synthetic biology, to enable the construction of programmable biological systems. Promoters and terminators play a pivotal role in this regulation, yet their availability and functionality are often limited to model organisms, hindering broader application in non-model bacterial hosts. Bacteriophages, with their diverse and efficient regulatory elements, offer a promising solution. Their promoters—either host-specific or phage-specific—provide both compatibility with native transcriptional systems and orthogonality for insulated gene expression. However, functional validation remains essential. This chapter presents a fluorescence-based workflow for characterizing phage-derived promoters and terminators in vivo, offering a scalable approach to expand the synthetic biology toolkit across diverse bacterial chassis. Using the SEVAtile DNA assembly method, modular constructs are generated to test host-specific and phage-specific promoters, as well as terminators, exemplified for *Pseudomonas chlororaphis*. Constructs are introduced via electroporation and assessed using plate reader-based fluorescence measurements of msfGFP and mCherry reporters. Promoter strength and terminator efficiency are quantified and normalized to cell growth, enabling comparative analysis across constructs. This scalable and adaptable workflow provides a robust framework for characterizing novel regulatory elements and expanding the synthetic biology toolkit for diverse bacterial systems.

Keywords Bacteriophages, Synthetic biology, Regulatory elements, Promotors, Terminators

1 Introduction

Precise control of gene expression is fundamental to synthetic biology, enabling the rational design of biological systems for diverse applications [1, 2]. Central to this control are promoters and terminators, which regulate the initiation and termination of transcription, respectively. Their coordinated function enables the fine-tuning of gene expression, thereby allowing for the controlled synthesis of proteins of interest. Given the modular nature of synthetic biology,

Huan Peng and S. Brian Kim (eds.), *Biological Macromolecule Engineering Technologies*,
Methods in Molecular Biology, vol. 3032, https://doi.org/10.1007/978-1-0716-5257-2_18,

there is a strong emphasis on the development and use of standardized, interchangeable genetic parts. Promoters and terminators, as key regulatory elements, can be systematically combined into complex genetic circuits that exhibit predictable and programmable behavior [3].

Despite their central role in synthetic biology, well-characterized promoters and terminators are not universally available for relevant bacterial species. Most regulatory elements have been developed and validated in model organisms such as *Escherichia coli* or *Bacillus subtilis*, limiting their direct applicability in nonmodel or newly emerging bacterial hosts [4, 5]. Differences in transcriptional machinery, sigma factor usage, and genomic context can significantly affect the functionality and predictability of these elements across species. As a result, the transferability of genetic parts remains a major challenge, necessitating host-specific characterization or the development of orthogonal regulatory systems to ensure reliable gene expression in diverse bacterial chassis [6].

Bacteriophages (phages), viruses that infect bacteria, represent a rich and underutilized source of such regulatory elements. Due to their coevolution with bacterial hosts, phage genomes often encode highly efficient promoters and terminators that are naturally adapted to bacterial transcriptional machinery [7]. These phage promoters can be broadly categorized into two types: host-specific phage promoters, which are recognized by the host's native RNA polymerase and typically drive early gene expression; and phage-specific promoters, which are recognized by phage-encoded RNA polymerases or modified host polymerases, and often control middle and late gene expression. This distinction is critical for synthetic biology, as phage-specific promoters offer orthogonality and insulation from host regulation, while host-specific promoters provide compatibility with native transcriptional systems [8]. Both types are valuable for expanding the regulatory toolkit, particularly in non-model bacteria where synthetic parts are scarce [4].

Recent advances in transcriptomic technologies, such as ONT-cappable-seq, have enabled the high-resolution mapping of transcription start and termination sites in phage genomes. This approach facilitates the systematic identification of candidate regulatory elements for building synthetic constructs [9]. However, the functional characterization of these elements in vivo remains essential. Fluorescent reporter proteins provide a robust and quantitative means to assess promoter activity and terminator efficiency across different host organisms (Fig. 1). These assays offer a rapid and scalable approach to evaluate the performance of phage-derived regulatory elements, serving as a critical first step in their validation for synthetic biology applications.

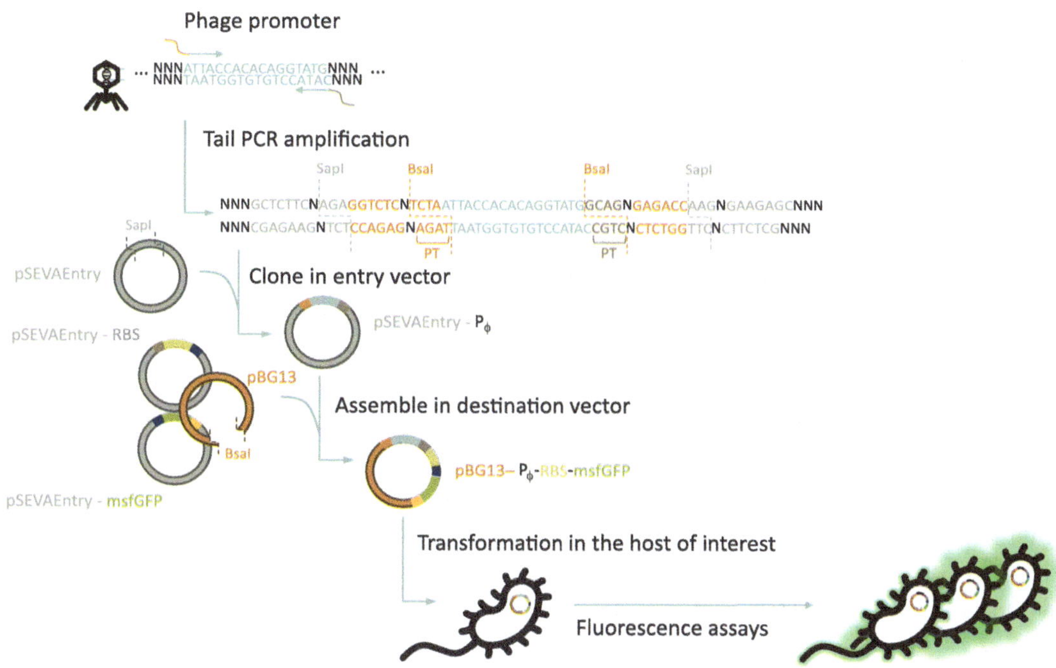

Fig. 1 Workflow for evaluating phage-derived regulatory elements using fluorescence-based assays. The sequence of the regulatory element of interest is first amplified via tail PCR and cloned into an entry vector (pSEVAEntry). These are then assembled into a destination vector (pBG13) containing a msfGFP reporter. The resulting constructs are transformed into the bacterial host, and promoter activity is quantified through fluorescence measurements, enabling functional characterization of phage regulatory elements. *PT*, position tag; *RBS*, ribosomal binding site; *msfGFP*, monomeric superfolder green fluorescent protein

2 Materials

2.1 Media, Buffers and Antibiotics

1. Lysogeny broth (LB): 10 g/L tryptone, 5 g/L yeast extract, 10 g/L NaCl (supplemented with 15 g/L agar to prepare solid medium) (*see* **Note 1**).
2. Super optimal broth with catabolite repression (SOC) medium: 20 g/L tryptone, 5 g/L yeast extract, 0.5 g/L NaCl, 10 mL/L 1 M $MgCl_2$, 10 mL/L 1 M $MgSO_4$, 20 mL/L 20% glucose.
3. M9 minimal medium: 100 mL/L 5% casein amino acid, 100 μL/L 1 M $CaCl_2$, 10 mL/L 20% citrate, 2 mL/L 1 M $MgSO_4$, 200 mL 5× M9 salts.
4. 10% sucrose; filter sterilized 0.22 μm.
5. 1 M 3-methylbenzoate.

2.2 Bacterial Strains, Plasmids and DNA

1. Chemically competent *E. coli* PIR2 cells (e.g., One Shot™ PIR2 Chemically Competent *E. coli*, Thermo Fisher Scientific).
2. Bacterial host (e.g., *Pseudomonas chlororaphis* NRRL-B-30761).

Table 1
Overview of the plasmid vectors for this method

Plasmid	Characteristics	Resource
pSTEntry–P_{EM7}	pRO1600 ori, ColE1 ori, oriT, Amp^R, SacB negative selection, P_{EM7} promoter	[10]
pSTEntry–BCD1	pRO1600 ori, ColE1 ori, oriT, Amp^R, SacB negative selection, BCD1 bicistronic design ribosomal binding site	[10]
pSTEntry–BCD2	pRO1600 ori, ColE1 ori, oriT, Amp^R, SacB negative selection, BCD2 bicistronic design ribosomal binding site	[10]
pSTEntry–mCherry	pRO1600 ori, ColE1 ori, oriT, Amp^R, SacB negative selection, mCherry fluorescent protein	[10]
pSTEntry–msfGFP	pRO1600 ori, ColE1 ori, oriT, Amp^R, SacB negative selection, msfGFP fluorescent protein	[10]
pBG13	*oriT*; *oriV(R6K)*; *P_{EM7}-BCD2-msfGFP fusion*; *Tn7L* and *Tn7R* extremes; Kan^R/Gm^R	[11]
pSTDesXa	pRO1600 ori, ColE1 ori, oriT, Kan^R, SacB negative selection, XylS-Pm expression; BsaI removed from XylS and RBS added after Pm	[10]
pTNS2	Helper plasmid; oriV(R6K); tnsABCD; Amp^R	[12]
pBGPromX	*oriT*; *oriV(R6K)*; *P_ϕ-BCD2-msfGFP fusion*; *Tn7L* and *Tn7R* extremes; Kan^R/Gm^R	This method
pBGTermX	*oriT*; *oriV(R6K)*; *P_{EM7}-BCD2-msfGFP-T_ϕ-BCD1-mCherry fusion*; *Tn7L* and *Tn7R* extremes; Kan^R/Gm^R	This method

3. Template phage DNA.
4. Plasmid vectors (Table 1).

2.3 Polymerase Chain Reaction

1. Phusion Polymerase and 5× HF Phusion buffer
2. DreamTaq polymerase and 10× Green DreamTaq buffer
3. dNTPs
4. Primers (100 μM)
   a) Tile primers (Sect. 3.1.1.)
   b) Plasmid primers:
      i) F-pSTDesXa: AATCTGGCTCCCCAACTAATGC
      ii) F-pBG13: GCGTTCGGTCAAGGTTCTGGAC
      iii) R-pBG13/pSTDesXb: GCGGCAACCGAGCGTTC
      iv) F-pBGProm-BsaI: GACGGTCTCTTAACGAATTCGAGCTCGGTAC

v) F-pBGTerm-BsaI: GACGGTCTCTAAGAGAATTCGAGCTCGGTAC

vi) R-pBG13-BsaI: GTCGGTCTCCTAGATTAATTAAGACGTCTTGAC

2.4 Gel Electrophoresis

1. Agarose
2. 50 μg/mL ethidium bromide
3. DNA ladder (e.g., GeneRuler 1 kb DNA ladder, GeneRuler 100 bp ladder)

2.5 Restriction Ligation

1. SapI restriction enzyme (5 U/μL)
2. BsaI restriction enzyme (20 U/μL)
3. T4 DNA ligase (5 U/μL)and 10× T4 DNA ligation buffer

2.6 Commercial Kits

1. PCR purification kit (e.g., GenJet PCR purification kit, Thermo Fisher Scientific)
2. Miniprep kit for plasmid extraction (e.g., GeneJet Plasmid Miniprep kit, Thermo Fisher Scientific)

2.7 Equipment

1. Incubators at 37°C and 30°C (static and shaking)
2. PCR thermocycler
3. Gel electrophoresis tray and combs
4. UV-transilluminator
5. Electroporator
6. 0.2 cm electroporation cuvette
7. Eppendorf tubes (0.5 and 1.5 mL)
8. A tabletop centrifuge
9. Glass test tubes
10. Erlenmeyer (250 mL)
11. Falcon tubes (50 mL)
12. Syringe filters (0.22 μm)
13. Water bath
14. Plate reader (e.g., Clariostar)
15. Microvolume spectrophotometer (e.g., simplinano)
16. Freezers (−20°C and −80°C)
17. Fridge (4°C)
18. COSTAR-96 well plate
19. 10 mm petri dishes
20. Micropipettes (single and multichannel)
21. Sterile pipette tips

3 Methods

3.1 Vector Creation, Propagation, and Assessment

All necessary plasmid vectors can be generated flexibly using the SEVAtile DNA assembly method. This is a modular DNA assembly technique based on Type IIs restriction enzymes, enabling the ordered assembly of genetic parts. Each SEVAtile is flanked by a unique four-nucleotide position tag (PT) that guides their ligation towards a predefined sequence using Golden Gate cloning. The method first uses tail-PCR to introduce the position tag and restriction sites to the tile of interest. Next, the separate tiles are assembled in a plasmid backbone. Finally, the assembled constructs are transformed into *E. coli* for propagation and sequence verification [10].

3.1.1 Tile Creation

1. Design two primers to amplify the tile of interest [e.g., promoter, terminator, ribosomal binding site (RBS), reporter, vector backbone]. Each primer starts with three random bp to allow annealing of the restriction enzymes, followed by a SapI and BsaI restriction site to clone the tile in the entry and destination vector, respectively (*see* **Note 2**). The restriction site is then followed by a position tag (PT) (*see* **Note 3**) to enable assembly of multiple parts in a one pot reaction. Template oligos (for oligo's to amplify the plasmid backbone (pBG13) see Sect. 2.3):
   a. Forward: 5′- NNN GCTCTTC N AGA GGTCTC PT(5 bp) sequence - 3′
   b. Reverse: 5′ - NNN GCTCTTC N CTT GGTCTC PT(5 bp) sequence - 3′
2. Amplify the tiles using tail PCR (*see* **Notes 4 and 5**):
   a. PCR mix: 1 μL of genomic/plasmid DNA (10 ng/μL), 1 μL of dNTPs, 1.5 μL of each primer (20 μM), 10 μL of 10× HF Phusion buffer, and 0.7 μL of Phusion Polymerase in a total volume of 50 μL.
   b. Cycling conditions: initial denaturation for 3 min at 98°C; 30 cycles of 10 s denaturation at 98°C, 10 s primer annealing at 2° below the melting temperature of your primers and 2 kb/min extension at 72°C; perform the final extension for 10 min at 72°C and hold at 12°C.
3. Verify the length of the PCR product by performing a gel electrophoresis using a 1% agarose gel. A DNA ladder can be used to assess the size of the PCR product.
4. Purify the PCR product using a PCR purification kit, eluting in 30 μl (*see* **Note 6**).
5. Measure the DNA concentration using a microvolume DNA spectrophotometer.
6. Store at −20°C.

3.1.2 Shuffle Tiles into the Entry and Destination Vectors

The type of regulatory element being tested (promoter or terminator) determines which specific tiles are needed (Fig. 2) (*see* **Note 7**). These tiles are first cloned into an entry vector to enable flexible combinations with other parts (e.g., different RBSs, fluorescent reporters, plasmid backbones) and then assembled into a destination vector to generate the final expression constructs.

1. Shuffle the tiles in the entry vector (*see* **Note 8**):
   a. Reaction mixture: 100 ng of plasmid backbone, 50 ng of the tile of interest (e.g., promoter, RBS, etc.), 2 μL of 10× T4 DNA ligation buffer, 3 μL of T4 DNA ligase (1 U/μL), and 1 μL of SapI (10 U/μL) in a total volume of 20 μL.
   b. Cycling conditions: 50 cycles of restriction (37°C, 2 min) and ligation (16°C, 3 min); followed by ligase (50°C, 5 min) and restriction enzyme (80°C, 5 min) inactivation.
2. Shuffle the tiles in the destination vector:
   a. Reaction mixture: 100 ng of plasmid backbone, 50 ng of each tile (Fig. 2), 2 μL of 10× T4 DNA ligation buffer, 3 μL of T4 DNA ligase (1 U/μL), and 1 μL of BsaI (10 U/μL) in a total volume of 20 μL (*see* **Note 9**).
   b. Cycling conditions: 50 cycles of restriction (37°C, 2 min) and ligation (16°C, 3 min); followed by ligase (50°C, 5 min) and restriction enzyme (80°C, 5 min) inactivation.

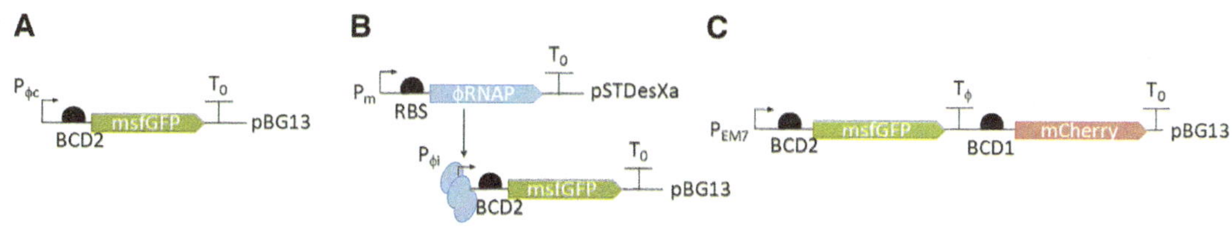

	P_ϕ	P_{EM7}	BCD1	BCD2	ϕRNAP	msfGFP	mCherry	T_ϕ	Plasmid backbones
A. Host-specific phage promoter	X			X		X			pBG13
B. Phage-specific phage promoter	X			X	X	X			pBG13; pSTDesXa
C. Phage terminator		X	X	X		X	X	X	pBG13

Fig. 2 Constructs used to evaluate different types of phage-derived regulatory elements. (**A**) Host-specific phage promoter construct: A host-specific phage promoter ($P_{\phi c}$) is cloned into the pBG13 plasmid backbone, upstream of a BCD2 ribosomal binding site (RBS) and an msfGFP reporter gene. (**B**) Phage-specific promoter construct: This setup includes a phage-specific promoter ($P_{\phi i}$) driving msfGFP expression (with BCD2 RBS) on the pBG13 backbone. Additionally, a second plasmid (pSTDesXa) is required to express the phage RNA polymerase (ϕRNAP) under the control of the XylS/Pm inducible system. (**C**) Phage terminator construct: The phage terminator (T_ϕ) is inserted between two reporter genes—msfGFP and mCherry—on the pBG13 backbone. Each reporter is preceded by its own RBS (BCD2 for msfGFP and BCD1 for mCherry), and both are transcribed from a constitutive promoter (P_{EM7}). The accompanying table summarizes the presence of each regulatory element and the plasmid backbones utilized in each construct

3.1.3 Vector Propagation

1. Transformation of the vector to *E. coli* chemical competent PIR2 cells (heat-shock) (*see* **Note 10**):
   a. Add 5 µL of shuffling reaction to the competent cells and incubate on ice for 30 min.
   b. Heat a water bath to 42°C and incubate the cells at 42°C for 30 s, followed by an incubation on ice for 5 min.
   c. Add 450 µL of SOC medium to the cells.
   d. Recover the cells for 1 h at 37°C, shaking (150 rpm).
   e. Plate the cells, divide the cells over three plates containing the antibiotic of interest (*see* **Note 1**) (5–50 µL—rest) and incubate overnight a 37°C.

3.1.4 Vector Assessment

1. Perform a colony PCR (cPCR) on eight random colonies to analyze the results. Use primers that bind to the plasmid backbone facilitate high-throughput screening of a variety of promoter and terminator elements (*see* **Note 11**).
   a. Reaction mixture: 1 colony, 0.5 µL of dNTPs, 0.5 µL of each plasmid primer (20 µM), 2.5 µL of Green Dream Taq buffer, and 0.5 µL of Dream Taq polymerase in a total volume of 25 µL.
   b. Cycling conditions: initial denaturation for 10 min at 95°C; 30 cycles of 30 s denaturation at 95°C, 30 s primer annealing at 2° below the melting temperature of your primers and 1 kb/min extension at 72°C; perform the final extension for 10 min at 72°C and hold at 12°C.
2. Verify the length of the PCR product by performing a gel electrophoresis using a 1% agarose gel. A DNA ladder can be used to assess the size of the PCR product.
3. Check your samples for correct, mutation-free assembly by Sanger Sequencing.

3.2 Electroporation to Pseudomonas chlororaphis NRRL-B-30761

To functionally characterize phage-derived regulatory elements in *Pseudomonas* species, the reporter constructs are introduced into bacterial cells via electroporation. Following electroporation, cells are recovered in rich medium and plated on selective agar to isolate transformants carrying the promoter construct. This approach enables the rapid and efficient introduction of genetic material for downstream expression analysis (*see* **Note 12**).

1. Two times 1.5 mL of *P. chlororaphis* culture is transferred into 2 mL Eppendorf tubes
2. Centrifuge 4 min (7000 × *g*) in tabletop centrifuge
3. Remove the supernatants and dissolve in 1 mL of 10% sucrose solution

4. Centrifuge 4 min (7000 × *g*)
5. Repeat Steps 3 and 4 two more times
6. Remove the remnant supernatants with a pipette
7. Resuspend the pellets in 100 μL 10% sucrose (100 μL per two tubes)
8. Add 300 ng of plasmid pBGPromX/pBGTermX plasmid and 500 ng pTNS2 (*see* **Note 13**)
9. Incubate for 5 min (room temperature)
10. Transfer the mixture to a 0.2 cm cuvette
11. Perform the electroporation (2.5 kV, 200 Ω, 25 μF, 1 pulse) (*see* **Note 14**)
12. Add 450 μL SOC medium
13. Incubate 1.20 h at 30°C (shaking)
14. Plate on LB containing Gm^{10}
15. Incubate overnight at 30°C

3.3 Fluorescence Assessment

Plate readers are used to assess the promoter strength and terminator activity. They are ideal for high-throughput, bulk measurements of fluorescence across many samples, providing average signal intensity per well (*see* **Note 15**).

1. Inoculate the bacteria of interest in triplicate in 4 mL M9 medium with the appropriate antibiotic and incubate overnight (150 rpm, 30°C) (see **Notes 1 and 16**).
2. Dilute the bacteria to an OD_{600} of 0.1 in 4 mL of fresh M9 medium with the appropriate antibiotic and incubate until an OD_{600} of 0.3 is reached (*see* **Note 1**).
3. Add the cultures to the COSTAR 96-well plate and measured every 15 min for 12 h using a plate reader (e.g., Clariostar) (*see* **Note 17**).
   a. Settings: Incubation at 30°C; OD_{600}; msfGFP, 485 nm (ex)–588 nm (em); mCherry, 485 nm (ex)–528 nm (em); shaking 300 rpm
4. Data analysis
   a. Export the data in an excel file.
   b. Perform a blank correction to get rid of any background related to the medium.
   c. To normalize for bacterial growth, calculate the ratio of fluorescent intensity over $OD_{600.}$
   d. To evaluate the strength of a terminator, calculate its termination activity. This metric compares the expression ratio of msfGFP to mCherry in a construct with the terminator

of interest to a construct without terminator. The following formula can be used for this purpose:

$$\text{Termination activity } (\%) = \left(1 - \frac{\frac{msfGFP_{\text{no terminator}}}{mCherry_{\text{no terminator}}}}{\frac{msfGFP}{mCherry}} \right) * 100\%$$

e. Use statistical software to determine significant differences in promoter strength (*see* **Note 18**).

4 Notes

1. To ensure plasmid selection, media were supplemented with the appropriate antibiotic: gentamycin (Gm; 10 µg/mL), kanamycin (Kan; 50 µg/mL), or ampicillin (Amp; 100 µg/mL), depending on the resistance marker encoded by the plasmid. The specific antibiotic associated with each plasmid is listed in Table 1.
2. If BsaI or SapI restriction sites are present in the sequence of interest. Other type IIs restriction enzymes with similar size overhangs can be used for this purpose. Alternatively, other cloning methods such as Gibson assembly can be used.
3. Position tags can be varied depending on the application. To enable flexible combinations with other parts, it is recommended to always use the same overhangs for each type of tile (e.g., backbone, promoters, RBSs, reporter proteins, terminators, etc.). Examples of position tags are given in Table 2.
4. For short sequences, oligo annealing can be used instead to obtain dsDNA tiles. In this case, 1 µL of each oligo (100 µM) is mixed with 18 µL of mQ. The sample is incubated at 95°C for 5 min and subsequently gradually cooled down to 12°C.
5. Any high fidelity DNA polymerase will do (e.g., Kappa, Q5, etc.).
6. Alternatively, if multiple bands are present GenJet Gel extraction kit can be used to purify the band of interest from the gel.
7. If expression is too high, this might result in toxicity for the cells. Upon transformation to *E. coli* this can result in a lack of colonies or mutations in regulatory elements. To overcome his challenge, a weaker RBS (e.g., BCD22) can be used.
8. Cloning in an entry vector is optional to allow flexible combinations with other parts. However, parts can also be cloned immediately into a destination vector.

Table 2
Position tags (PTs) assigned to each genetic tile (5′ to 3′ Orientation)

Tile	pBGProm	pBGTerm	PT—left	PT—right
Backbone–pBGProm	x		NTAAC	NTCTA
Backbone–pBGTerm		x	TAAGA	NTCTA
Promoter	x	x	NTCTA	NGCAG
BCD2	x	x	NGCAG	NAATG
msfGFP	x	x	NAATG	NTAAC
Terminator		x	NTAAC	NGGCT
BCD1		x	NGGCT	NTAAT
mCherry		x	NTAAT	TAAGA

This table lists the left and right position tags (PTs) for each genetic tile in the construct, all oriented from the 5′ to 3′ direction

9. When shuffling multiple genetic tiles, it is possible that certain tiles may be underrepresented or entirely absent in the final construct. This issue can be mitigated by increasing the relative concentration of the affected tiles in the assembly mixture, thereby enhancing their likelihood of successful incorporation.
10. Different chemo competent *E. coli* cells can be used depending on the compatibility of the plasmid (e.g., pBG13–PIR2 cells; pSTDesXa–TOP10 cells).
11. To keep track of which colonies are screened using cPCR, the colonies that are picked can be plated on a gritted agar plate where each square in the grid refers to the position in the PCR plate.
12. In case of a phage specific-phage promoter, the electroporation protocol needs to be repeated to introduce a plasmid containing the corresponding RNA polymerase gene for expression in the cells.
13. To control plasmid copy number and reduce the burden on the bacterial cells, the plasmid is integrated in the genome. However, a different SEVA tile compatible backbone can be used to maintain a plasmid system.
14. Time constants above 3.5 are considered a benchmark for successful electroporation.
15. Alternatively, flow cytometry can be used to analyze fluorescence at the single-cell level, offering detailed insights into population heterogeneity and cell-to-cell variability.

16. Recommended controls include a blank to correct for background fluorescence caused by the medium, wild-type bacteria to measure background fluorescence of the bacterial cells (e.g., due to pyoverdine production), construct without promoter or terminator (negative control), construct with regulatory element described in literature (positive control; e.g., strong constitutive promoter P_{EM7}, terminator T_0, pET system).
17. In case an inducible promoter is used (e.g., to induce expression of the phage RNA polymerase in case of a phage-specific phage promoter), the culture can be measured both in presence and absence of inducer to determine the level of leaky expression. We recommend testing different concentrations of inducers to determine the optimal concentration.
18. A Shapiro–Wilk test can be used to verify if the data are normally distributed across different promoters at a certain point in time. If the data are normally distributed, a Tuckey HSD test can be used to compare significant difference between regulatory elements. If the data are not normally distributed, a Steel-Dwass test can be used instead.

References

1. De Wannemaeker L, Mey F, Bervoets I et al (2023) Standardization of fluorescent reporter assays in synthetic biology across the visible light spectrum. ACS Synth Biol 12:3591–3607
2. Smanski MJ, Zhou H, Claesen J et al (2016) Synthetic biology to access and expand nature's chemical diversity. Nat Rev Microbiol 14:135–149
3. Lucks JB, Qi L, Whitaker WR et al (2008) Toward scalable parts families for predictable design of biological circuits. Curr Opin Microbiol 11:567–573
4. Adams BL (2016) The next generation of synthetic biology chassis: moving synthetic biology from the laboratory to the field. ACS Synth Biol 5:1328–1330
5. Pontrelli S, Chiu T-Y, Lan EI et al (2018) *Escherichia coli* as a host for metabolic engineering. Metab Eng 50:16–46
6. Hwang S, Joung C, Kim W et al (2023) Recent advances in non-model bacterial chassis construction. Curr Opin Syst Biol 36:100471
7. Lammens E-M, Nikel PI, Lavigne R (2020) Exploring the synthetic biology potential of bacteriophages for engineering non-model bacteria. Nat Commun 11:5294
8. Lammens E-M, Feyaerts N, Kerremans A et al (2023) Assessing the orthogonality of phage-encoded RNA polymerases for tailored synthetic biology applications in Pseudomonas species. Int. J. Mol. Sci. 24:7175
9. Putzeys L, Intizar D, Lavigne R et al (2024) Obtaining detailed phage transcriptomes using ONT-cappable-seq. In: Peng H, Liu J, Chen IA (eds) Phage engineering and analysis: methods and protocols. Springer, US, New York, NY, pp 207–235
10. Lammens E-M, Boon M, Grimon D et al (2022) SEVAtile: a standardised DNA assembly method optimised for Pseudomonas. 15:370–386
11. Zobel S, Benedetti I, Eisenbach L et al (2015) Tn7-based device for calibrated heterologous gene expression in *Pseudomonas putida*. ACS Synth Biol 4:1341–1351
12. Choi K-H, Gaynor JB, White KG et al (2005) A Tn7-based broad-range bacterial cloning and expression system. Nat Methods 2:443–448

Check for updates

Chapter 19

Using peCHYRON to Record Mammalian Cells' History of Chemical Exposure into Their DNA

Matt C. Demelo and Theresa B. Loveless

Abstract

Vertebrate development involves progressive differentiation guided by intrinsic and extrinsic cues; at the population level this yields reproducible outcomes even though single-cell lineage choices can vary between individuals. Variability between individual animals, along with opaque bodies and large numbers of cells, complicates the study of development in vertebrate models. DNA recording technology allows developmental biologists to follow each cell's history, without needing to average across multiple organisms, by stably recording lineage and historical cell signaling as small changes to each cell's genome, which can be read out by DNA sequencing at the end of the experiment [1,2]. Recently, we developed a continuous DNA recording platform, peCHYRON, which allows long-term, multiplexed recording of events in mammalian cells [3]. Here, we describe the design, implementation, and analysis of peCHYRON experiments, demonstrating an example experiment in which constitutive, doxycycline-inducible, and IPTG-inducible gene expression events are recorded concurrently into the DNA of populations of cells. This protocol covers the following steps: the molecular cloning of peCHYRON component plasmids, construction of peCHYRON cell lines, execution of peCHYRON experiments, long-read amplicon sequencing of DNA records, and analysis.

Keywords DNA memory, DNA recording, Molecular recorder, Cell fate history, Prime editing, Golden Gate toolkit, Nanopore sequencing, Amplicon sequencing, PiggyBac integration, Lineage tracing

1 Introduction

DNA recording is used to track transient cellular events, such as gene expression, by making durable records in each cell's genome throughout an experiment [1]. DNA recording technology has expanded rapidly due to CRISPR innovations, but it has until recently been unable to record multiple events in temporal order [2]. Pairing DNA polymerases with CRISPR nucleases has allowed iterative recording [3, 4], but these tools are still limited in the number of sequential edits they can make before losing activity, in part due to the unpredictable repair of the double-strand breaks that are an intermediate in the recording process [3].

Huan Peng and S. Brian Kim (eds.), *Biological Macromolecule Engineering Technologies*,
Methods in Molecular Biology, vol. 3032, https://doi.org/10.1007/978-1-0716-5257-2_19,

Moreover, the ability to multiplex, or record multiple types of events to the same "DNA tape" or "recording locus," has been very limited [5].

The invention of prime editing (PE) has enabled DNA recording technologies that support both many temporally-ordered rounds of recording and substantial multiplexing: DNA Typewriter [6] and peCHYRON [7]. In both approaches, each prime edit is an insertion mutation made at a "DNA tape" recording target that includes both a "signature" sequence that identifies the particular prime editor guide (peg)RNA that made the record and a "propagator" sequence that creates a target site for the next edit (see Fig. 1 and Table 1 for peCHYRON pegRNA design rules). To enable multiplexed recording of cellular events, such as the activation of a specific promoter, pegRNAs with different signatures (but the same propagator) can be put under the control of different promoters of interest. Then, the frequency of insertion of signatures linked to each promoter at the DNA tape reflects the duration or strength of activation of the

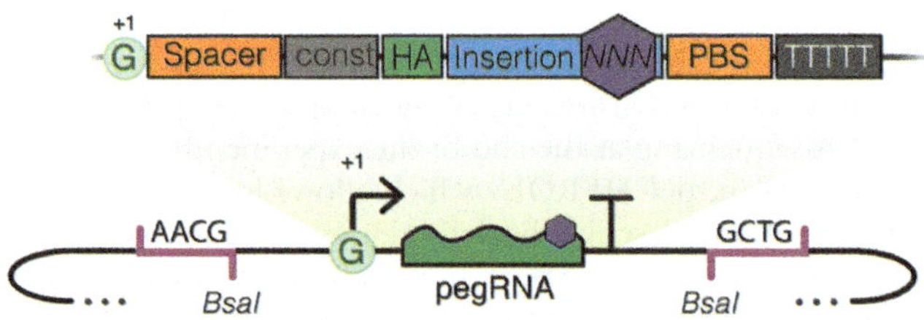

Fig. 1 Schematic of a part plasmid encoding a peCHYRON pegRNA driven by an RNAP3 promoter. The highlighted G is the first nucleotide transcribed, the Spacer targets the peCHYRON DNA tape, the constant region binds to prime editor, and the homology arm (HA) matches the DNA downstream of the edit on the tape. The *NNN* outlined in a purple hexagon is the signature, while the remainder of the insertion in blue is the propagator. Poly-T is the terminator for RNAP3. This construct is flanked by BsaI overhangs compatible with cloning into a transcriptional unit as part of the Mammalian Toolkit

Table 1
Sequences for construction of peCHYRON pegRNAs

	Sequence (5′ to 3′)	
pegRNA domain	**A-to-B pegRNA**	**B-to-A pegRNA**
Spacer	GCCATCATCACCATCATTGA	GACTGGGCCATCTATCATTGA
Constant Cas9-binding	GTTTAAGAGCTAAGCTGGAAACAGCATAGCAAGTTTAAATAA GGCTAGTCCGTTATCAACTTGAAAAAGTGGCACC GAGTCGGTGC	
Homology arm	TGGAGGAAGCAGGGCTTCCTTTCCTCTGCCATCA	
Insertion	ATGATAGATGGCCCAGT*NNN*	ATGATGGTGATGATGGC*NNN*
PBS	ATGATGGTGAT	ATGATAGATGGCCC
Terminator	TTTTT	

NNN signifies variable signature mutations

corresponding promoter; the position relative to other signatures on the DNA tape reflects the relative timing of activation of the corresponding promoter. DNA Typewriter is particularly effective in short-term recording experiments, as its use of short (5–6 bp) insertions allows it to quickly accumulate information, at the cost of eventual exhaustion of the synthetic, genomically-integrated DNA tape with, typically, five target sites. Conversely, peCHYRON uses 20-bp insertions at each edit, meaning that recording can start at a genomic site, with the DNA tape being built up as recording proceeds. This design requires alternating propagators, such that an edit by an "A-to-B" pegRNA is required to facilitate an edit by a "B-to-A" pegRNA, which will then allow an edit by an "A-to-B" pegRNA, etc.; each of these edits deposits an information-containing signature. An additional component of peCHYRON is a "protective nicking sgRNA," which directs prime editor to nick adjacent to and on the same strand as the target sites present in the loci encoding the pegRNAs, which would otherwise be "cannibalized." peCHYRON is most useful in situations that meet one or more of the following conditions: (1) records longer than five are especially informative, (2) recording must continue for >1 month, and (3) recording at endogenous loci is especially valuable.

Recording gene expression from many conditionally-active promoters is complicated by their requirement for RNA Polymerase II (RNAP2)-mediated transcription, which is unable to natively transcribe functional pegRNAs [8]. Traditionally, effective transcription of pegRNAs must proceed from RNAP3 promoters; endogenous RNAP3 promoters generally control the expression of small non-coding RNAs, such as tRNAs, and endogenous inducible promoters are not available, but synthetic repressible RNAP3 promoters have been constructed. Alternatively, use of endoribonucleases such as Csy4 allows RNA Polymerase II-mediated transcription of pegRNAs through post-transcriptional cleavage of transcripts to yield functional pegRNAs [9]. While Csy4-mediated expression of pegRNAs allows expanded possibilities for recording events in cells, it is limited by inherent toxicity of the Csy4 endoribonuclease [8, 10] and potential cross-talk between orthogonal Csy4-processed genetic circuits.

Here, we provide a detailed protocol for the design of peCHYRON pegRNAs for recording events of interest within cells, as well as a strategy for hierarchical Golden-Gate assembly of peCHYRON genetic circuits for stable integration in mammalian cells. Additionally, we provide guidelines for purification of high-quality, transfection-grade plasmids for transfection and recommendations for transfection of peCHYRON component plasmids into cells. Finally, we describe the engineering of cells for peCHYRON recording, the execution of a peCHYRON experiment within these engineered cells, and the pipeline for obtaining and interpreting recording data from a peCHYRON recording locus via amplicon-nanopore sequencing. For demonstration purposes, we outline an experiment, similar to Figs. 5

and 6 in [7], in which a peCHYRON cell line capable of recording treatment with doxycycline and/or IPTG is made and used. All plasmids needed for this experiment can be obtained from Addgene, but we additionally provide a detailed protocol for the construction of integrating vectors expressing pegRNAs under the control of TetO-hU6 promoters, TetR, and a selectable marker. In these constructs, the pegRNA will be expressed only in the presence of doxycycline, which removes TetR from the promoter. This protocol can be customized by the reader to put TetR, and thus pegRNA expression, under the control of a degron [11, 12, 13] or promoter of interest.

2 Materials

2.1 Molecular Cloning

1. Nuclease-free water.
2. Esp3I, 10 U/μL (ThermoFisher Scientific).
3. BsaI-HFv2, 20 U/μL (New England Biolabs).
4. Hi-T4 Ligase (New England Biolabs).
5. Thermal cycler.
6. Fluorescence transilluminator.
7. 0.2 mL PCR tubes.
8. *E. coli* DH10β competent cells.
9. Mammalian Toolkit Modular Cloning part plasmid collection (available from Addgene [14]).
10. peCHYRON plasmid collection (available from Addgene [7]).
11. Chloramphenicol, 25 mg/mL in sterile 100% ethanol.
12. Kanamycin, 50 mg/mL in sterile water.
13. Carbenicillin, 100 mg/mL in sterile 50% ethanol.

2.2 Transfection-Grade Plasmid Preparation and Transfection

1. ZymoPURE Miniprep Kit (Zymo Research).
2. LB_{MG}: Lysogeny Broth with 10 mM Magnesium Chloride.
3. Chloramphenicol, 25 mg/mL in sterile 100% ethanol.
4. Kanamycin, 50 mg/mL in sterile water.
5. Carbenicillin, 100 mg/mL in sterile 50% ethanol.
6. Sterile nuclease-free water.
7. Sterile 1.5 mL microcentrifuge tubes.
8. Gibco OptiMEM (Thermo Fisher Scientific).
9. Complete DMEM (cDMEM): 10% Fetal Bovine Serum (v/v) in e.g., Gibco DMEM, high glucose, GlutaMAX (Thermo Fisher Scientific).

10. Transfection reagent.
11. Tissue culture plates.
12. Biological Safety Cabinet.
13. Tissue Culture CO_2 incubator.
14. Fluorescent microscope or fluorescent cell imaging device.
15. Agarose.
16. Horizontal electrophoresis system with power supply.

2.3 Preparation of peCHYRON Cell Lines and Execution of peCHYRON Experiments

1. Sterile 1.5 mL microcentrifuge tubes.
2. Gibco OptiMEM (Thermo Fisher Scientific).
3. Complete DMEM (cDMEM): 10% Fetal Bovine Serum (v/v) in e.g., Gibco DMEM, high glucose, GlutaMAX (Thermo Fisher Scientific).
4. Super PiggyBac Transposase expression vector (System Biosciences or equivalent).
5. Fluorescence-Assisted Cell Sorting-capable Flow Cytometer (e.g., Sony MA900).
6. Transfection reagent (we use FuGene HD for HEK293T cells and Lipofectamine-3000 for immortalized mouse embryonic fibroblasts).
7. Tissue culture-treated plates.
8. Fluorescent microscope or fluorescent cell imaging device.
9. Biological Safety Cabinet.
10. Tissue Culture CO_2 incubator.
11. Blasticidin Hydrochloride, 10 mg/mL (Invivogen).
12. Puromycin, 10 mg/mL (Invivogen).
13. Hygromycin B gold, 100 mg/mL (Invivogen).

2.4 Cell Collection and Genomic DNA Extraction

1. Biological Safety Cabinet.
2. Tissue Culture CO_2 incubator.
3. Trypsin-EDTA (0.05%), phenol red.
4. Thermal cycler or dry bath.
5. 1.5 mL microcentrifuge tubes.
6. 0.2 mL PCR tubes.
7. QuickExtract DNA Extraction Solution (Lucigen).

2.5 Library Preparation and Sequencing

1. Thermal cycler.
2. QuickExtract DNA Extraction Solution (Lucigen).
3. QuantaBio repliQa 2× HiFi ToughMix PCR master mix (QuantaBio).

Table 2
Primers for amplicon-nanopore sequencing

Primer name	Amplicon	Sequence
site3outside-outsideF	HEK3	TAGCATACCGGTnnnnnnnnnn GCTTACAGCAGGAGTCCCA
site3outside-outsideR	HEK3	TAGCATACCGGTnnnnnnnnnn GGCGACTGTCCCTCTCAAAT
pTBL3328-F	pTBL3328 target site	TAGCATACCGGTnnnnnnnnnn AGCACGCAAAGCGTTCTGTC
pTBL3328-F	pTBL3328 target site	TAGCATACCGGTnnnnnnnnnn AGAGCCCACCGCTGTAAGG

The sequence "nnnnnnnnnn" is a barcode that enables multiplexing

Table 3
Barcodes for amplicon-nanopore sequencing

Barcode	Sequence
Barcode1	AAACAAACAC
Barcode2	AACCTTAGAT
Barcode3	ACCCGTTCGT
Barcode4	AGCATAAGTA
Barcode5	AAACCCTGAC
Barcode6	AAGGTAAACT
Barcode7	ACGTAAGAAA
Barcode8	AGTAACACCC

These sequences are designed to have a hamming distance of at least four from all other barcodes, and are compatible with the "barcodes_subset.fasta file within the MAPLE "ref" folder (Step 3.6.3)

4. 0.2 mL PCR tubes.
5. 1.5 mL microcentrifuge tubes.
6. Nuclease-free water.
7. Molecular grade ethanol.
8. Nanopore amplicon sequencing primers collection (see Table 2).
9. Agarose.
10. Horizontal electrophoresis system with power supply.
11. "Nuclease-Away" Decontamination solution: 10% bleach (v/v), 0.25 N Sodium Hydroxide, 1% Sparkleen detergent (or equivalent) (w/v).

12. AMPure XP beads (Beckman-Coulter).
13. NEBNext® Companion Module v2 for Oxford Nanopore Technologies® Ligation Sequencing (New England Biolabs).
14. Bovine Serum Albumin (BSA), 50 mg/mL.
15. Oxford Nanopore MinION sequencing device, Mk1b or later model (Oxford Nanopore Technologies).
16. MinION Flow Cell (R10.4.1) (Oxford Nanopore Technologies).
17. Sequencing management and analysis computer (we use a professional/workstation GPU for local real-time basecalling Table 3).

2.6 Sequencing Analysis: Demultiplexing and Preliminary Analysis

1. Linux-operated computer.
2. Miniconda installation of Anaconda Python distribution (Anaconda, Inc.).
3. Mutational Analysis for ParalleL Evolution (MAPLE) snakemake pipeline and associated dependencies [15].

3 Methods

3.1 Assembly of TetR-Controlled pegRNA Integrating Plasmids Using Hierarchical Golden Gate Assembly (see Note 1)

1. Assemble necessary Mammalian Toolkit (MTK) [14]-compatible part plasmids to create 4 transcriptional units (TUs):
   a. TU1 is the depressible RNAP3 promoter hU6-TetO (*see* **Note 2**) driving an A-to-B pegRNA (Table 4).
   b. TU2 is hU6-TetO driving a B-to-A pegRNA (Table 5).
   c. TU3 is the strong constitutive promoter pCAG driving TetR (Table 6 and *see* **Note 3**).
   d. TU4 is the strong constitutive promoter full-length human EF1alpha driving mCherry (Table 7).

Table 4
Part plasmids for creation of TU1: TetR-controlled A-to-B pegRNA expression

Plasmid ID	Bacterial resistance marker	MTK part	Addgene ID
MTK1_008 ConLS-cHS4 insulator	*camR*	MTK1	123671
pTBL3331 MTK234-TetO-AtoBpegRNA	*camR*	MTK234	226668
MTK5_010 ConR7	*camR*	MTK5	123854
pTBL2327 TU backbone	*ampR*	MTK678	254858

Table 5
Part plasmids for creation of TU2: TetR-controlled B-to-A pegRNA expression

Plasmid ID	Bacterial resistance marker	MTK part	Addgene ID
MTK1_008 ConLS-cHS4 insulator	*camR*	MTK1	123671
pTBL3331 MTK234-TetO-BtoApegRNA	*camR*	MTK234	254860
MTK5_010 ConR7	*camR*	MTK5	123854
pTBL2327 TU backbone	*ampR*	MTK678	254858

Table 6
Part plasmids for creation of TU3: TetR expression

Plasmid ID	Bacterial resistance marker	MTK part	Addgene ID
MTK1_029 ConL7-B2 insulator	*camR*	MTK1	123691
MTK2_005 pCAG	*camR*	MTK2	123700
pTBL2620 MTK3-TetR	*camR*	MTK3	226670
MTK4_003 rabbit beta-globin pA	*camR*	MTK4	123812
MTK5_009 ConR6	*camR*	MTK5	123853
pTBL2327 TU backbone	*ampR*	MTK678	254858

Table 7
Part plasmids for creation of TU4: mCherry expression

Plasmid ID	Bacterial resistance marker	MTK part	Addgene ID
MTK1_028 ConL6-F1 insulator	*camR*	MTK1	123690
MTK2_007 human pEF1alpha	*camR*	MTK2	123702
pTBL3481 MTK3-mCherry	*camR*	MTK3	226672
MTK4_001 BGH pA	*camR*	MTK4	123810
pTBL2436 MTK5-ConRE	*kanR*	MTK5	254859
pTBL2327 TU backbone	*ampR*	MTK678	254858

2. Clone TUs 1–4 via BsaI-mediated Golden Gate assembly, by mixing together equimolar amounts of the component plasmids listed in each table.
3. Transform TU assemblies into *E. coli* DH10β competent cells and select successful assemblies by plating transformants on LB agar + 100 μg/mL carbenicillin and selection of GFP-negative colonies under illumination with blue light.

4. Purify assembled TU plasmids via plasmid miniprep purification and validate successful assembly by plasmid size and whole-plasmid sequencing.
5. Assemble 2 "destination vectors" for stable integration of transcriptional unit cassettes by mixing equimolar amounts of the listed TU plasmids for Esp3I-mediated Golden Gate assembly: DV1 includes TU1–TU3–TU4; DV2 includes TU2–TU3–TU4 (*see* **Note 4**).
6. Transform destination vector assemblies into *E. coli* DH10β competent cells and select successful assemblies by plating transformants on LB agar + 50 μg/mL kanamycin and selection of GFP-negative colonies under fluorescent illumination.
7. Purify assembled destination vector plasmids via plasmid miniprep purification to validate successful assembly by plasmid size and whole-plasmid sequencing.

3.2 Transfection-Grade Plasmid Preparation

1. Inoculate overnight cultures of *E. coli* DH10β containing plasmids of interest into LBMG with appropriate selective antibiotics and incubate overnight (14–18 h) in a shaking incubator at 37°C and 250 rpm (*see* **Note 5**).
2. Pellet up to 4.5 mL of culture in a sterile microcentrifuge tube and purify transfection grade plasmids using the ZymoPURE Miniprep Kit, per the manufacturer's protocol, performing all steps aseptically to reduce potential of contamination of the plasmid product (*see* **Note 6**).
3. Working aseptically, transfer ~90% of the volume of purified plasmid to a new, sterile microcentrifuge tube, using the remaining volume in the original tube to quantify plasmid concentration and evaluate plasmid size and quality on an agarose gel (*see* **Note 7**).

3.3 Preparation of peCHYRON Cell Lines and Execution of peCHYRON Experiments

1. Prepare cell lines with an integrated initial peCHYRON target site by one of two strategies:
   a. For human cells, the initial target site can be knocked in to the HEK3 locus by transiently transfecting with PE and the pegRNA whose sequence is shown in Table 8. If it is important that every cell has the site inserted, clones can be isolated and propagated. A single transient transfection is likely to yield target sites at ~40% of HEK3 loci, so unselected populations of cells can also be used.
   b. For any mammalian cell, a target site cassette (Table 8) can be integrated by cotransfection with PiggyBac transposase (see below for detailed protocol) and then selected for. Cells must be passaged for at least 2 weeks after transfection before attempting to sequence this integrated locus, to allow for adequate dilution of unincorporated plasmids.

Table 8
Methods to introduce initial peCHYRON target site.

pegRNA sequence that will direct the insertion of an initial peCHYRON target site at human *HEK3*				
GGCCCAGACTGAGCACGTGAGTTTTAGAGCTAGAAATAGCAAGTTAAAATAAGGCTAGT CCGTTATCAACTTGAAAAAGTGGCACCGAGTCGGTGCTGGAGGAAGCAGGGCTTCCTT TCCTCTGCCATCAATGATGGTGATGATGGCTTGCGTGCTCAGTCTG				
Plasmid ID	Bacterial resistance marker	Tissue culture selective marker	Notes	Addgene ID
pTBL3328 PB-sfGFP-WPRE-target	*kanR*	sfGFP	PB-integratable peCHYRON recording locus	226676

Table 9
Core plasmids for implementation of peCHYRON

Plasmid ID	Bacterial resistance marker	Tissue culture selective marker	Notes	Addgene ID
pTBL3342 PB-PEmax-puro	*kanR*	*puroR*	PB-integratable pCAG-driven PEmax expression unit	226674
pTBL3478 PB-PEmax-mTagBFP2	*kanR*	*hygR*, mTagBFP	PB-integratable pCAG-driven PEmax expression unit	226664
pTBL3346 PB-nicking guide-blast	*kanR*	*blastR*	PB-integratable protective nicking guide RNA	226665
pTBL3241 PB-AtoBpeg RNA-mCherry	*kanR*	mCherry	PB-integratable A to B pegRNA expression unit	226666
pTBL3401 BtoApegRNA-sfGFP	*kanR*	sfGFP	PB-integratable B to A pegRNA expression unit	226667
pTBL3439 lacO-hU6-AtoBpegRNA-LacI-mCherry	*kanR*	mCherry	PB-integratable IPTG-controlled AtoB pegRNA expression unit	254861
pTBL3440 lacO-hU6-BtoApegRNA-LacI-mCherry	*kanR*	mCherry	PB-integratable IPTG-controlled BtoA pegRNA expression unit	254862

2. Prior to transfection of target site-containing cells with additional peCHYRON components, seed cells to be ~20% confluent at the time of transfection in an appropriately sized well/plate. Ensure cells are fully adherent (if applicable), at sufficient confluency, and healthy before proceeding with transfection.
3. Acquire transfection-grade minipreps of all necessary PiggyBac integration vectors. Each experiment must include a separate vector expressing each of the following (detailed listing in Table 9):
   a. Prime editor (we have used PE2 [16], PEmax [17], and PE7 [18] successfully), under the control of a strong, constitutive, intron-containing promoter such as pCAG or full-length human EF1alpha promoter, with an antibiotic marker such as *puroR* and, optionally, a fluorescent marker to track silencing.
   b. A-to-B pegRNA, with a fluorescent marker.
   c. B-to-A pegRNA, with a fluorescent marker.
   d. Protective nicking guide RNA, with an antibiotic marker such as *blastR*.
   e. Additional pegRNAs controlled by doxycycline or IPTG, marked with fluorescent markers, can be added if desired.
4. Thoroughly decontaminate the sterile miniprep tubes with 70% ethanol and transfer them to a sterile tissue culture workspace (e.g., a sterile biosafety cabinet in a separate tissue culture room).
5. Prepare transfections in OptiMEM with molarity-normalized ratios of plasmids using a desired transfection method (e.g., 50 fmol PE, 20 fmol pegRNA A to B, 20 fmol pegRNA B to A, 20 fmol protective nicking sgRNA, 1/20 mass of transient SuperPiggyBac expression plasmid).
6. Transfect cells per manufacturer's protocol of chosen transfection method and allow cells to incubate overnight in transfection mixtures. The peCHYRON experiment should be considered initiated from the moment of transfection of both Prime Editor and suitable pegRNAs.
7. Observe transfectants 1 day post-transfection to evaluate transfection efficiency and maintain cells as needed.
8. 2 days post-transfection, add appropriate selection antibiotics. Maintain treatment, while passaging cells as needed, for at least 1 week.
9. Optionally, sort for cells with multiple pegRNA integrations by FACS. In general, SuperPiggyBac transposase is efficient

enough that any cells surviving antibiotic selection will have at least two pegRNA integrations even without FACS.

10. Treat with doxycycline and IPTG as desired.
11. Proceed to cell collection as desired.

3.4 Cell Collection and Genomic DNA Extraction

1. Trypsinize transfected cells and neutralize with cDMEM to collect.
2. Pellet collected cells at 1000*g* for 5 min, and aspirate supernatant completely.
3. Add an appropriate volume of QuickExtract DNA Extraction solution (see Table 10) to the pelleted cells and resuspend by pipetting.
4. Vortex the resuspend cell pellet and incubate the resuspension in a dry bath or thermal cycler as follows:
   Dry bath: 65°C for 25 min, 98°C for 10 min.
   Thermal cycler: 65°C for 10 min, 68°C for 15 min, 95°C for 10 min.
5. Transfer extractions to low-retention tubes for storage. Extracted genomic DNA should be stored at 4°C for <4 weeks or stored at −80°C for long-term storage.

3.5 Library Preparation and Sequencing

1. Thaw an appropriate volume of 2× repliQa HiFi Toughmix PCR master mix at on ice or at 4°C.
2. Thoroughly decontaminate the workspace with "Nuclease-Away" and allow the workspace to sit for at least 10 min after treatment (*see* **Note 8**).
3. Wipe the workspace down with 70% ethanol and allow ethanol to evaporate completely.
4. Assign unique primer index pairs to each unique sample processed, such that each sample has a unique combination of forward and reverse primers (Tables 2 and 3).

Table 10
Volumes for QuickExtract DNA Extraction Solution-mediated genomic DNA extraction from mammalian cells

Culture plate	Approx. cell count	Recommended volume (μL)
6-Well plate	1.2×10^6 cells	500
12-Well plate	0.5×10^6 cells	300
24-Well plate	0.24×10^6 cells	100
48-Well plate	0.12×10^6 cells	50
96-Well plate	0.04×10^6 cells	20

5. Prepare PCR reaction mixtures with unique primer pairs for each unique sample in 0.2 mL PCR tubes, using 5 μL of the genomic DNA extract obtained in Sect. 3.4 (Table 11).
6. Gently pipette mix each reaction to ensure distribution of reaction components, followed a quickly centrifuging the reactions.
7. Load the reactions into a thermal cycler and run the thermal cycler program "Nanopore-iPCR-rQ" (Table 12).
8. After PCR completion, run the reactions on a 1% agarose gel for 30 min at 100 V and then image the gel to assess PCR success.
9. Pool 5 μL of each iPCR in a single 1.5 mL microcentrifuge tube and quickly spin the tube down to collect all reactions in the bottom of the tube.
10. Bring Ampure XP beads (AXP) to room temperature and mix well by vortexing to completely resuspend the beads.
11. Add 90% of the total volume of the PCR pool of AXP to the pool (e.g., a 20-reaction pool at 100 μL will use 90 μL AXP).

Table 11
Reaction mixture for index PCR setup

Component	Quantity
2× repliQa HiFi toughmix	25 μL
Forward primer (10 μM)	2.5 μL
Reverse primer (10 μM)	2.5 μL
Genomic DNA extract	5 μL
Nuclease-free water	15 μL

Table 12
Thermal cycling protocol "Nanopore-iPCR-rQ" for index PCR reaction

Reaction step	Temperature (°C)	Length (mm:ss)	Notes
1—Initial denaturation	98.0	05:00	
2—Cycle denaturation	98.0	00:10	Repeat for 30 cycles
3—Cycle annealing	62.0[a]	00:10	
4—Cycle extension	72.0	00:30[b]	
5—Final extension	72.0	02:00	
6—Hold	12.0	∞	

[a]Annealing temperature may require optimization for new primer sequences. [b]Extension time should be approximately 10 s per kb. This example uses a 2.2 kb amplicon.

Gently flick the mixture and allow the tube to sit at room temperature for 5 min.

12. Quickly spin the tube down and transfer it to a magnetic separation rack for 5 min to pellet beads.
13. During this period, prepare 1.2 mL of 80% molecular-grade ethanol in nuclease free water (v/v).
14. Aspirate supernatant from the tube, avoiding the pelleted beads.
15. Add 500 μL of 80% Ethanol to the tube while it remains on the magnetic separator and wait 30 s.
16. Aspirate the ethanol and repeat the ethanol wash.
17. Completely aspirate the ethanol, quickly spin the tube down, and place back on the magnetic separator.
18. Using a 10 μL pipette tip, remove residual ethanol from the tube, and let the tube air dry for ~30 s.
19. Do not let the pellet dry excessively, resuspend the pellet before it completely dries or cracks.
20. Resuspend the pellet in 50–100 μL of nuclease free water, and let the resuspended pellet sit at room temperature for 5 min.
21. For <36 samples, 50 μL is recommended. For 37–72 samples, 75 μL is recommended. For pools with >72 samples, 100 μL is recommended.
22. Spin down the resuspension and place on a magnetic separator for 5 min.
23. Collect 95% of the volume of added water of the eluate and quantify DNA concentration by NanoDrop (or similar method).
24. Prepare sequencing libraries via the Oxford Nanopore LSK-114 Ligation Sequencing protocol, using 100 fmol of pooled index PCRs as an input.
25. Load libraries onto an Oxford Nanopore MinION sequencing instrument and begin sequencing per the manufacturer's protocol.

3.6 Sequencing Analysis: Demultiplexing and Preliminary Analysis

1. Install the MAPLE python library as a conda environment on an analysis computer run on a Linux Operating System per the developer's instructions.
2. After sequencing completion, create a new directory for sequencing run analysis titled after the run name (e.g., 20250505_SEQ001-nanopore1).
3. Copy and paste the "config.yaml" file, "ref" directory, and "demux" from the REPOSITORY into the newly-created analysis directory.

4. Open the "config.yaml" file and edit the value of "runname" (line 123) to match that of the directory (e.g., runname: -20250505_SEQ001-nanopore1). Save the file and exit.
5. Within the enclosed "ref" directory, open the "barcodeGroups.csv" file and change the following match the desired values:
   a. "tag": reference sequence name (i.e., "refseq" from "refseq-ref.fasta").
   b. "barcodeGroup": Name of each individual sample.
   c. "fwd": Index ID for the forward primer as "BarcodeN" used for a given sample, where "N" is the number assigned to a given index.
   d. "rev": Index ID for reverse primer as "BarcodeN" for a given sample, where "N" is the number assigned to a given index.
6. Using the command line (and performing subsequent steps from command line), copy the "/fastq_pass" directory within the sequencing run data directory into the analysis directory created above.
7. Set the analysis directory as the working directory and activate an instance of the MAPLE conda environment.
8. Execute the MAPLE analysis with the command "snakemake --snakefile ~/maple/Snakefile -j 30 targets."
9. Upon completion of the MAPLE analysis, deactivate the MAPLE conda environment and set the "/demux" directory as the working directory and execute the batchProc_bedtools.sh" script to convert the demultiplexed.bam sequences files to.fastq format.
10. Set the "/fastqanalysis" directory as the working directory and execute the "executeinsAna.sh" script to batch analyze insertions in the demultiplexed.fastq files.
11. Examine the "/fastqanalysis" directory to observe four new data files created for every unique.fastq file input:
    a. "SAMPLEID_indel_histogram.csv": contains read counts delineated by total insertion size in the read.
    b. "SAMPLEID_entropy_dataframe.csv": contains counts of occurrences of each possible insertion signature for each possible "insertional polarity" (i.e., whether inserted by an A-to-B or B-to-A pegRNA) across reads.
    c. "SAMPLEID_insertions_dataframe.csv": a list of every read processed by the insertional analysis pipeline.

d. "SAMPLEID_signatures_data.csv": contains read counts of each unique insertional pattern identified (insertions converted to single-letter representations).

12. Proceed to downstream analysis as desired. Notable strategies include evaluation of general insertional efficiency (see Fig. 3 in [7]), analysis of bigram frequency (see Fig. 5 in [7]), or analysis of signature proportion at each position in a string of peCHYRON insertions (see Fig. 6 and Extended Data Fig. 10 in [7]).

4 Notes

1. The Mammalian Toolkit (MTK) is a hierarchical Golden Gate cloning toolkit for modular assembly of expression and integration vectors for mammalian systems. MTK part vectors carry expression vector parts that are assembled into functional Transcriptional Units (TUs) via BsaI-mediated Golden Gate assembly. The order of the parts in the TU is enforced the unique overhang sequences created by each BsaI cut, which only match to allow ligation when the parts are assembled in the proper order. (See example overhangs for an "MTK234" part in Fig. 1.) TUs are assembled with upstream and downstream "connector" sequences to allow assembly of multiple TUs in ordered fashion in a Destination Vector (DV). DVs are assembled via Esp3I-mediated Golden Gate assembly of one or multiple TUs into a backbone carrying sequences for stable integration (e.g., BxB1 sites, PiggyBac inverted terminal repeats (ITRs), etc.) and selectable markers of interest.
2. Conditional RNAP3 promoters include tetO-hU6 and lacO-U6, which have insertions of *tetO* or *lacO* sites, respectively, flanking the U6 TATA box, allowing TetR or LacI-mediated repression of transcription. Such promoters can be regulated by doxycycline or IPTG, or by regulating the expression or stability of TetR or LacI. We find the tetO-hU6 promoter to have far less leaky expression than lacO-hU6, at the cost of a lower maximum expression level [7].
3. To add control of TetR levels, two modifications are possible to MTK parts: the MTK2 part containing the promoter controlling TetR can be an inducible promoter, or the MTK3 part encoding TetR can be modified to include a degron.
4. It is recommended that conditionally-expressed TUs be positioned upstream of constitutively-expressed TUs to avoid potential "read-through" transcription from an upstream constitutively-expressed TU. (e.g., assembly of a hU6-pegRNA-AtoB

with pEF1a-mCherry in a PiggyBac transposon-integratable destination vector backbone). In addition, highly-expressed genes can reduce expression of genes downstream, due to altered supercoiling [19]. Finally, we avoid as much as possible any repeated sequences within one DV by varying insulators, promoters, and poly-adenylation sequences, and putting related sequences such as different pegRNAs on separate DVs. It is especially important to avoid repeats of sequences that will be targeted by prime editor (such as our pegRNAs, which are nicked by prime editor even though actual editing is blocked by protective nicking).

5. *endA*- (Endonuclease I deficient) *E. coli* K12 cells should be used to prevent nuclease degradation of plasmids during purification. *E. coli* DH-10β-derived strains (e.g., NEB-10β , TOP10, etc.) are ideal for propagation and purification of peCHYRON component plasmids due to their improved ability in handling large (>5 kb) plasmids. For cloning repetitive sequences, a "stable" *E. coli* strain (e.g., NEBStable, Stbl4, etc.) may be desirable to prevent unwanted recombination.
6. Using a high-purity plasmid miniprep kit is strongly recommended to ensure sufficient removal of bacterial endotoxins from the purified plasmid and to minimize plasmid degradation during purification. The ZymoPURE miniprep kit is ideal for transfection-grade spin minipreps, but similar miniprep kits with dedicated endotoxin removal wash steps may be suitable alternatives.
7. Both plasmid size and supercoiled proportion should be evaluated by agarose gel electrophoresis. Supercoiled Plasmid DNA ladder should be used to evaluate plasmid size, and supercoiled plasmid should be the predominant isoform of plasmid present (supercoiled plasmid will present at the size indicated by the ladder, with a trailing linearized plasmid band, and a further trailing "nicked" plasmid band). Notably, duplication of the cHS4 insulator in TU and DV plasmids may result in a 200 bp shift in plasmid size which may be diagnosed by prior whole-plasmid sequencing.
8. It is crucially important to avoid the introduction of homologous amplified PCR products from previous experiments into gDNA or amplicon sequencing PCR reactions being set up. Once cells have been lysed, we carry out gDNA extraction and sequencing PCR setup in PCR hoods in a positive-pressure room separate from the main lab space in which amplified products are analyzed.

References

1. Wagner DE, Klein AM (2020) Lineage tracing meets single-cell omics: opportunities and challenges. Nat Rev Genet 21:410–427
2. Askary A, Chen W, Choi J, Du LY, Elowitz MB, Gagnon JA, Schier AF, Seidel S, Shendure J, Stadler T, Tran M (2025) The lives of cells, recorded. Nat Rev Genet 26:203–222
3. Loveless TB, Grotts JH, Schechter MW, Forouzmand E, Carlson CK, Agahi BS, Liang G, Ficht M, Liu B, Xie X, Liu CC (2021) Lineage tracing and analog recording in mammalian cells by single-site DNA writing. Nat Chem Biol 17:739–747
4. Bhan N, Callisto A, Strutz J, Glaser J, Kalhor R, Boyden ES, Church G, Kording K, Tyo KEJ (2021) Recording temporal signals with minutes resolution using enzymatic DNA synthesis. J Am Chem Soc 143:16630–16640
5. Carlson CK, Loveless TB, Milisavljevic M, Kelly PI, Mills JH, Tyo KEJ, Liu CC (2024) A massively parallel in vivo assay of TdT mutants yields variants with altered nucleotide insertion biases. ACS Synth Biol 13:3326–3343
6. Choi J, Chen W, Minkina A, Chardon FM, Suiter CC, Regalado SG, Domcke S, Hamazaki N, Lee C, Martin B, Daza RM, Shendure J (2021) A time-resolved, multi-symbol molecular recorder via sequential genome editing. Nature 608:98–107
7. Loveless TB, Carlson CK, Helmy CAD, Hu VJ, Ross SK, Demelo MC, Murtaza A, Liang G, Ficht M, Singhai A, Pajoh-Casco MJ, Liu CC (2025) Open-ended molecular recording of sequential cellular events into DNA. Nat Chem Biol 21:512-521. https://doi.org/10.1038/s41589-024-01764-5
8. Nissim L, Perli SD, Fridkin A, Perez-Pinera P, Lu TK (2014) Multiplexed and programmable regulation of gene networks with an integrated RNA and CRISPR/Cas toolkit in human cells. Mol Cell 54:698–710
9. Chen W, Choi J, Li X, Nathans JF, Martin B, Yang W, Hamazaki N, Qiu C, Lalanne J-B, Regalado S, Kim H, Agarwal V, Nichols E, Leith A, Lee C, Shendure J (2024) Symbolic recording of signalling and cis-regulatory element activity to DNA. Nature 632:1073–1081. https://doi.org/10.1038/s41586-024-07706-4
10. Zhang L, Qiu X, Zhou Y, Luo Z, Zhu L, Shao J, Xie M, Wang H (2025) A trigger-inducible split-Csy4 architecture for programmable RNA modulation. Nucleic Acids Res 53:gkae1319
11. Zeng Y, Bhagyashree B, Zhao W, Nguyen T, Segatori L (2019) Hysteretic genetic circuit for detection of proteasomal degradation in mammalian cells. ACS Synth Biol 8:2025–2035
12. Zeng Y, Jones AM, Thomas EE, Nassif B, Silberg JJ, Segatori L (2018) A Split transcriptional repressor that links protein solubility to an orthogonal genetic circuit. ACS Synth Biol 7:2126–2138
13. Zhao W, Pferdehirt L, Segatori L (2018) Quantitatively predictable control of cellular protein levels through proteasomal degradation. ACS Synth Biol 7:540–552
14. Fonseca JP, Bonny AR, Kumar GR, Ng AH, Town J, Wu QC, Aslankoohi E, Chen SY, Dods G, Harrigan P, Osimiri LC, Kistler AL, El-Samad H (2019) A toolkit for rapid modular construction of biological circuits in mammalian cells. ACS Synth Biol 8:2593–2606
15. Rix G, Williams RL, Hu VJ, Spinner A, Pisera A(Olek), Marks DS, Liu CC (2024) Continuous evolution of user-defined genes at 1 million times the genomic mutation rate. Science 386:eadm9073
16. Anzalone AV, Randolph PB, Davis JR, Sousa AA, Koblan LW, Levy JM, Chen PJ, Wilson C, Newby GA, Raguram A, Liu DR (2019) Search-and-replace genome editing without double-strand breaks or donor DNA. Nature 576:149–157
17. Chen PJ, Hussmann JA, Yan J, Knipping F, Ravisankar P, Chen P-F, Chen C, Nelson JW, Newby GA, Sahin M, Osborn MJ, Weissman JS, Adamson B, Liu DR (2021) Enhanced prime editing systems by manipulating cellular determinants of editing outcomes. Cell 184:5635–5652.e29
18. Yan J, Oyler-Castrillo P, Ravisankar P, Ward CC, Levesque S, Jing Y, Simpson D, Zhao A, Li H, Yan W, Goudy L, Schmidt R, Solley SC, Gilbert LA, Chan MM, Bauer DE, Marson A, Parsons LR, Adamson B (2024) Improving prime editing with an endogenous small RNA-binding protein. Nature 628:639–647. https://doi.org/10.1038/s41586-024-07259-6
19. Johnstone CP, Galloway KE (2022) Supercoiling-mediated feedback rapidly couples and tunes transcription. Cell Rep 41:111492

Index

Huan Peng and S. Brian Kim (eds.), *Biological Macromolecule Engineering Technologies*,
Methods in Molecular Biology, vol. 3032, https://doi.org/10.1007/978-1-0716-5257-2,

The manufacturer's authorised representative in the EU is Springer Nature Customer Service Centre GmbH, Europaplatz 3, 69115 Heidelberg, Germany. If you have any concerns regarding our products, please contact ProductSafety@springernature.com

Printed and bound by CPI Group (UK) Ltd, Croydon, CR0 4YY
15/09/2026
02217000-0007